Intelligent Nanomaterials for Drug Delivery Applications

Intelligent Nanomaterials for Drug Delivery Applications

Edited by

NABEEL AHMAD
Associate Professor and Head
School of Biotechnology
IFTM University Moradabad
Uttar Pradesh, India

P. GOPINATH
Associate Professor
Nanobiotechnology Laboratory
Centre for Nanotechnology
Department of Biotechnology
Indian Institute of Technology Roorkee
Uttarakhand, India

ELSEVIER

Elsevier
Radarweg 29, PO Box 211, 1000 AE Amsterdam, Netherlands
The Boulevard, Langford Lane, Kidlington, Oxford OX5 1GB, United Kingdom
50 Hampshire Street, 5th Floor, Cambridge, MA 02139, United States

Library of Congress Cataloging-in-Publication Data
A catalog record for this book is available from the Library of Congress

British Library Cataloguing-in-Publication Data
A catalogue record for this book is available from the British Library

ISBN: 978-0-12-817830-0

For information on all Elsevier publications visit our website at
https://www.elsevier.com/books-and-journals

Publisher: Matthew Deans
Acquisitions Editor: Simon Holt
Editorial Project Manager: John Leonard
Production Project Manager: Sreejith Viswanathan
Cover Designer: Alan Studholme

Typeset by TNQ Technologies

List of Contributors

Jyoti Ahlawat, MS
Chemistry and Biochemistry Department
The University of Texas at El Paso
El Paso, TX, United States

Kanwal Akhtar, MS
Magnetic Materials Laboratory
Department of Physics
University of Agriculture Faisalabad
Faisalabad, Pakistan

Rakhi Chaudhary, M.Tech
School of Biotechnology
Gautam Buddha University
Greater Noida, Uttar Pradesh, India

Helon Guimarães Cordeiro, MSc
Department of Biochemistry and Tissue Biology
IB/UNICAMP
Campinas, Brazil

Núbia Alexandre de Melo Nunes, MSc
Department of Biochemistry and Immunology
ICB/UFMG
Belo Horizonte, Brazil

Deepanjalee Dutta, PhD
Centre for Nanotechnology
Indian Institute of Technology Guwahati
Guwahati, Assam, India

Karine Emanuelle da Silva, BSc
Department of Biochemistry
FMRP/USP
Ribeirão Preto, Brazil

Danilo Roberto Carvalho Ferreira, BSc
Department of Biochemistry
UFSJ/CCO
Divinópolis, Brazil

Deepa Garg, M.Tech
Ubiquitous Analytical Techniques Division
CSIR-Central Scientific Instruments Organization
Chandigarh, India
Academy of Scientific and Innovative Research (AcSIR)
Ghaziabad, Uttar Pradesh, India

Upasana Gupta, B.Tech
Ubiquitous Analytical Techniques Division
CSIR-Central Scientific Instruments Organization
Chandigarh, India

Marlia Mohd Hanafiah, PhD
Center for Earth Sciences and Environment
Faculty of Science and Technology
Universiti Kebangsaan Malaysia
Bangi, Selangor, Malaysia
Centre for Tropical Climate Change System
Institute of Climate Change
Universiti Kebangsaan Malaysia
Bangi, Selangor, Malaysia

Jaidip M. Jagtap, PhD
Department of Biomedical Engineering
Medical College of Wisconsin
Milwaukee, WI, United States

Yasir Javed, PhD
Magnetic Materials Laboratory
Department of Physics
University of Agriculture Faisalabad
Faisalabad, Pakistan

Mohammad Nadeem Khan, PhD
Assistant Professor
School of Studies in Biotechnology
Bastar University
Jagdalpur, Chattisgarh, India

Vinay Kumar, PhD
Department for Management of Science and Technology Development
Ton Duc Thang University
Ho Chi Minh City, Vietnam
Faculty of Environment and Labour Safety
Ton Duc Thang University
Ho Chi Minh City, Vietnam

Sivarama Krishna Lakkaboyana, PhD
Department of Chemical Technology
Chulalongkorn University
Bangkok, Thailand

Subhrangsu Sunder Maitra, PhD
School of Biotechnology
Jawaharlal Nehru University
New Delhi, India

Ishita Matai, PhD
Ubiquitous Analytical Techniques Division
CSIR-Central Scientific Instruments Organization
Chandigarh, India
Academy of Scientific and Innovative Research (AcSIR)
Ghaziabad, Uttar Pradesh, India

Mukesh Kumar Meher, M.Tech
Department of Biotechnology
Indian Institute of Technology Roorkee (IIT-Roorkee)
Roorkee, Uttarakhand, India

Mahesh Narayan, PhD
Chemistry and Biochemistry Department
The University of Texas at El Paso
El Paso, TX, United States

Amanda Tomie Ouchida, PhD
Department of Physiology and Pharmacology
Karolinska Institutet
Stockholm, Sweden

Abdul K. Parchur, PhD
Department of Biomedical Engineering
Medical College of Wisconsin
Milwaukee, WI, United States

Krishna Mohan Poluri, PhD
Department of Biotechnology
Indian Institute of Technology Roorkee (IIT-Roorkee)
Roorkee, Uttarakhand, India
Centre for Nanotechnology
Indian Institute of Technology Roorkee (IIT-Roorkee)
Roorkee, Uttarakhand, India

Abhay Sachdev, PhD
Ubiquitous Analytical Techniques Division
CSIR-Central Scientific Instruments Organization
Chandigarh, India
Academy of Scientific and Innovative Research (AcSIR)
Ghaziabad, Uttar Pradesh, India

Sunil Kumar Sailapu, PhD
Centre for Nanotechnology
Indian Institute of Technology Guwahati
Guwahati, Assam, India

Naveed Akhtar Shad, PhD
Department of Physics
Government College University Faisalabad
Faisalabad, Punjab, Pakistan

Gayatri Sharma, PhD
Department of Biomedical Engineering
Medical College of Wisconsin
Milwaukee, WI, United States

Neha Sharma, MSc
Department of Biochemistry
Chulalongkorn University
Bangkok, Thailand

Barkha Singhal, PhD
Assistant Professor
School of Biotechnology
Gautam Buddha University
Greater Noida, Uttar Pradesh, India

Fernanda Maria Policarpo Tonelli, PhD
Department of Morphology
ICB/UFMG
Belo Horizonte, Brazil

Flávia Cristina Policarpo Tonelli, MSc
Department of Pharmacy
UFSJ/CCO
Divinópolis, Brazil

Contents

CHAPTER 1

Introduction to Active, Smart, and Intelligent Nanomaterials for Biomedical Application

JYOTI AHLAWAT, MS • MAHESH NARAYAN, PHD

1 INTRODUCTION

Nanomaterials are materials that have the smallest dimension in the nanometer range. These nanosized systems have a size range from a few nanometers such as micelles to several hundreds of nanometers like liposomes [1]. Comparatively, the size of a water molecule is ~0.16 nm, benzene is ~0.43 nm, the diameter of DNA is ~2 nm, the size of hemoglobin is ~6.4 nm, and the thickness of human hair is 80,000 nm [2]. Interactions between these nanomaterials and biomolecules can be considered for both in vitro as well as in vivo diagnosis [1]. In vitro, these nanomaterials allow recognition and capture of the biological molecules while in vivo, these particulate assemblies are used for imaging purposes [1]. The properties and structure of these nanomaterials differ significantly from the bulk materials and the atoms and molecules from which they are synthesized [3]. Therefore, nanoparticulate systems that can release a drug at an appropriate rate in response to a specific physiological stimulus and at the destined target site are termed as active, smart, and intelligent nanomaterials [4] (Fig. 1.1).

It is noteworthy that there is explosive scientific growth in the field of nanoscience and technology due to the introduction of new methods of nanomaterial synthesis and advancement in the analytical techniques for the characterization and manipulation of material in the nanometer range [3]. For instance, gene targeting/drug targeting, chips for chemical/biochemical assays, DNA chips and arrays, nanosensors, and nanocomposites are some of the applications based on nanomaterials.

Nanotechnology is expanding rapidly due to support from researchers from diverse backgrounds such as academia, clinicians, industry, and federal sectors. As a result, National Nanotechnology Initiative was launched in 2001, which supports research, development, and commercialization of nanotechnology. Furthermore, in 2003 "Forward look on nanomedicine" was launched by the European Science Foundation that aimed at biomedical applications of nanotechnology and nanoscience. Later, in 2004 the high-level group European technology was established by the European Commission with the aim to improve the quality of life and healthcare for the European citizens. As a follow-up, eight national nanomedicine development centers were established by NIH in 2005 and 2006 with the aim to determine how the cellular machinery at nanoscale works and then develop devices based on these principles for the detection, treatment, and prevention of various diseases. There are active scientific community and medical industrial partners in countries such as France, Germany, Spain, and the United Kingdom where development in biomedical nanotechnology is significant although the transfer of technology to the industries is still not as effective as in North America [1].

In summary, nanotechnology refers to the study of material at the ultrasmall scale. The nanomaterials operate at the nanoscale at which the biomolecules inside the living body work. At this scale, the surface-to-volume ratio of the particle is large, which results in the huge surface for interactions with the biological molecules. Furthermore, this results in shorter reaction time, and therefore the miniaturized devices are more sensitive, faster, and portable.

2 TYPES OF NANOMATERIALS

In simple and medium terms, the nanomaterials are divided into three generations [1]:

- First generation: these include nanocapsules and nanospheres

Intelligent Nanomaterials for Drug Delivery Applications. https://doi.org/10.1016/B978-0-12-817830-0.00001-1

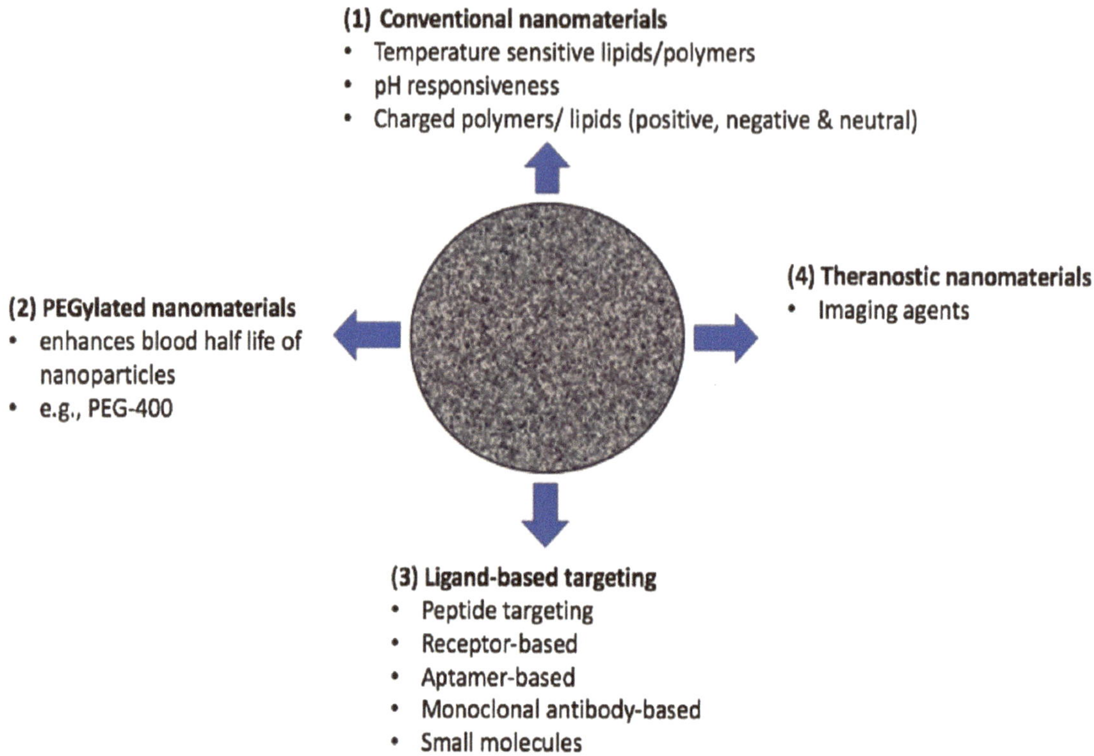

FIG. 1.1 Schematic illustration of various strategic designs for nanomaterials.

- Second generation: these include nanoparticles coated with hydrophilic polymeric envelope such as polyethylene glycol (PEG).
- Third generation: it involves attachment of ligand on the surface of a biodegradable core coated with a polymer.

This section of the chapter sheds light on some of the nanoparticulate systems (Tables 1.1 and 1.2) used for biomedical applications. The main advantages of the nanoparticulate system include the protection of drugs from degradation inside the body before reaching the destined location, increased circulation time, enhanced drug delivery, lowered dose, controlled and sustained release of drug, and reduced toxicity.

2.1 Inorganic Nanoparticles

The inorganic nanoparticles can be defined as particles of metallic composition exhibiting at least one dimension in the nanometer range [5]. These nanoparticles act as excellent drug-delivery vehicles, are biocompatible, provide enhanced bioavailability of drugs, promote targeted drug delivery, reduce the therapeutic dose, and are stable over a wide range of temperature and pH [6,7]. The inorganic nanoparticles can be synthesized using physical and chemical methods. The physical method involves the deposition of vapor and relies on the principle of breaking down the bulk precursor into smaller nanoparticles. The second method, chemical synthesis, requires the breakdown of the metal ion into atoms in the presence of a stabilizing agent followed by controlled aggregation of these atoms [7]. The latter is generally preferred over the physical mode of synthesis of nanoparticles. It is necessary to design inorganic nanoparticles with enhanced stability, increased circulation time without reticuloendothelial clearance, and neutrophil activation without compromising the therapeutic efficacies. As most of the inorganic nanoparticles have to be chemically or biologically modified before use for cellular delivery, concerns such as toxicity, carcinogenesis, and immunogenicity arising due to the use of these nanoparticles must be addressed [6,8,9].

TABLE 1.1
Properties and Various Applications of Selected Nanoparticles.

Sr. No.	Nanoparticle	Properties	Applications	References
1	Gold	Optical and photothermal properties due to surface plasmon resonance, thermal, electrical, antiangiogenic, catalytic, magnetic properties	Antibacterial and antiviral applications, biosensing, MRI, cancer diagnosis, and photothermal cancer therapy	[7,21–41]
2	Silver	Optical properties due to surface plasmon resonance (SPR), antiangiogenic, structural, thermal, electrical, and catalytic properties	Antibacterial and antifungal applications in water purification systems, paints and household products, antiviral applications against HIV-I and monkey pox virus, biosensing	[21,42–65]
3	Magnetic	Magnetic, caloric, physical, and hydrodynamic properties	Biomedicine, cancer treatment, MRI, drug delivery, removal of toxic metal ions and antimicrobial applications	[66–81]
4	Mesoporous silica-based	Porosity, controlled particle size, stable over a wide range of pH and temperature, chemical stability	Drug delivery, diagnostic catalysis, biosensing, and separation	[5]

TABLE 1.2
Drug-Delivery Application of Biodegradable Polymers: Synthetic and Natural.

SYNTHETIC BIODEGRADABLE POLYMERS		NATURAL BIODEGRADABLE POLYMERS	
Polyesters	Polyoxalates	Starch	Dextran
Polyorthoesters	Polyiminocarbonates	Hyaluronic acid	Chitosan
Polyanhydrides	Polyurethanes	Heparin	–
Polydioxanones	Polyphosphazenes	Gelatin	
Poly(a-cyanoacrylates)	–	Albumin	

Reprinted with permission from Coulembier O, Degée P, Hedrick JL, Dubois P. From controlled ring-opening polymerization to biodegradable aliphatic polyester: especially poly (β-malic acid) derivatives. Progress in Polymer Science. 2006;31(8):723–747.

2.1.1 Metal nanoparticles

Biomedical applications like the visualization of cellular components using electron microscopy, targeted and nontargeted delivery of the drug-loaded nanoparticles, detection, and diagnosis of various diseases involve the use of metal nanoparticles [10–12].

In a study, Mirkin et al. showed that gold nanoparticles conjugated with a specific oligonucleotide can detect (due to color change) complementary DNA strands [13]. In another study, Loo et al. found out that gold nanoparticles when tagged with antibodies, enzymes, or nucleotides had applications in several biosensor assays, drug- and gene-delivery systems, and biomedical-based imaging systems [14].

Silver is known for its antimicrobial activity since ancient times. In a study, Mahapatra and Karak showed the use of silver particles, in the nanometer range, on protheses, catheters, vascular grafts, and skin grafts due to their broad spectrum antimicrobial activity but was also found to be little toxic to mammalian cells [15].

2.1.2 Magnetic nanoparticles

These nanoparticles have gained popularity in recent years due to their special magnetic properties and ability to act as a contrast agent for magnetic resonance imaging [16] and as a drug carrier [17]. However, the safety of using these nanoparticles remains unexplored.

A study carried out by Lu et al. showed that the biocompatibility of carbon-coated iron nanoparticles is dependent on the surface chemistry of nanoparticle and on the cell type studied [18].

2.1.3 Mesoporous silica system

The mesoporous silica system has gained intensive attention around the world in the past decade because of its highly ordered structure, large pore size, worm-like interior, and huge surface area [19]. These properties make it an ideal candidate for the hosting and delivery of a variety of molecules such as drugs and protein [9].

As these microspheres have a size in the range of bacteria and can trigger immune response, to ensure effortless endocytosis by cells without any cell toxicity, the size of mesoporous silica nanoparticles can be modulated between 50 and 300 nm [9].

The mesoporous silica-based nanoparticle MCM-41 is one of the important synthesized materials [20]. Although biocompatibility is not so strong, it has been employed for drug delivery because of their bioactivity. Other mesoporous silica materials include SBA: SBA-15, SBA-16, SBA-1, SBA-3, HMS, and MSU [20].

2.2 Micelles

Micelles are core–shell structure formed by the self-assembly of amphiphiles in the aqueous environment when the concentration of these amphiphiles exceeds the critical micelles concentration [82]. Most of the micellar drug-delivery systems are composed of a hydrophilic outer corona that consists of PEG and an inner low molecular weight hydrophobic core-forming block. Such nanoparticulate systems have a size below 100 nm [83–85]. The hydrophobic interactions are the major driving force for the formation of such thermodynamically stable structures [3]. Moreover, these micellar systems possess reduced toxicity and have a polarity gradient (from the outer hydrophilic corona to the inner hydrophobic core) that helps in the solubilization of the poorly soluble hydrophobic compounds. This system is further classified into phospholipid, pluronic, poly(L-amino acid), and polyester micelles. Finally, the pharmacokinetics, biodistribution, and circulation time of drug, either in the core (nonpolar) or corona (polar molecule) or in between the core and shell (intermediate polarity), are increased compared to the free drug.

2.3 Carbon Nanotubes

Carbon nanotubes are self-assembling atoms of carbon arranged in the form of a tube [86]. These tubes have good electrical, mechanical, and surface properties making them an excellent candidate for drug delivery, biosensors, and biomedical devices. However, CNTs have solubility issues in most of the solvents and are cytotoxic without any functionalization [5]. Therefore, functionalization could be achieved by (1) additional reactions on the tips and the sidewalls or (2) oxidation followed by carboxyl-based coupling. Furthermore, functionalization improves the biocompatibility of the CNTs and makes them soluble [5]. Such functionalization also helps in the penetration of the CNTs across the cell membrane due to its hydrophilicity [5].

2.4 Liposomes

Liposomes are defined as bilayered vesicles composed of inner aqueous volume entirely surrounded by a lipid outer membrane [86]. These nanoparticulate systems can vary in size from few nanometers to several micrometers [87,88]. Furthermore, properties of the liposome have been found to depend on vesicle size, surface charge, lipid composition, and preparation method employed [88]. For instance, liposomal vesicles with a size greater than few nanometers have been investigated to play a role in Mononuclear Phagocyte System (MPS) clearance and complement activation and hence requires additional methods for the prevention of opsonization [89]. Therefore, surface modification using approaches such as coating with linear dextran, PEG, and polyvinylalcohol can help in protection from MPS uptake, increasing their circulation time and hence sustained drug release [90–92]. These vesicular systems are considered successful therapeutic drug carriers and have been used as carriers for many anticancer agents such as cisplatin [93] and paclitaxel [94] and antibiotics such as ciprofloxacin [95] and amikacin [96]. Furthermore, liposomes can be surface-modified with a ligand for achieving targeted drug delivery [97].

2.5 Polymer-Based Nanoparticles

These nanoparticles possess a core–shell structure and are made up of biodegradable polymers and copolymers to increase circulation time in the body and reduce MPS uptake [5,98]. For example, some of the well-studied nanoparticles are poly(glycolic acid), poly(lactic acid), poly(methylmethacrylate), and poly(lactic-co-glycolic acid) [98–100]. These are FDA approved and are biodegradable in nature. Furthermore, the corona consists of hydrophilic polymers, such as PEG and PVP and the core is a polymeric matrix for the encapsulation of hydrophobic drugs [5]. The drug molecule can be either physically adsorbed or chemically linked to the surface or entrapped inside the nanoparticle. Moreover, these polymeric nanoparticles have a size below 1000 nm [86].

3 CHARACTERISTICS OF NANOMATERIALS

3.1 Particle Size

The particle size can be determined using photon-correlation spectroscopy or dynamic light scattering. Photon-correlation spectroscopy measures the diameter by light scattering and using Brownian motion. Scanning electron microscopy and transmission electron microscopy are used to verify the results obtained using photon-correlation microscopy.

Particle size and size distribution determine the stability of nanoparticles, targeting ability, toxicity, in vivo fate, drug release profile, and loading of drugs. For example, smaller particles have a large surface area-to-volume ratio, implying the presence of drugs near or at the surface of nanoparticles allowing the faster release of drugs following zero-order kinetics [102]. Furthermore, the degradation of polymer also depends on the particle size, for example, poly lactic-co-glycolic acid (PLGA) [103].

3.2 Surface Properties

Nanomaterials, when administered through the intravenous route, can be recognized and cleared by the phagocytes [104]. In addition, hydrophobicity of the nanoparticles determines opsonization and clearance by MPS. Therefore, to enhance the lifetime of these nanomaterials in vivo and to minimize opsonization, surface modification of nanoparticles with hydrophilic polymers such as PEG, polyethylene oxide (PEO), and polysorbate 80 (Tween 80) is done. Furthermore, the zeta potential can be used to determine the charge on the surface of the nanoparticles [105]. For instance, a charge greater than ± 30 mV on nanoparticles provides stability and prevents aggregation.

3.3 Drug Loading

Drug loading can be done using two methods:

- Incorporation method: it involves the incorporation of drugs during the time of nanoparticle development.
- Adsorption methods: it requires the adsorption of drugs after nanoparticle development.

Loading of the drug depends on the solubility of the drug in the solid polymer or liquid dispersion agents (matrix) that in turn depends on the composition of matrix, drug–polymer interactions, and molecular weight [107]. It has been previously reported that active substances such as drugs and peptides show enhanced entrapment efficiency at or near their pI (isoelectric point) [107].

3.4 Drug Release

The drug release rate depends on the following:

- solubility of the drug
- desorption of surface-bound drug
- diffusion of drugs from the nanoparticle core
- matrix erosion
- Combination of erosion and diffusion processes.

Rate of diffusion or erosion determines the release of drugs in the case of nanospheres. For instance, if the rate of diffusion is faster than the erosion of the matrix, then the release mechanism is diffusion dependent. Moreover, the burst release of drug from nanoparticles is due to the weak interaction between the drug and the matrix [108]. For instance, if the surface of nanoparticle is coated with polymers such as PEO and PEG, then the release profile is diffusion dependent from the polymeric barrier due to the ionic interaction between the drug and the auxiliary ingredients.

4 TYPES OF TARGETING

4.1 Passive Targeting

To begin with, passive targeting is the basic targeting strategy employed by the smart nanoparticulate system [109]. The enhanced permeability and retention (EPR) effect or passive targeting refers to the preferential localization of nanomaterials in the pathological tissues described by leaky vasculature and poor lymphatic draining [110–113]. Nanoparticles such as polymers, liposomes, and micelles take advantage of the leaky tumor vasculature allowing entrapment and accumulation inside the tumors [113,114]. Moreover, a direct relationship between passive targeting and leaky vascularization has been documented in a different study [115]. Furthermore, prolonged circulation in the bloodstream can be achieved by conjugating PEG onto the surface of drug-loaded nanomaterials resulting in enhanced therapeutic efficiency of these nanomaterials compared to free drug molecules [109]. For example, Doxil (doxorubicin-loaded PEGylated liposome) has prolonged circulation half-life, avoids high plasma level, and has minimized reticuloendothelial system uptake and clearance [116].

4.2 Active Targeting

Active targeting, also referred as receptor-mediated targeting or ligand-based targeting, involves the use of a variety of targeting ligands such as peptides and antibodies to allow targeted delivery and uptake by the target cells expressing high levels of the specific biomarkers (e.g., cell surface receptors or cell adhesion molecules). These targeting moieties can be either physically or chemically conjugated on the surface of the

TABLE 1.3
Various Examples of Ligands, Targets, and Nanocarrier Systems Used for Active Drug Targeting.

Targeting Ligands	Targets	Examples of Nanomaterials	References
Folate	Folate receptor	Liposomes Albumin nanoparticles	[123–125]
Transferrin	Transferrin receptor	Liposomes Nanogels	[126–129]
Insulin	Insulin receptor	Magnetic nanoparticles	[130]
Antibodies and their fragments			
Anti-HER2 and its fragments	HER-2 receptor	Liposomes	[131]
OX26 (anti-transferrin)	Transferrin receptor	Liposomes	[132]
323/A3	Epithelial glycoproteins	Liposomes	[133]
Anti-Flk-1	Vascular endothelial growth factor receptor 2 (Flk-1)	Lipid nanoparticles	[134]
Anti-CD19	CD-19 epitope	Liposomes	[135]
Peptides			
Luteinizing hormone-releasing hormone	Luteinizing hormone-releasing hormone receptor aminopeptidase (CD13)	Polymeric nanoparticles	[136]
NGR peptide	Aminopeptidase N (CD13)	Liposomes	[137]
Gelatinase inhibitory peptide CTTHWGFTLC	MMP-2 and MMP-9 gelatinase	Liposomes Albumin nanoparticles	[138]

Adapted with permission from O.M. Koo et al., 2005.

nanomaterials [117,118]. Table 1.3 summarizes the various targeting ligands used to functionalize nanomaterials. However, there is still a lot of debate going on whether ligand functionalized nanomaterials have the ability to provide enhanced nanomaterial accumulation at the targeted site (e.g., malignant cells) compared to nontargeted nanomaterials [119]. For example, a similar level of accumulation in breast cancer xenograft models, overexpressing HER2 antigen, was observed for both HER2 monoclonal antibody fragments conjugated liposomes and nonfunctionalized liposomes [120]. Pharmacodynamic differences were suggested as the reason for the improved antitumor effect and enhanced drug delivery of the targeted nanoparticulate nanomedicine formulation in vivo [121]. So far, denileukin diftitox was the only FDA-approved nanoparticles in the market, which too was clinically discontinued in 2016, for the treatment of cancer using active targeting [121]. Reasons such as unspecificity of the targeting moiety, dose loss due to lysosomal digestion, and immune clearance are considered as key factors in blood clearance resulting in a lack of clinical application of such ligand functionalized nanoparticulate systems [122] (Fig. 1.2).

4.3 Stimuli-Responsive Targeting

A third type of targeting termed as stimuli-responsive targeting is currently receiving attention around the globe as it allows triggered drug release [114]. In this, the nanoparticulate system can undergo physicochemical changes in its structure resulting in release of drug at a particular location and time when subjected to endogenous stimuli such as low pH, presence of redox gradients or certain enzymes in tumor microenvironment, or exogenous stimuli such as light, heat, electric/magnetic fields, and ultrasound [114]. For example, Thermodox (thermosensitive liposomal doxorubicin) has temperature-sensitive lipids and polymers. Such combination allows the nanomaterial to remain stable at physiologic temperature and later undergo phase change increasing their permeability upon heating hence triggering the release of the drug

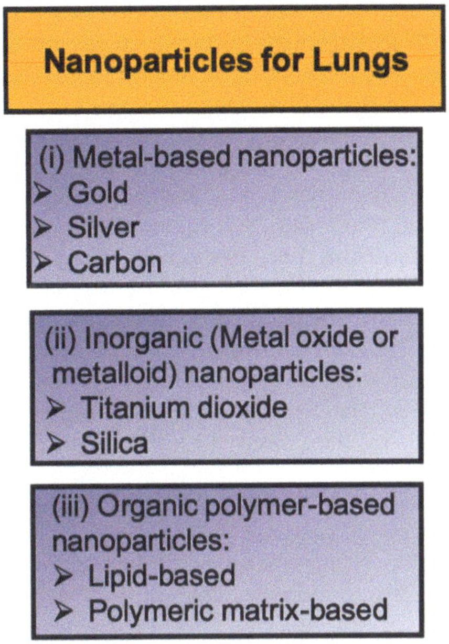

FIG. 1.2 Different types of nanoparticles for the improvement of therapy for pulmonary diseases.

[139]. Furthermore, nanomaterials that have pendant acidic or basic groups are termed "pH-sensitive" and can accept or lose protons depending on the surrounding pH and thus are believed to take advantage of the low pH of the tumor microenvironment (6.4–6.7) [140,141]. Such pH sensitivity of the nanomaterials is considered beneficial for the delivery of thermolabile drugs [114].

5 NANOMATERIALS IN MEDICINE: THE IDEAL SCALE

In the past few years, nanomedicine has gained the trust of scientists around the globe for the local delivery of drugs while addressing pharmacokinetic and pharmacodynamic drawbacks associated with the use of conventional free drug formulations [142,143]. Various nanoparticulate drug-delivery systems for enhancing therapeutic efficacy include liposomes, solid lipid nanoparticles, dendrimers, polymeric nanoparticles, micelles, and solid metal-containing nanoparticles. This section of the chapter describes the major fields of interest for drug-delivery applications using nanomaterials.

5.1 Pulmonology

Lung diseases such as lung carcinoma, asthma, and chronic obstructive pulmonary disease have a huge occurrence and are considered life threating. Nanoparticles such as liposomes, dendrimers, lipid- and polymer-based micelles, and polymeric nanoparticles are a choice for the improvement of therapy of such life-threatening diseases. For example, nanoparticulate systems such as polyamidoamine (PAMAM) dendrimers were evaluated for the delivery of poorly soluble antiasthma drug beclomethasone dipropionate [142]. In another study, inorganic nanoparticles, organic nanoparticles, and metallic nanoparticles were synthesized and tested for pulmonary immune hemostasis [143]. Furthermore, polymeric drug-delivery vehicles are of considerable interest to the scientist, as these can be surface-modified and copolymerized for application in the treatment of lung diseases. These include gelatin, chitosan, PLGA, PEG, and alginate [145].

5.2 Ophthalmology

Eye is a tiny intricate organ, and drug delivery following this route is of keen interest for the pharmaceutical scientists around the globe [142]. Conditions such as corneal disorder, for example, glaucoma and eye infection, for example, conjunctivitis requires ocular administration of the drug. Nanoparticulate systems such as liposomes and pharmacosomes have been observed to improve residence time, corneal permeability, and bioavailability of drug molecules [146,147]. For instance, PLGA-loaded pranoprofen nanoparticles, in the form of a hydrogel, for ophthalmic administration were evaluated. As a result, improved pharmacokinetic in comparison to the free drug was observed [148]. In a different study, curcumin-loaded polyethylene glycol-poly caprolactone (MePEG-PCL) nanoparticles were evaluated for ophthalmic administration of curcumin and were reported to show enhanced retention of drugs in the cornea with improved efficiency [149]. Furthermore, DC-TMCNS (diclofenac-loaded N-trimethyl chitosan nanoparticles) were evaluated for improved bioavailability of drugs [150].

5.3 Cardiovascular System

A huge number of people are affected by the cardiovascular disorder in developing countries, with reported morbidity close to 80% and is reported to occur almost equally in both men and women [151]. Nanoparticulate systems such as sirolimus-loaded chitosan-based liposomes were evaluated for restenosis treatment and were found to be effective [152]. In a different study, rat myocardial ischemic model was tested with vascular endothelial growth factor-loaded polymeric nanoparticles for the delivery of cytokines [153].

Moreover, CoQ10 (Coenzyme Q10) can be used in the treatment of myocardial ischemia owing to its role in the mitochondrial electron transport chain. However, its poor pharmaceutical properties limit its use. Therefore, CoQ10-loaded polymeric nanoparticles were developed to improve the poor pharmacokinetic properties of CoQ10 alone.

5.4 Oncology

Cancer is of the leading cause of death around the globe. Cancer patients often face the cytotoxic effect of chemotherapeutic drugs and therefore nanoparticulate drug-delivery systems can help in increasing the life span and quality of patient's life by minimizing systemic toxicity of the chemotherapeutics and by increasing the tumor specificity [154]. For instance, Doxil (doxorubicin-loaded liposome) is an FDA-approved nanoparticle for the treatment of ovarian cancer. The half-life of this nanoparticulate system is 100 times more than the free doxorubicin with overall reduced cardiotoxicity [155,156]. Furthermore, nab-Ptx (Abraxane) is an albumin-bound Ptx used in the treatment of lung, breast, and pancreatic cancer [157]. Nab-Ptx in combination with gemcitabine (Gem) was FDA-approved for pancreatic ductal adenocarcinoma treatment in 2013 [158]. It acts by targeting cancer cells in the G2/M stage of the cell cycle. In a study, Liu et al. developed a phospholipid-modified cationic PAMAM-siMDR1 complex for the treatment of human breast cancer cells [159]. This new complex was reported to show increased cellular uptake of siRNA targeting the MDR1 gene and hence enhanced MDR1 gene silencing [157]. In a different study, Li et al. showed that HSPC (hR3-siMDR1-PAMAM complex) could reverse MDR in adriamycin-resistant Michigan Cancer Foundation-7 cells (MCF/ADR) breast cancer cell line [160]. This complex showed reduced toxicity and improved endosomal escape with enhanced cellular uptake.

5.5 Brain—The Ultimate Target for Drug-Delivery Application

Specific targeting to the brain is challenging for drug-delivery applications, as free drugs including many small molecules cannot cross the brain–blood barrier. In a study, the passage of nanoparticle (about 200 nm) was reported [161]. In a different study carried out by Kreuter et al. in 2003, the same effect was not found for a different type of nanoparticle [162]. Furthermore, Lockman et al. in 2004 reported that nanoparticles with a neutral charge and low concentration of anionic nanoparticles were considered less toxic to blood–brain barrier integrity compared to cationic nanoparticles and high concentration of anionic nanoparticles [163]. Related to this, Kreuter et al. in 2004 showed that coating nanoparticle surfaces with a surfactant such as polysorbate (Tween) enhanced the penetration of drug-loaded nanoparticles across the blood–brain barrier [164]. Furthermore, binding of nanoparticle with lipoprotein receptors (apolipoprotein-E and low-density lipoprotein) on brain capillary endothelial cells was suggested as the reason for this uptake of drug by the brain [165]. In addition, PEG-PLA nanoparticles surface functionalized with lectin were developed for translocation to the brain via inhalation route [166]. In a different study, Gao et al. in 2006 showed twofold brain uptake of the nanoparticle. Wheat germ agglutinin, which binds to N-acetyl-D-glucosamine and sialic acid present in the nasal cavity, was used [167].

In summary, nanoparticles can be surface-modified with ligands to provide targeted drug delivery and reduced systemic toxicity [142]. These drug-delivery nanoforms can provide enhanced therapeutic efficacy of the drug. Advancement in the analytical techniques has led to the manufacturing of nanomaterials for the delivery of therapeutics for the treatment of pulmonary, ophthalmological, cardiovascular diseases, and cancer therapy and crossing blood–brain barrier. Therefore, the future of the nanoparticulate nanomedicine is propitious.

6 TOXICOLOGICAL ISSUES OF NANOMATERIALS

In the past few years, the special properties of nanomaterials are being utilized for designing the toxicological test for checking the hazardous or toxic effect of the engineered nanomaterials. Recent studies with cationic nanoparticles such as gold and polystyrene showed hemolysis and blood clotting, whereas anionic nanoparticles were observed to be comparatively less cytotoxic. Moreover, studies with carbon-based nanoparticles such as SWCNT and MWCNT showed platelet aggregation whereas no such aggregation was observed in the case of C60-fullerenes (a building block for these CNTs) [168]. In addition, reactive oxygen species generation, mitochondrial dysfunction, and lipid peroxidation were observed in keratinocytes and bronchial epithelium when cells were incubated with a high dose of single-walled carbon nanotubes [169,170]. In another study, neutral and low concentration of anionic nanoparticles was observed to cause less damage to the blood–brain barrier integrity compared to the cationic and high concentration of anionic nanoparticles. For instance, Net et al. in a study showed that nanoparticles induce oxidative stress by the production of reactive oxygen species in the brain that has been implicated in the pathogenesis of Alzheimer's and Parkinson's disease [171]. Furthermore, gold nanorods were found to be considerably cytotoxic, and this was attributed to the stabilizer Cetyltrimethylammonium Bromide (CTAB) used during its synthesis. For silica nanoparticles, increased dose and exposure time caused increased toxicity in vitro. It was observed to be due to the generation of reactive oxygen species and reduced glutathione levels resulting in the onset of oxidative stress inside the cell [172].

For the application of nanotechnology in medicine, knowledge of the safety and toxicological issues of the free drug, different nanoparticulate delivery systems, effect of rate of drug release at target and nontarget sites, and interaction of these nanomaterials with the biological components inside the body needs full attention [173]. For instance, increased complexity of nanomaterials such as surface coating and ligand attachment can affect biodistribution, bioavailability, and biocompatibility of nanomaterials following in vivo administration [174,175]. Moreover, comprehensive bioassays to assess absorption, distribution, metabolism, excretion, cytotoxicity, immunotoxicity, and genotoxicity before preclinical evaluations for toxicological responses to nanomaterials are essential [176]. For example, rats administered with Doxil showed decreased bacterial blood clearance due to macrophage suppression [177].

Some of the testing approaches (under development) for comparative and simultaneous risk potential analysis of many nanomaterials include computational modeling, high-throughput screening method, and high content screening [178]. In addition, special emphasis on the discrimination in toxicity of drug-loaded and "empty" nanoparticles should be taken into account, as the slow or nondegrading nanomaterials may persist and accumulate at the site causing chronic inflammatory reactions.

In summary, there is a tendency of nanosized (small) particles to be more toxic than the larger particles having the same chemical composition. Various studies focus on the reduction in the toxicity of the loaded drug but do not describe anything regarding the toxic effect of the carrier molecule itself. Costigan et al. in 2006 reviewed the various evidence of the potentially hazardous effect of nanomaterials in healthcare products. However, it is believed that not all the toxic effect for the use of nanomaterials is known at this moment [179–188].

7 CONCLUSION AND OUTLOOK

Recent advances in the nanoparticulate drug-delivery system are set to spread rapidly. For decades, nanoparticles are used for drug-delivery applications to reduce toxicity and side effects associated with the use of free drugs. In addition, these can help in overcoming various physiological barriers such as blood–brain barrier. Smart-drug delivery involves enhancing the therapeutic index while minimizing the cytotoxic effect using multifunctional nanomaterials. Moreover, the physicochemical properties of the nanoparticles such as size and structure can be modified according to the severity of disease for attaining maximal therapeutic benefits. For instance, FA-DABA-SMA polymers employ three levels of targeting different from older formulations that reply to the EPR effect [114]. Therefore, optimizing the drug release profile and employing different targeting strategies are of keen interest and importance in this drug-delivery field but the cost of production, manufacturing, regulatory standards, and technical difficulties must be addressed simultaneously. In addition, a key point that was overlooked for so many years was the risks of these carrier platforms to patients. Overall, a scientist should consider minimizing the complexity of the nanoparticulate system for generating clinically translatable nanosized therapeutics. So far, we have little knowledge about the basics of the interaction of these nanomaterials with the biological components such as living cells and organs inside the organism. Therefore, the collaboration between those working in particle toxicology and the development of various drug-delivery nanomaterials are necessary; then, we will see how nanotechnology will bring real added values in a cost-effective way.

REFERENCES

[1] Boisseau P, Loubaton B. Nanomedicine, nanotechnology in medicine. Physical Rendering 2011;12:620–36.

[2] Faraji AH, Wipf P. Nanoparticles in cellular drug delivery. Bioorganic and Medicinal Chemistry 2009; 17(8):2950–62.

[3] Rao CN, et al. Nanomaterials-An introduction. In: The chemistry of nanomaterials: synthesis, properties and applications. Wiley; 2006. p. 1–11.

[4] James HP, John R, Alex A, Anoop KR. Smart polymers for the controlled delivery of drugs–a concise overview. Acta Pharmaceutica Sinica B 2014;4(2):120–7.

[5] Safari J, Zarnegar Z. Advanced drug delivery systems: nanotechnology of health design A review. Journal of Saudi Chemical Society 2014;18(2):85–99.

[6] Xu ZP, Zeng QH, Lu GQ, Yu AB. Inorganic nanoparticles as carriers for efficient cellular delivery. Chemical Engineering Science 2006;61(3):1027–40.

[7] Kango S, Kalia S, Celli A, Njuguna J, Habibi Y, Kumar R. Surface modification of inorganic nanoparticles for development of organic–inorganic nanocomposites—a review. Progress in Polymer Science 2013;38(8): 1232–61.
[8] Cho K, Wang XU, Nie S, Shin DM. Therapeutic nanoparticles for drug delivery in cancer. Clinical Cancer Research 2008;14(5):1310–6.
[9] Ensign LM, Cone R, Hanes J. Oral drug delivery with polymeric nanoparticles: the gastrointestinal mucus barriers. Advanced Drug Delivery Reviews 2012;64(6): 557–70.
[10] Ranjan AP, Mukerjee A, Helson L, Vishwanatha JK. Scale up, optimization and stability analysis of Curcumin C3 complex-loaded nanoparticles for cancer therapy. Journal of Nanobiotechnology 2012;10(1):38.
[11] Sankari Z, Adeli H, Adeli A. Intrahemispheric, interhemispheric, and distal EEG coherence in Alzheimer's disease. Clinical Neurophysiology 2011;122(5):897–906.
[12] Alivisatos AP, Gu W, Larabell C. Quantum dots as cellular probes. Annual Review of Biomedical Engineering 2005;7:55–76.
[13] Rosi NL, Giljohann DA, Thaxton CS, Lytton-Jean AK, Han MS, Mirkin CA. Oligonucleotide-modified gold nanoparticles for intracellular gene regulation. Science 2006;312(5776):1027–30.
[14] Loo C, Lin A, Hirsch L, Lee MH, Barton J, Halas N, et al. Nanoshell-enabled photonics-based imaging and therapy of cancer. Technology in Cancer Research & Treatment 2004;3(1):33–40.
[15] Mahapatra SS, Karak N. Hyperbranched aromatic polyamines with s-triazine rings. Journal of Applied Polymer Science 2007;106(1):95–102.
[16] Idée JM, Port M, Raynal I, Schaefer M, Le Greneur S, Corot C. Clinical and biological consequences of transmetallation induced by contrast agents for magnetic resonance imaging: a review. Fundamental and Clinical Pharmacology 2006;20(6):563–76.
[17] Chen H, Khemtong C, Yang X, Chang X, Gao J. Nanonization strategies for poorly water-soluble drugs. Drug Discovery Today 2011;16(7–8):354–60.
[18] Lu AH, Salabas EE, Schüth F. Magnetic nanoparticles: synthesis, protection, functionalization, and application. Angewandte Chemie International Edition 2007;46(8):1222–44.
[19] Meek ST, Greathouse JA, Allendorf MD. Metal-organic frameworks: a rapidly growing class of versatile nanoporous materials. Advanced Materials 2011; 23(2):249–67.
[20] Kresge CT, Leonowicz ME, Roth WJ, Vartuli JC, Beck JS. Ordered mesoporous molecular sieves synthesized by a liquid-crystal template mechanism. Nature 1992; 359(6397):710.
[21] Frederix F, Friedt JM, Choi KH, Laureyn W, Campitelli A, Mondelaers D, et al. Biosensing based on light absorption of nanoscaled gold and silver particles. Analytical Chemistry 2003;75(24):6894–900.
[22] Frens G. Controlled nucleation for the regulation of the particle size in monodisperse gold suspensions. Nature Physical Science 1973;241(105):20.
[23] Sun Y, Mayers B, Xia Y. Metal nanostructures with hollow interiors. Advanced Materials 2003;15(7-8): 641–6.
[24] Kundu S, Panigrahi S, Praharaj S, Basu S, Ghosh SK, Pal A, Pal T. Anisotropic growth of gold clusters to gold nanocubes under UV irradiation. Nanotechnology 2007;18(7):075712.
[25] Kundu S, Peng L, Liang H. A new route to obtain high-yield multiple-shaped gold nanoparticles in aqueous solution using microwave irradiation. Inorganic Chemistry 2008;47(14):6344–52.
[26] Elghanian R, Storhoff JJ, Mucic RC, Letsinger RL, Mirkin CA. Selective colorimetric detection of polynucleotides based on the distance-dependent optical properties of gold nanoparticles. Science 1997;277(5329): 1078–81.
[27] Jain PK, Huang X, El-Sayed IH, El-Sayed MA. Noble metals on the nanoscale: optical and photothermal properties and some applications in imaging, sensing, biology, and medicine. Accounts of Chemical Research 2008;41(12):1578–86.
[28] Zhang Y, Schwartzberg AM, Xu K, Gu C, Zhang JZ. Electrical and thermal conductivities of gold and silver nanoparticles in solutions and films and electrical field enhanced Surface-Enhanced Raman Scattering (SERS). In: Physical chemistry of interfaces and nanomaterials IV, vol. 5929. International Society for Optics and Photonics; August 2005. p. 592912.
[29] Arvizo RR, Rana S, Miranda OR, Bhattacharya R, Rotello VM, Mukherjee P. Mechanism of anti-angiogenic property of gold nanoparticles: role of nanoparticle size and surface charge. Nanomedicine: Nanotechnology, Biology and Medicine 2011;7(5):580–7.
[30] Chen X, Zhao D, Zhao L, An Y, Ma R, Shi L, et al. Optic and catalytic properties of gold nanoparticles tuned by homopolymers. Science in China, Series B: Chemistry 2009;52(9):1372–81.
[31] Gréget R, Nealon GL, Vileno B, Turek P, Mény C, Ott F, et al. Magnetic properties of gold nanoparticles: a room-temperature quantum effect. ChemPhysChem 2012; 13(13):3092–7.
[32] Baffou G, Quidant R. Thermo-plasmonics: using metallic nanostructures as nano-sources of heat. Laser and Photonics Reviews 2013;7(2):171–87.
[33] Ortega MA, Rodriguez L, Castillo J, Piscitelli V, Fernandez A, Echevarria L. Thermo-optical properties of gold nanoparticles in colloidal systems. Journal of Optics A: Pure and Applied Optics 2008;10(10): 104024.
[34] Gil-Tomás J, Tubby S, Parkin IP, Narband N, Dekker L, Nair SP, Street C. Lethal photosensitisation of *Staphylococcus aureus* using a toluidine blue O–tiopronin–gold nanoparticle conjugate. Journal of Materials Chemistry 2007;17(35):3739–46.

[35] Perni S, Piccirillo C, Pratten J, Prokopovich P, Chrzanowski W, Parkin IP, Wilson M. The antimicrobial properties of light-activated polymers containing methylene blue and gold nanoparticles. Biomaterials 2009;30(1):89–93.
[36] Gu H, Ho PL, Tong E, Wang L, Xu B. Presenting vancomycin on nanoparticles to enhance antimicrobial activities. Nano Letters 2003;3(9):1261–3.
[37] Di Gianvincenzo P, Marradi M, Martínez-Ávila OM, Bedoya LM, Alcamí J, Penadés S. Gold nanoparticles capped with sulfate-ended ligands as anti-HIV agents. Bioorganic and Medicinal Chemistry Letters 2010;20(9):2718–21.
[38] Debouttière PJ, Roux S, Vocanson F, Billotey C, Beuf O, Favre-Réguillon A, et al. Design of gold nanoparticles for magnetic resonance imaging. Advanced Functional Materials 2006;16(18):2330–9.
[39] Huang X, Qian W, El-Sayed IH, El-Sayed MA. The potential use of the enhanced nonlinear properties of gold nanospheres in photothermal cancer therapy. Lasers in Surgery and Medicine: The Official Journal of the American Society for Laser Medicine and Surgery 2007;39(9):747–53.
[40] Huang X, El-Sayed IH, Qian W, El-Sayed MA. Cancer cell imaging and photothermal therapy in the near-infrared region by using gold nanorods. Journal of the American Chemical Society 2006;128(6):2115–20.
[41] Agarwal A, Huang SW, O'donnell M, Day KC, Day M, Kotov N, Ashkenazi S. Targeted gold nanorod contrast agent for prostate cancer detection by photoacoustic imaging. Journal of Applied Physics 2007;102(6):064701.
[42] Iravani S, Korbekandi H, Mirmohammadi SV, Zolfaghari B. Synthesis of silver nanoparticles: chemical, physical and biological methods. Research in Pharmaceutical Sciences 2014;9(6):385.
[43] Pingali KC, Rockstraw DA, Deng S. Silver nanoparticles from ultrasonic spray pyrolysis of aqueous silver nitrate. Aerosol Science and Technology 2005;39(10):1010–4.
[44] Lee I, Han SW, Kim K. Simultaneous preparation of SERS-active metal colloids and plates by laser ablation. Journal of Raman Spectroscopy 2001;32(11):947–52.
[45] Long D, Wu G, Chen S. Preparation of oligochitosan stabilized silver nanoparticles by gamma irradiation. Radiation Physics and Chemistry 2007;76(7):1126–31.
[46] Bönnemann H, Richards RM. Nanoscopic metal particles– synthetic methods and potential applications. European Journal of Inorganic Chemistry 2001;2001(10):2455–80.
[47] Mallick K, Witcomb MJ, Scurrell MS. Polymer stabilized silver nanoparticles: a photochemical synthesis route. Journal of Materials Science 2004;39(14):4459–63.
[48] Navaladian S, Viswanathan B, Viswanath RP, Varadarajan TK. Thermal decomposition as route for silver nanoparticles. Nanoscale Research Letters 2007;2(1):44.
[49] Rodriguez-Sanchez L, Blanco MC, Lopez-Quintela MA. Electrochemical synthesis of silver nanoparticles. The Journal of Physical Chemistry B 2000;104(41):9683–8.
[50] Haes AJ, Van Duyne RP. A nanoscale optical biosensor: sensitivity and selectivity of an approach based on the localized surface plasmon resonance spectroscopy of triangular silver nanoparticles. Journal of the American Chemical Society 2002;124(35):10596–604.
[51] Gurunathan S, Lee KJ, Kalishwaralal K, Sheikpranbabu S, Vaidyanathan R, Eom SH. Antiangiogenic properties of silver nanoparticles. Biomaterials 2009;30(31):6341–50.
[52] Khan MAM, Kumar S, Ahamed M, Alrokayan SA, AlSalhi MS. Structural and thermal studies of silver nanoparticles and electrical transport study of their thin films. Nanoscale Research Letters 2011;6(1):434.
[53] Chen D, Qiao X, Qiu X, Chen J. Synthesis and electrical properties of uniform silver nanoparticles for electronic applications. Journal of Materials Science 2009;44(4):1076–81.
[54] Xu R, Wang D, Zhang J, Li Y. Shape-dependent catalytic activity of silver nanoparticles for the oxidation of styrene. Chemistry–An Asian Journal 2006;1(6):888–93.
[55] Naderi S, Ghaderi A, Solaymani S, Golzan MM. Structural, optical and thermal properties of silver colloidal nanoparticles. The European Physical Journal-Applied Physics 2012;58(2).
[56] Feng QL, Wu J, Chen GQ, Cui FZ, Kim TN, Kim JO. A mechanistic study of the antibacterial effect of silver ions on *Escherichia coli* and *Staphylococcus aureus*. Journal of Biomedical Materials Research 2000;52(4):662–8.
[57] Smetana AB, Klabunde KJ, Marchin GR, Sorensen CM. Biocidal activity of nanocrystalline silver powders and particles. Langmuir 2008;24(14):7457–64.
[58] Nanda A, Saravanan M. Biosynthesis of silver nanoparticles from *Staphylococcus aureus* and its antimicrobial activity against MRSA and MRSE. Nanomedicine: Nanotechnology, Biology and Medicine 2009;5(4):452–6.
[59] Kim KJ, Sung WS, Suh BK, Moon SK, Choi JS, Kim JG, Lee DG. Antifungal activity and mode of action of silver nano-particles on Candida albicans. Biometals 2009;22(2):235–42.
[60] Baram-Pinto D, Shukla S, Perkas N, Gedanken A, Sarid R. Inhibition of herpes simplex virus type 1 infection by silver nanoparticles capped with mercaptoethane sulfonate. Bioconjugate Chemistry 2009;20(8):1497–502.
[61] Lara HH, Ayala-Nuñez NV, Ixtepan-Turrent L, Rodriguez-Padilla C. Mode of antiviral action of silver nanoparticles against HIV-1. Journal of Nanobiotechnology 2010;8(1):1.
[62] Lu L, Sun RW, Chen R, Hui CK, Ho CM, Luk JM, Che CM. Silver nanoparticles inhibit hepatitis B virus replication. Antiviral Therapy 2008;13(2):253.
[63] Sun L, Singh AK, Vig K, Pillai SR, Singh SR. Silver nanoparticles inhibit replication of respiratory syncytial virus. Journal of Biomedical Nanotechnology 2008;4(2):149–58.

[64] Elechiguerra JL, Burt JL, Morones JR, Camacho-Bragado A, Gao X, Lara HH, Yacaman MJ. Interaction of silver nanoparticles with HIV-1. Journal of Nanobiotechnology 2005;3(1):6.
[65] Rogers JV, Parkinson CV, Choi YW, Speshock JL, Hussain SM. A preliminary assessment of silver nanoparticle inhibition of monkeypox virus plaque formation. Nanoscale Research Letters 2008;3(4):129.
[66] Jolivet JP, Chanéac C, Tronc E. Iron oxide chemistry. From molecular clusters to extended solid networks. Chemical Communications 2004;(5):481–3.
[67] Wan S, Huang J, Yan H, Liu K. Size-controlled preparation of magnetite nanoparticles in the presence of graft copolymers. Journal of Materials Chemistry 2006;16(3):298–303.
[68] Albornoz C, Jacobo SE. Preparation of a biocompatible magnetic film from an aqueous ferrofluid. Journal of Magnetism and Magnetic Materials 2006;305(1):12–5.
[69] Hou Y, Yu J, Gao S. Solvothermal reduction synthesis and characterization of superparamagnetic magnetite nanoparticles. Journal of Materials Chemistry 2003;13(8):1983–7.
[70] Pascal C, Pascal JL, Favier F, Elidrissi Moubtassim ML, Payen C. Electrochemical synthesis for the control of γ-Fe2O3 nanoparticle size. Morphology, microstructure, and magnetic behavior. Chemistry of Materials 1999;11(1):141–7.
[71] Franzel L, Bertino MF, Huba ZJ, Carpenter EE. Synthesis of magnetic nanoparticles by pulsed laser ablation. Applied Surface Science 2012;261:332–6.
[72] Vijayakumar R, Koltypin Y, Felner I, Gedanken A. Sonochemical synthesis and characterization of pure nanometer-sized Fe_3O_4 particles. Materials Science and Engineering: A 2000;286(1):101–5.
[73] Bandyopadhyay M, Bhattacharya J. Magnetic and caloric properties of magnetic nanoparticles: an equilibrium study. Journal of Physics: Condensed Matter 2006;18(49):11309.
[74] Akbarzadeh A, Samiei M, Davaran S. Magnetic nanoparticles: preparation, physical properties, and applications in biomedicine. Nanoscale Research Letters 2012;7(1):144.
[75] Martchenko I, Dietsch H, Moitzi C, Schurtenberger P. Hydrodynamic properties of magnetic nanoparticles with tunable shape anisotropy: prediction and experimental verification. The Journal of Physical Chemistry B 2011;115(49):14838–45.
[76] Roca AG, Costo R, Rebolledo AF, Veintemillas-Verdaguer S, Tartaj P, Gonzalez-Carreno T, et al. Progress in the preparation of magnetic nanoparticles for applications in biomedicine. Journal of Physics D: Applied Physics 2009;42(22):224002.
[77] Chan DC, Kirpotin DB, Bunn Jr PA. Synthesis and evaluation of colloidal magnetic iron oxides for the site-specific radiofrequency-induced hyperthermia of cancer. Journal of Magnetism and Magnetic Materials 1993;122(1–3):374–8.
[78] Babes L, Denizot B, Tanguy G, Le Jeune JJ, Jallet P. Synthesis of iron oxide nanoparticles used as MRI contrast agents: a parametric study. Journal of Colloid and Interface Science 1999;212(2):474–82.
[79] Chertok B, Moffat BA, David AE, Yu F, Bergemann C, Ross BD, Yang VC. Iron oxide nanoparticles as a drug delivery vehicle for MRI monitored magnetic targeting of brain tumors. Biomaterials 2008;29(4):487–96.
[80] Khodabakhshi A, Amin MM, Mozaffari M. Synthesis of magnetite nanoparticles and evaluation of its efficiency for arsenic removal from simulated industrial wastewater. Iranian Journal of Environmental Health Science and Engineering 2011;8(3):189–200.
[81] Tran N, Mir A, Mallik D, Sinha A, Nayar S, Webster TJ. Bactericidal effect of iron oxide nanoparticles on *Staphylococcus aureus*. International Journal of Nanomedicine 2010;5:277.
[82] Ball V, Maechling C. Isothermal microcalorimetry to investigate non specific interactions in biophysical chemistry. International Journal of Molecular Sciences 2009;10(8):3283–315.
[83] Kwon GS. Polymeric micelles for delivery of poorly water-soluble compounds. Critical Reviews in Therapeutic Drug Carrier Systems 2003;20(5).
[84] Torchilin VP. PEG-based micelles as carriers of contrast agents for different imaging modalities. Advanced Drug Delivery Reviews 2002;54(2):235–52.
[85] Ashok B, Arleth L, Hjelm RP, Rubinstein I, Önyüksel H. In vitro characterization of PEGylated phospholipid micelles for improved drug solubilization: effects of PEG chain length and PC incorporation. Journal of Pharmaceutical Sciences 2004;93(10):2476–87.
[86] Ahlawat J, Henriquez G, Narayan M. Enhancing the delivery of chemotherapeutics: role of biodegradable polymeric nanoparticles. Molecules 2018;23(9):2157.
[87] New RR, editor. Liposomes. IRL at Oxford University Press; 1990.
[88] Sethi V, Önyüksel H, Rubinstein I. Liposomal vasoactive intestinal peptide. In: Methods in enzymology, vol. 391. Academic Press; 2005. p. 377–95.
[89] Allen TM, Chonn A. Large unilamellar liposomes with low uptake into the reticuloendothelial system. FEBS Letters 1987;223(1):42–6.
[90] Papahadjopoulos D, Allen TM, Gabizon A, Mayhew E, Matthay K, Huang SK, et al. Sterically stabilized liposomes: improvements in pharmacokinetics and antitumor therapeutic efficacy. Proceedings of the National Academy of Sciences 1991;88(24):11460–4.
[91] Lasic DD, Martin FJ, Gabizon A, Huang SK, Papahadjopoulos D. Sterically stabilized liposomes: a hypothesis on the molecular origin of the extended circulation times. Biochimica et Biophysica Acta (BBA)-Biomembranes 1991;1070(1):187–92.
[92] Takeuchi H, Kojima H, Yamamoto H, Kawashima Y. Evaluation of circulation profiles of liposomes coated with hydrophilic polymers having different molecular weights in rats. Journal of Controlled Release 2001;75(1–2):83–91.

[93] Zamboni WC, Gervais AC, Egorin MJ, Schellens JH, Zuhowski EG, Pluim D, et al. Systemic and tumor disposition of platinum after administration of cisplatin or STEALTH liposomal-cisplatin formulations (SPI-077 and SPI-077 B103) in a preclinical tumor model of melanoma. Cancer Chemotherapy and Pharmacology 2004;53(4):329–36.
[94] Zhang JA, Anyarambhatla G, Ma L, Ugwu S, Xuan T, Sardone T, Ahmad I. Development and characterization of a novel Cremophor® EL free liposome-based paclitaxel (LEP-ETU) formulation. European Journal of Pharmaceutics and Biopharmaceutics 2005;59(1):177–87.
[95] Kadry AA, Al-Suwayeh SA, Abd-Allah AR, Bayomi MA. Treatment of experimental osteomyelitis by liposomal antibiotics. Journal of Antimicrobial Chemotherapy 2004;54(6):1103–8.
[96] Donald PR, Sirgel FA, Venter A, Smit E, Parkin DP, Van de Wal BW, Mitchison DA. The early bactericidal activity of a low-clearance liposomal amikacin in pulmonary tuberculosis. Journal of Antimicrobial Chemotherapy 2001;48(6):877–80.
[97] Dagar S, Krishnadas A, Rubinstein I, Blend MJ, Önyüksel H. VIP grafted sterically stabilized liposomes for targeted imaging of breast cancer: in vivo studies. Journal of Controlled Release 2003;91(1–2):123–33.
[98] Gessner A, Waicz R, Lieske A, Paulke BR, Mäder K, Müller RH. Nanoparticles with decreasing surface hydrophobicities: influence on plasma protein adsorption. International Journal of Pharmaceutics 2000;196(2):245–9.
[99] Zweers ML, Engbers GH, Grijpma DW, Feijen J. In vitro degradation of nanoparticles prepared from polymers based on DL-lactide, glycolide and poly (ethylene oxide). Journal of Controlled Release 2004;100(3):347–56.
[100] Radwan MA, Zaghloul IY, Aly ZH. In vivo performance of parenteral theophylline-loaded polyisobutylcyanoacrylate nanoparticles in rats. European Journal of Pharmaceutical Sciences 1999;8(2):95–8.
[101] Deleted in review
[102] Redhead HM, Davis SS, Illum L. Drug delivery in poly (lactide-co-glycolide) nanoparticles surface modified with poloxamer 407 and poloxamine 908: in vitro characterisation and in vivo evaluation. Journal of Controlled Release 2001;70(3):353–63.
[103] Dunne M, Corrigan OI, Ramtoola Z. Influence of particle size and dissolution conditions on the degradation properties of polylactide-co-glycolide particles. Biomaterials 2000;21(16):1659–68.
[104] Müller R, Maaben S, Weyhers H, Mehnert W. Phagocytic uptake and cytotoxicity of solid lipid nanoparticles (SLN) sterically stabilized with poloxamine 908 and poloxamer 407. Journal of Drug Targeting 1996;4(3):161–70.
[105] Couvreur P, Barratt G, Fattal E, Vauthier C. Nanocapsule technology: a review. Critical Reviews in Therapeutic Drug Carrier Systems 2002;19(2).
[106] Deleted in review
[107] Calvo P, Remunan-Lopez C, Vila-Jato JL, Alonso MJ. Novel hydrophilic chitosan-polyethylene oxide nanoparticles as protein carriers. Journal of Applied Polymer Science 1997;63(1):125–32.
[108] Magenheim B, Levy MY, Benita S. A new in vitro technique for the evaluation of drug release profile from colloidal carriers-ultrafiltration technique at low pressure. International Journal of Pharmaceutics 1993;94(1–3):115–23.
[109] Matsumura Y, Maeda H. A new concept for macromolecular therapeutics in cancer chemotherapy: mechanism of tumoritropic accumulation of proteins and the antitumor agent smancs. Cancer Research 1986;46(12 Part 1):6387–92.
[110] Bae YH. Drug targeting and tumor heterogeneity. Journal of Controlled Release: Official Journal of the Controlled Release Society 2009;133(1):2.
[111] Bae YH, Park K. Targeted drug delivery to tumors: myths, reality and possibility. Journal of Controlled Release 2011;153(3):198.
[112] Jain RK, Stylianopoulos T. Delivering nanomedicine to solid tumors. Nature Reviews Clinical Oncology 2010;7(11):653.
[113] Wang M, Thanou M. Targeting nanoparticles to cancer. Pharmacological Research 2010;62(2):90–9.
[114] Kalaydina RV, Bajwa K, Qorri B, Decarlo A, Szewczuk MR. Recent advances in "smart" delivery systems for extended drug release in cancer therapy. International Journal of Nanomedicine 2018;13:4727.
[115] Theek B, Gremse F, Kunjachan S, Fokong S, Pola R, Pechar M, et al. Characterizing EPR-mediated passive drug targeting using contrast-enhanced functional ultrasound imaging. Journal of Controlled Release 2014;182:83–9.
[116] Nichols JW, Bae YH. EPR: evidence and fallacy. Journal of Controlled Release 2014;190:451–64.
[117] Li X, Szewczuk MR, Malardier-Jugroot C. Folic acid-conjugated amphiphilic alternating copolymer as a new active tumor targeting drug delivery platform. Drug Design, Development and Therapy 2016;10:4101.
[118] Byrne JD, Betancourt T, Brannon-Peppas L. Active targeting schemes for nanoparticle systems in cancer therapeutics. Advanced Drug Delivery Reviews 2008;60(15):1615–26.
[119] Ferrari M. Nanovector therapeutics. Current Opinion in Chemical Biology 2005;9(4):343–6.
[120] Kirpotin D, Park JW, Hong K, Zalipsky S, Li WL, Carter P, et al. Sterically stabilized anti-HER2 immunoliposomes: design and targeting to human breast cancer cells in vitro. Biochemistry 1997;36(1):66–75.
[121] Kirpotin DB, Drummond DC, Shao Y, Shalaby MR, Hong K, Nielsen UB, et al. Antibody targeting of long-circulating lipidic nanoparticles does not increase tumor localization but does increase internalization in animal models. Cancer Research 2006;66(13):6732–40.
[122] Chen WC, Zhang AX, Li SD. Limitations and niches of the active targeting approach for nanoparticle drug

delivery. European Journal of Nanomedicine 2012; 4(2–4):89–93.
[123] Ni S, Stephenson SM, Lee RJ. Folate receptor targeted delivery of liposomal daunorubicin into tumor cells. Anticancer Research 2002;22(4):2131–5.
[124] Koo OM, Rubinstein I, Onyuksel H. Role of nanotechnology in targeted drug delivery and imaging: a concise review. Nanomedicine: Nanotechnology, Biology and Medicine 2005;1(3):193–212.
[125] Zhang L, Hou S, Mao S, Wei D, Song X, Lu Y. Uptake of folate-conjugated albumin nanoparticles to the SKOV3 cells. International Journal of Pharmaceutics 2004; 287(1–2):155–62.
[126] Maruyama K, Ishida O, Kasaoka S, Takizawa T, Utoguchi N, Shinohara A, et al. Intracellular targeting of sodium mercaptoundecahydrododecaborate (BSH) to solid tumors by transferrin-PEG liposomes, for boron neutron-capture therapy (BNCT). Journal of Controlled Release 2004;98(2):195–207.
[127] Omori N, Maruyama K, Jin G, Wang SJ, Hamakawa Y, Sato K, et al. Targeting of post-ischemic cerebral endothelium in rat by liposomes bearing polyethylene glycol-coupled transferrin. Neurological Research 2003;25(3):275–9.
[128] Ishida O, Maruyama K, Tanahashi H, Iwatsuru M, Sasaki K, Eriguchi M, Yanagie H. Liposomes bearing polyethyleneglycol-coupled transferrin with intracellular targeting property to the solid tumors in vivo. Pharmaceutical Research 2001;18(7):1042–8.
[129] Vinogradov SV, Batrakova EV, Kabanov AV. Nanogels for oligonucleotide delivery to the brain. Bioconjugate Chemistry 2004;15(1):50–60.
[130] Gupta AK, Berry C, Gupta M, Curtis A. Receptor-mediated targeting of magnetic nanoparticles using insulin as a surface ligand to prevent endocytosis. IEEE Transactions on Nanobioscience 2003;2(4):255–61.
[131] Park JW, Hong K, Kirpotin DB, Colbern G, Shalaby R, Baselga J, et al. Anti-HER2 immunoliposomes: enhanced efficacy attributable to targeted delivery. Clinical Cancer Research 2002;8(4):1172–81.
[132] Gosk S, Vermehren C, Storm G, Moos T. Targeting anti—transferrin receptor antibody (OX26) and OX26-conjugated liposomes to brain capillary endothelial cells using in situ perfusion. Journal of Cerebral Blood Flow and Metabolism 2004;24(11):1193–204.
[133] Fonseca MJ, Jagtenberg JC, Haisma HJ, Storm G. Liposome-mediated targeting of enzymes to cancer cells for site-specific activation of prodrugs: comparison with the corresponding antibody-enzyme conjugate. Pharmaceutical Research 2003;20(3):423–8.
[134] Li L, Wartchow CA, Danthi SN, Shen Z, Dechene N, Pease J, et al. A novel antiangiogenesis therapy using an integrin antagonist or anti–Flk-1 antibody coated 90Y-labeled nanoparticles. International Journal of Radiation Oncology Biology Physics 2004;58(4):1215–27.
[135] Sapra P, Allen TM. Ligand-targeted liposomal anticancer drugs. Progress in Lipid Research 2003;42(5): 439–62.
[136] Dharap SS, Qiu B, Williams GC, Sinko P, Stein S, Minko T. Molecular targeting of drug delivery systems to ovarian cancer by BH3 and LHRH peptides. Journal of Controlled Release 2003;91(1–2):61–73.
[137] Pastorino F, Brignole C, Marimpietri D, Cilli M, Gambini C, Ribatti D, et al. Vascular damage and anti-angiogenic effects of tumor vessel-targeted liposomal chemotherapy. Cancer Research 2003;63(21): 7400–9.
[138] Medina OP, Kairemo K, Valtanen H, Kangasniemi A, Kaukinen S, Ahonen I, et al. Radionuclide imaging of tumor xenografts in mice using a gelatinase-targeting peptide. Anticancer Research 2005;25(1A):33–42.
[139] Kono K, Ozawa T, Yoshida T, Ozaki F, Ishizaka Y, Maruyama K, et al. Highly temperature-sensitive liposomes based on a thermosensitive block copolymer for tumor-specific chemotherapy. Biomaterials 2010; 31(27):7096–105.
[140] Adochite RC, Moshnikova A, Golijanin J, Andreev OA, Katenka NV, Reshetnyak YK. Comparative study of tumor targeting and biodistribution of pH (Low) Insertion Peptides (pHLIP® peptides) conjugated with different fluorescent dyes. Molecular Imaging and Biology 2016;18(5):686–96.
[141] Weerakkody D, Moshnikova A, El-Sayed NS, Adochite RC, Slaybaugh G, Golijanin J, et al. Novel pH-sensitive cyclic peptides. Scientific Reports 2016;6: 31322.
[142] Mir M, Ishtiaq S, Rabia S, Khatoon M, Zeb A, Khan GM, et al. Nanotechnology: from in vivo imaging system to controlled drug delivery. Nanoscale Research Letters 2017;12(1):500.
[143] Blanco E, Shen H, Ferrari M. Principles of nanoparticle design for overcoming biological barriers to drug delivery. Nature Biotechnology 2015;33(9):941.
[144] Deleted in review
[145] Sung JC, Pulliam BL, Edwards DA. Nanoparticles for drug delivery to the lungs. Trends in Biotechnology 2007;25(12):563–70.
[146] Wadhwa S, Paliwal R, Paliwal SR, Vyas SP. Nanocarriers in ocular drug delivery: an update review. Current Pharmaceutical Design 2009;15(23):2724–50.
[147] Puglia C, Offerta A, Carbone C, Bonina F, Pignatello R, Puglisi G. Lipid nanocarriers (LNC) and their applications in ocular drug delivery. Current Medicinal Chemistry 2015;22(13):1589–602.
[148] Andrei G, Peptu CA, Popa M, Desbrieres J, Peptu C, Gardikiotis F, et al. Formulation and evaluation of cefuroxim loaded submicron particles for ophthalmic delivery. International Journal of Pharmaceutics 2015; 493(1–2):16–29.
[149] Pradhan N, Guha R, Chowdhury S, Nandi S, Konar A, Hazra S. Curcumin nanoparticles inhibit corneal neovascularization. Journal of Molecular Medicine 2015;93(10):1095–106.
[150] Asasutjarit R, Theerachayanan T, Kewsuwan P, Veeranodha S, Fuongfuchat A, Ritthidej GC. Development and evaluation of diclofenac sodium loaded-N-

Trimethyl chitosan nanoparticles for ophthalmic use. AAPS PharmSciTech 2015;16(5):1013–24.
[151] Singh B, Garg T, Goyal AK, Rath G. Recent advancements in the cardiovascular drug carriers. Artificial Cells, Nanomedicine, and Biotechnology 2016;44(1): 216–25.
[152] Haeri A, Sadeghian S, Rabbani S, Anvari MS, Ghassemi S, Radfar F, Dadashzadeh S. Effective attenuation of vascular restenosis following local delivery of chitosan decorated sirolimus liposomes. Carbohydrate Polymers 2017;157:1461–9.
[153] Formiga FR, Pelacho B, Garbayo E, Abizanda G, Gavira JJ, Simon-Yarza T, et al. Sustained release of VEGF through PLGA microparticles improves vasculogenesis and tissue remodeling in an acute myocardial ischemia–reperfusion model. Journal of Controlled Release 2010;147(1):30–7.
[154] Danhier F, Feron O, Préat V. To exploit the tumor microenvironment: passive and active tumor targeting of nanocarriers for anti-cancer drug delivery. Journal of Controlled Release 2010;148(2):135–46.
[155] Rahman AM, Yusuf SW, Ewer MS. Anthracycline-induced cardiotoxicity and the cardiac-sparing effect of liposomal formulation. International Journal of Nanomedicine 2007;2(4):567.
[156] Batist G. Cardiac safety of liposomal anthracyclines. Cardiovascular Toxicology 2007;7(2):72–4.
[157] Kundranda MN, Niu J. Albumin-bound paclitaxel in solid tumors: clinical development and future directions. Drug Design, Development and Therapy 2015;9:3767.
[158] Bhaw-Luximon A, Jhurry D. New avenues for improving pancreatic ductal adenocarcinoma (PDAC) treatment: selective stroma depletion combined with nano drug delivery. Cancer Letters 2015;369(2):266–73.
[159] Zhou L, Wang H, Li Y. Stimuli-responsive nanomedicines for overcoming cancer multidrug resistance. Theranostics 2018;8(4):1059.
[160] Li J, Liu J, Guo N, Zhang X. Reversal of multidrug resistance in breast cancer MCF-7/ADR cells by h-R3-siMDR1-PAMAM complexes. International Journal of Pharmaceutics 2016;511(1):436–45.
[161] Olivier JC, Fenart L, Chauvet R, Pariat C, Cecchelli R, Couet W. Indirect evidence that drug brain targeting using polysorbate 80-coated polybutylcyanoacrylate nanoparticles is related to toxicity. Pharmaceutical Research 1999;16(12):1836–42.
[162] Kreuter J, Ramge P, Petrov V, Hamm S, Gelperina SE, Engelhardt B, et al. Direct evidence that polysorbate-80-coated poly (butylcyanoacrylate) nanoparticles deliver drugs to the CNS via specific mechanisms requiring prior binding of drug to the nanoparticles. Pharmaceutical Research 2003;20(3):409–16.
[163] Lockman PR, Koziara JM, Mumper RJ, Allen DD. Nanoparticle surface charges alter blood–brain barrier integrity and permeability. Journal of Drug Targeting 2004; 12(9–10):635–41.
[164] Kreuter J. Influence of the surface properties on nanoparticle-mediated transport of drugs to the brain. Journal of Nanoscience and Nanotechnology 2004; 4(5):484–8.
[165] Michaelis K, Hoffmann MM, Dreis S, Herbert E, Alyautdin RN, Michaelis M, et al. Covalent linkage of apolipoprotein e to albumin nanoparticles strongly enhances drug transport into the brain. Journal of Pharmacology and Experimental Therapeutics 2006;317(3): 1246–53.
[166] De Jong WH, Borm PJ. Drug delivery and nanoparticles: applications and hazards. International Journal of Nanomedicine 2008;3(2):133.
[167] Gao X, Tao W, Lu W, Zhang Q, Zhang Y, Jiang X, Fu S. Lectin-conjugated PEG–PLA nanoparticles: preparation and brain delivery after intranasal administration. Biomaterials 2006;27(18):3482–90.
[168] Radomski A, Jurasz P, Alonso-Escolano D, Drews M, Morandi M, Malinski T, Radomski MW. Nanoparticle-induced platelet aggregation and vascular thrombosis. British Journal of Pharmacology 2005;146(6):882–93.
[169] Sayes CM, Liang F, Hudson JL, Mendez J, Guo W, Beach JM, et al. Functionalization density dependence of single-walled carbon nanotubes cytotoxicity in vitro. Toxicology Letters 2006;161(2):135–42.
[170] Shvedova A, Castranova V, Kisin E, Schwegler-Berry D, Murray A, Gandelsman V, et al. Exposure to carbon nanotube material: assessment of nanotube cytotoxicity using human keratinocyte cells. Journal of Toxicology and Environmental Health Part A 2003;66(20): 1909–26.
[171] Nel A, Xia T, Mädler L, Li N. Toxic potential of materials at the nanolevel. Science 2006;311(5761):622–7.
[172] Lin W, Huang YW, Zhou XD, Ma Y. In vitro toxicity of silica nanoparticles in human lung cancer cells. Toxicology and Applied Pharmacology 2006;217(3):252–9.
[173] Hare JI, Lammers T, Ashford MB, Puri S, Storm G, Barry ST. Challenges and strategies in anti-cancer nanomedicine development: an industry perspective. Advanced Drug Delivery Reviews 2017;108:25–38.
[174] Allen TM, Cullis PR. Drug delivery systems: entering the mainstream. Science 2004;303(5665):1818–22.
[175] Tinkle S, McNeil SE, Mühlebach S, Bawa R, Borchard G, Barenholz Y, et al. Nanomedicines: addressing the scientific and regulatory gap. Annals of the New York Academy of Sciences 2014;1313(1):35–56.
[176] Gaspar R. Regulatory issues surrounding nanomedicines: setting the scene for the next generation of nanopharmaceuticals. Nanomedicine 2007;2(2):143–7.
[177] Storm G, Ten Kate MT, Working PK, Bakker-Woudenberg IA. Doxorubicin entrapped in sterically stabilized liposomes: effects on bacterial blood clearance capacity of the mononuclear phagocyte system. Clinical Cancer Research 1998;4(1):111–5.
[178] Oomen AG, Bos PM, Fernandes TF, Hund-Rinke K, Boraschi D, Byrne HJ, et al. Concern-driven integrated approaches to nanomaterial testing and assessment–

report of the NanoSafety Cluster Working Group 10. Nanotoxicology 2014;8(3):334–48.
[179] Costigan S. The toxicology of nanoparticles used in health care products. Available at: the website of the Medicines and Healthcare products Regulatory Agency, Department of Health, UK. Accessed, 20. 2006.
[180] Sahai N, Ahmad N, Gogoi M. Nanoparticles based drug delivery for tissue regeneration using biodegradable scaffolds: a review. Current Pathobiology Reports 2018;6(4):219–24.
[181] Khatoon A, Khan F, Ahmad N, Shaikh S, Rizvi SMD, Shakil S, et al. Silver nanoparticles from leaf extract of Mentha piperita: eco-friendly synthesis and effect on acetylcholinesterase activity. Life Sciences 2018;209: 430–4.
[182] Ahmad N, Bhatnagar S, Saxena R, Iqbal D, Ghosh AK, Dutta R. Biosynthesis and characterization of gold nanoparticles: kinetics, in vitro and in vivo study. Materials Science and Engineering: C 2017;78:553–64.
[183] Ahmad N, Rizvi S, Sahai N, Dutta R. Biosynthesis, characterization of gold nanoparticles using M. indica leaf extract and their anticancer activity. International Journal of Nanotechnology 2016;2(2):1–4.
[184] Ahmad N, Bhatnagar S, Ali SS, Dutta R. Phytofabrication of bioinduced silver nanoparticles for biomedical applications. International Journal of Nanomedicine 2015;10:7019.
[185] Ahmad N, Shree K, Srivastava M, Dutta R. Novel rapid biological approach for synthesis of silver nanoparticles and its characterization. International Journal of Pharmacology 2014;1(1):28–31.
[186] Ahmad N, Bhatnagar S, Dubey SD, Saxena R, Sharma S, Dutta R. Nanopackaging in food and electronics. In: Nanoscience in food and agriculture, vol. 4. Cham: Springer; 2017. p. 45–97.
[187] Ahlawat J, Kumar V, Gopinath P. Carica papaya loaded poly (vinyl alcohol)-gelatin nanofibrous scaffold for potential application in wound dressing. Materials Science and Engineering: C 2019:109834.
[188] Ahlawat J, Deemer EM, Narayan M. Chitosan nanoparticles rescue rotenone-mediated cell death. Materials 2019;12(7):1176.

CHAPTER 2

pH-Sensitive Nanomaterials for Smart Release of Drugs

MUKESH KUMAR MEHER, MTECH • KRISHNA MOHAN POLURI, PHD

1 INTRODUCTION

Over the decade, pH-sensitive nanomaterials have continuously gained attention and developed for biomedical applications as systemic smart drug delivery. Nanomaterials are usually in a size range of 1−100 nm, which are similar to the size range of biomolecules/biomolecular complexes and are smaller than the size of a cell. The architectures of nanomaterials used in biomedical applications are comparable to those of biomolecular structures. Henceforth, nanomaterials were encapsulated with bio- or biocompatible polymers to make them bioactive and biodegradable materials for the improvement of smart drug delivery and targeting, medical devices, diagnosis, and therapeutics [1−4]. Due to the greater surface-to-volume ratio, nanomaterials are having higher surface functionalization capacity to be used as potential therapeutic materials. For medical applications, several types of nanomedicines have been formulated; they include polymeric nanoparticles, lipid nanoparticles, metal nanoparticles, and various organic−inorganic hybrid nanoparticles. The high stability and greater drug-loading capacity of nanoparticles are subjected to longer body circulation and overcoming highly dense cellular barriers like blood−brain barrier [5,6]. Thus, nanomaterials are more likely to develop a potential drug carrier for smart drug release. Compared to the conventional drug-delivery system, the nanosystem offers various advantages including easily tunable properties, high stability, target specificity, enhanced cellular penetrations, and capability to deliver both hydrophobic and hydrophilic therapeutic agents [7,8]

The smart drug-release (SDR) process can be achieved by the surface functionalization of unique nanostructures. This SDR process basically required a mechanism to sense the targeted microenvironment by specific stimuli such as pH (acidic/basic), redox potential, light, ultrasound, temperature, analytes (glucose, glutathione, etc.), and proteins/enzymes concentrations [9,10]. Furthermore, the nanomaterials can be decorated with signaling molecules to achieve ligand-receptor binding over the cells [11,12]. Among the all stimuli response, pH is one of the stimuli that has been extensively explored for smart and sustainable drug-release applications due to a wide pH range body fluids found in different tissue and cellular compartments (Table 2.1). For example, the gastrointestinal (GI) tract is having a broad pH range (pH 4 in the stomach and pH 7.2 in the colon), which is different than the blood physiological pH 7.4 [13]. Furthermore, pH differences are found in different cell organelles such as lysosome (pH 4−5) and endosome (pH 5.5−6) that maintain acidic pH for cellular function. It is also well documented that pH change occurs in various disease conditions such as microbial infection, inflammation, stomach ulcer, tumorigenesis, and carcinogenesis [14]. In cancer or tumor cells, pH of the cells is 0.5−1 unit less pH than the surrounding cellular microenvironment due to the following: (a) metabolic glycolysis in both aerobic and anaerobic conditions, and (b) high amount of lactic acid production by Warburg's effect [15]. Therefore, varied pH in the prerequisite site can be a potential approach for pH-sensitive drug delivery, and nanomaterials are the best possible selection for the development of pH-sensing materials [8,16]. For this, various pH stimuli-responsive polymers that respond to a specific pH environment by undergoing structural transition/modification and/or properties such as surface charge, bond break, and solubility were being used to fabricate pH-sensing nanomaterials [17]. The value of the surrounding pH at which these variations occur is called as transition or critical pH. In general, these polymers contain ionizable acidic or basic functional residue that gets protonated or deprotonated in response to the surrounding pH [18,19]. For drug-delivery applications, pH-sensitive

Intelligent Nanomaterials for Drug Delivery Applications. https://doi.org/10.1016/B978-0-12-817830-0.00002-3

TABLE 2.1
Summary of Essential pH-Sensitive Organs/Sites in the Human Body.

Sites	pH	Function	References
Skin	4–6.5	Prevents microbial growth	[25]
Stomach	1.35–3.5	Food digestion	[26]
Duodenum	~6	Chemical digestion of food	[26]
Jejunum and ileum	6–7.5	Absorbs vitamin B_{12} and bile salts	[26]
Bile juice	7.6–8.8	Helps digestion, and absorption of fats	[30]
Pancreatic fluid	~8.8	Neutralizes gastric acid	[30]
Cerebrospinal fluid	7.3	Protects the brain	[30]
Intracellular fluid	6–7.2	Maintains solutes or electrolytes concentration in cells	[30]
Vaginal fluid	<4.7	Prevents microbial growth	[28,29]
Arterial blood serum	7.4	Transports oxygen and nutrients from the heart to body tissues	[31]
Venous blood serum	7.35	Transports CO_2 from the tissues to lungs	[31]
Urine	4.6–8	Contains metabolic by-products	[30]
Lysosome	4.5–5	Contains digestive enzymes	[8,20]
Endosome	5.5–6	Material internalization in cell	[8,20]
Endoplasmic reticulum	~7	Helps in protein folding	[20]
Peroxisomes	~7	Helps in breakdown of long-chain fatty acids	[20]
Secretory granules	5.5–6	Store hormones	[20]
Cytosol	7.4	Helps in cellular functioning	
Cancerous tissue	6.2–6.9/7.12-7.65		[32,33]
Gastric cancer	5.4 ± 2.1		[37]
Gastritis	3 ± 2.2		[37]
Gastric ulcer	2.4 ± 1.9		[37]

nanomaterials are composed of pH-responsive polymers with several structural elements such as liposome, silica nanoparticles, metal nanoparticles, quantum dots, and other inorganic nanomaterials. The pH-responsive polymers either alone or in combination gets assembled to produce a nanostructure or else incorporated in a nanoframework [18,20]. The stability of these materials depends upon the composition and the surface charge of the polymers.

For smart drug-delivery applications, along with pH-sensitive mechanism, nanomaterials are designed to sense various other stimuli (photo-, thermal-, magnetic, ultrasound, etc.). Cellular microenvironment-sensitive agents, receptor-specific ligands, imaging or diagnostic agents are incorporated into the nanocomposites for specificity. The addition of secondary sensitive materials will render a synergistic effect on the therapy. For example, Zheng et al. synthesized a dual pH and thermal reactive core–shell silica nanoparticles for anticancer drug-delivery applications. These dual temperature and pH-responsive nanoparticles were investigated to carry anticancer drug doxorubicin (DOX), and demonstrated a significant amount of drug release in the targeted cell at an elevated temperature of 40°C and at slightly acidic condition (pH 5), which is a typical cancer cell environment [21]. Similarly, magnetic or supermagnetic nanoparticles can be incorporated with the nanocarrier for theranostic approach and to monitor the sustainable drug-release process [22,23]. Fluorophore molecules are also incorporated with therapeutic drugs for bioimaging applications [24]. Depending on the pH and prerequisite site, multifunctional drug nanocarriers have been developed for the smart release of drugs. This chapter reviews the different types of pH-responsive nanomaterial and their mechanisms for smart drug-release applications.

2 PH-DEPENDENT CELLULAR MICROENVIRONMENTS FOR TARGETED DRUG DELIVERY

The pH level of blood is constricted to a slightly alkaline range of 7.35–7.45 for maintaining the proper functionality of the human body. However, the pH level of different organs varies considerably according to their specific functions (Table 2.1). Skin, which is the body's largest organ, maintains acidic pH in a range of 4–6.5 to prevent environmental microbial growth [25]. The pH of the gastric acid in the stomach is 1.35–3.5 to help digestion, activate digestive enzymes, and prevent unsolicited microbial growth. Following an ascending pH order, the lower part of the gastrointestinal (GI) tract such as duodenum (pH ~6), jejunum, and ileum (pH 6–7.5) are comparatively neutral and slightly alkaline to the stomach [26]. Therefore, the high pH-sensitive drug-delivery system can be a potential approach to get maximum stability in the stomach and sustainable release in the small intestine. The pH buffer system works efficiently in these organs, which are maintained by phosphate ions, carbonic acid, bicarbonate ions, and free hydrogen. However, various fluids present in body are possessing considerably different pH than blood serum such as bile (pH 7.6 to 8.8), pancreatic fluid (pH ~8.8), vaginal fluid (pH less than 4.7), cerebrospinal fluid (pH 7.3), and intracellular fluid (pH 6–7.2) [27–30]. The role of bile and pancreatic fluid is to neutralize the gastric acid in the stomach and induce the digestion process. The acidic nature of vaginal fluid inhibits the overgrowth of various microbial organisms [28]. However, the arterial and venous blood serums are the same in composition but maintain different arterial (pH 7.4) and venous (pH 7.35) serum pH, as venous blood serum transports deoxygenated blood from tissue to heart [31]. Depending on the all body fluid, the urine may have a diverse range of pH from acidic (pH ~4.6) to alkaline (pH ~8). At the cellular level, different cellular organelles such as lysosome (pH 4–5), endosome (pH 5.5–6), endoplasmic reticulum (pH ~7), peroxisomes (pH ~7), and secretory granules (pH 5.5) maintain lower pH than cytosol (pH 7.4) [20].

Apart from these normal physiological conditions, many diseases and complications create different pH microenvironments in the human body. The pH level of cancerous tissue is comparatively more acidic (pH 6.2–6.9) extracellularly and slightly alkaline (pH 7.12–7.65) intracellularly than healthy tissue due to the production of lactic acid in aerobic/anaerobic glycolysis [32–34]. Similarly, the acidic microenvironment is generated at the site of infected/inflamed tissues due to the chemotactic lactic acid secretion from the involved leukocytes [35,36]. The normal gastric pH is found to increase in several complications such as pH 5.4 ± 2.1 in gastric cancer, pH 3 ± 2.2 in gastritis, and pH 2.4 ± 1.9 in gastric ulcer [37]. Considering other inflammatory bowel disorders in the gastrointestinal tract, Crohn's disease and ulcerative colitis can be potential targets for low pH-sensitive drug delivery [38].

3 MECHANISM OF PH-SENSING NANOMATERIAL FOR SMART DRUG RELEASE

Nanotechnology bestows an innovative and outstanding platform for smart drug-delivery strategy through a favorable nanocarrier for drug encapsulation/loading and stimuli-sensitive drug delivery. The application of nanocarrier materials leads to overcome multiple problems associated with traditional drug delivery such as nonspecific drug release, deficiency of tumor specificity, insufficient drug administrations, and development of multidrug resistance (MDR) in various cancer cells [39–41]. pH-sensitive nanomaterials are potential therapeutic solutions for targeting and delivering the drug to a differential pH microenvironment such as tumor cells, GI tract, and infected or inflamed cells [42,43]. Generally, the drug-loaded nanocarriers encapsulate and sustain the drug at normal pH conditions and are expected to release drugs at physiologically different pH conditions by undergoing structural modifications of nanocarriers. The surface ionizable functional groups (i.e., amines, phosphoric acids, and carboxylic acids etc.) receive or offer protons and undergo pH-dependent physicochemical modification such as an increase in particle size or pore size resulting for drug release [20,44]. The physicochemical modification leads to three types of structural modifications in nanocarrier in three methods: (i) dissociation, (ii) disruption (through bursting or swelling), and (iii) alterations of distribution-coefficient between the therapeutic agent and nanocarrier [7,26].

For precise cell targeting, drug-loaded pH-sensitive nanocarriers are designed with cell-specific ligand to bind the targeted cells/cell receptors only (Fig. 2.1). The nanocarrier is engulfed by the invagination of the plasma membrane with the activation of receptor-mediated endocytosis. The pit over the cell forms a vesicular structure to produce early endosomes for the

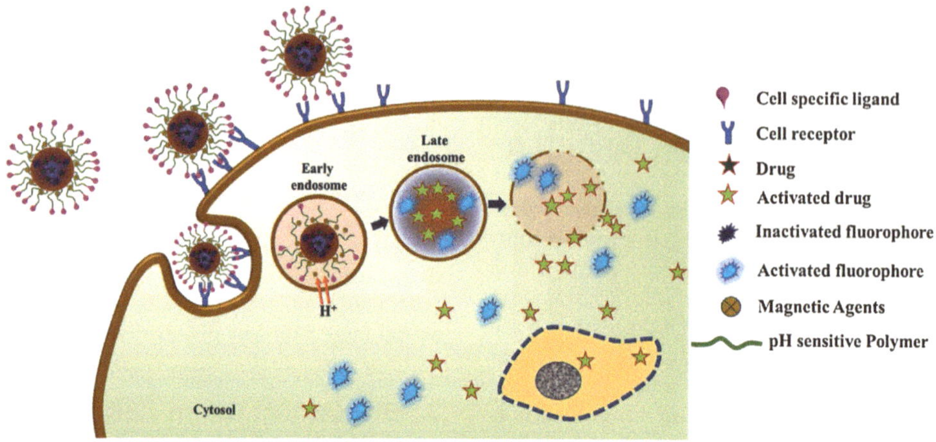

FIG. 2.1 Schematic representation of cellular uptake of drug-loaded pH-sensitive nanocarrier via receptor-mediated endocytosis.

endocytic uptake pathway. An ATP-driven proton (H^+) pump activates on the endosomal membrane to pull cytosolic H^+ inside, which causes an acidic environment by a decrease in endosomal pH. In response to the acidic environment, a pH-sensitive nanocarrier undergoes conformational modification and releases the drug into the late endosome, which later releases to the cell cytoplasm by endosomolysis [45–47]. Then, antitumor drugs activate different pathways to kill tumor cells [48]. In addition to the drugs, nanocarriers can be accompanied with magnetic nanoparticles and fluorophore molecules, which also be released to the cytoplasm to assist the bioimaging processes as well [49,50]. For diseased/inflamed tissues, the nanocarrier responds to the different microenvironment's pH and releases single or multiple drugs to the external microenvironment [43,47]. The multiple drug-release systems may produce a synergistic effect on the therapeutics. Various types of nanocarriers for controlled pH-sensitive smart drug release have been discussed in the following sections.

4 TYPES OF NANOMATERIALS USED FOR PH-SENSITIVE DRUG DELIVERY

pH-sensitive nanomaterials have been fabricated for the smart drug-delivery system. The highly sensitive nanocarriers can be constructed by using different organic, inorganic, and both constituents such as natural/synthetic polymers, phospho-/lipids, silica, ceramic, and metal nanocomposites. For smart drug delivery, the nanocomposites materials have been combined with fluorescent tags, magnetic nanoparticles, and ligands for the molecular marker to acquire high accuracy of site-specific drug delivery. In general, the nanocarrier materials are divided into three main categories according to the materials used for their fabrications: (1) polymeric nanocomposites, (2) liposomes, and (3) inorganic–organic nanocomposites [7,8].

4.1 Polymeric Nanocomposites

The pH-responsive polymers are the key ingredients to construct pH-responsive polymeric nanocomposites. These stimuli-responsive polymers in different pH solutions undergo structural and functional alterations such as surface charge alteration, aggregation or disaggregation, chain formations, change in water solubility, and hydrophobicity. The pH-responsive polymers are composed of ionizable (acidic or basic) monomeric units, which can either accept or donate H^+ in response to external pH. This mechanism can modulate the surface charge of the polymer, which leads to conformational changes over the nanostructure. These polymers can have linear, branched, or network-like structures. The solubility and self-assembly behavior of these polymers depend upon the side chain functional groups such as acidic (carboxyl, phosphate, sulfonic) or basic (amines, pyridines) groups [51]. The external pH change condition may cause protonation/deprotonation of the polymer chain, which regulates the cluster formation or separation of homopolymers in nanostructures. According to the pH-responsive behavior of a particular polymer, the drug nanocarrier materials can be designed for drug entrapment and release system [17]. Owing to these unique characteristics, the pH-responsive polymers can be used as a potential material for different biomedical applications, for example, gene/drug delivery, constructing nanosensors, biomedical imaging, and chemotherapy [14,52,53]. For drug delivery, different nanostructures have been prepared from the pH-responsive materials such as (A) polymeric nanospheres, (B) dendrimers, (C) core–shell nanoparticles, (D) polymer-drug conjugates, (E) nanocomposite hydrogels, and (F) micelles.

4.1.1 Polymeric nanospheres

Polymeric nanospheres or nanobeads are a homogenous assembly of polymers into a nanoscale structure with a permissible range of 10–200 nm in diameter (Fig. 2.2A). In this colloidal system, therapeutic agents or drugs are entrapped or non/covalently bound to the polymer. Mostly, these nanospheres are fabricated with biocompatible, biodegradable, and resorbable pH-sensitive polymers to target-specific disease (Table 2.2). Several techniques are used to prepare pH-sensitive polymeric nanospheres such as polymerization, emulsion evaporation method, solvent evaporation method, and phase inversion temperature method [54]. Moreover, these nanospheres acquire sustained and control drug release on the specific site. The surface of the polymeric nanospheres/beads has also been improved by using some fluorescent tags, compounds, or site-specific ligands.

Tang et al. constructed two new amphiphilic self-assembled polymers such as polyethyleneimine-block-poly[(1,4-butanediol)-diacrylate-b-5-hydroxyamylamine] (PEI-PDHA) and polyethylene glycol-block-poly[(1,4-butanediol)-diacrylate-b-5-hydroxyamylamine] (PEG-PDHA) for drug-delivery applications. The PEI-PDHA and PEG-PDHA were complexed with Snail siRNA (siSna) and Twist siRNA (siTwi) as anticancer drug nanocarrier for paclitaxel (PTX). At pH 5, both siRNA and PTX released, which resulted in the inhibition of cancer cell motility with extracellular matrix (ECM) degradation [55]. Beloqui et al. fabricated polymeric nanospheres by linking two poly(lactide-co-glycolide) acid (PLGA) and a polymethacrylate (Eudragit S100) polymers for curcumin (CC) delivery in inflammatory bowel disease.

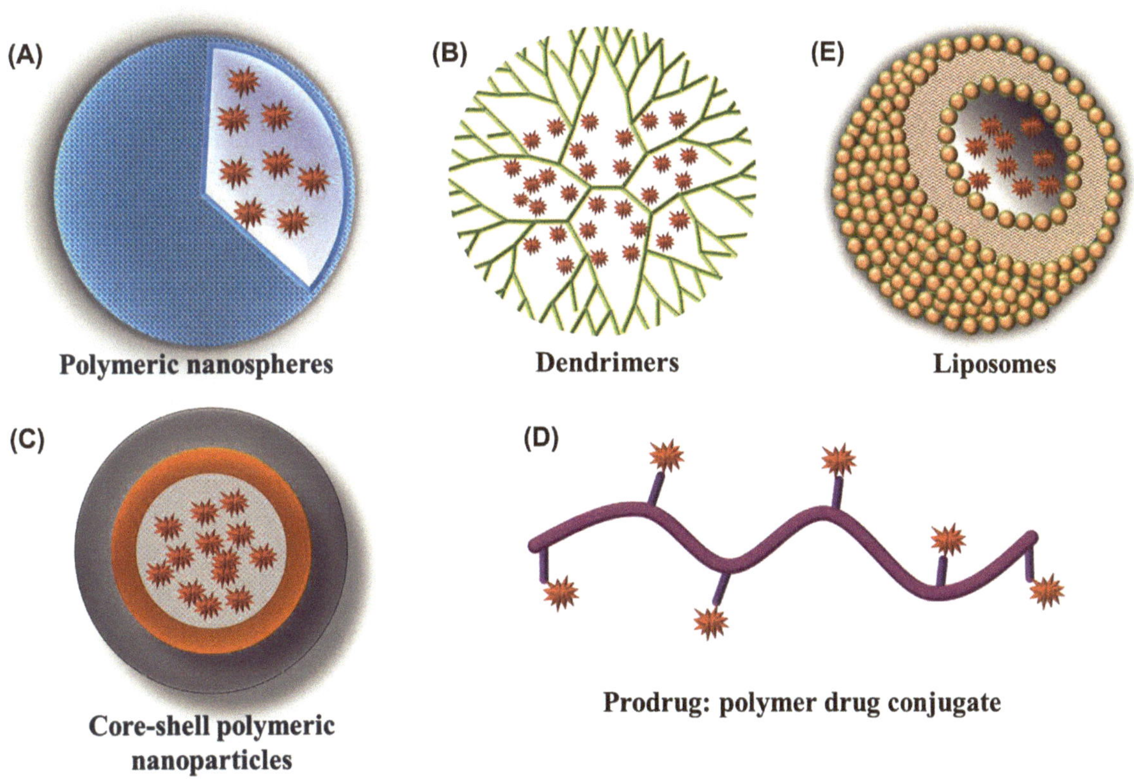

FIG. 2.2 Schematic of various polymeric nanocarrier for pH-sensitive smart drug release: **(A)** polymeric nanoparticle; **(B)** dendrimers; **(C)** core–shell polymeric nanoparticles; **(D)** prodrug: polymer-drug conjugate; **(E)** liposomes.

TABLE 2.2
Summary of Polymeric Nanospheres for pH-Sensitive Drug-Release Applications.

Polymeric Nanospheres	pH-Sensitive	Drug	Application	References
PEI-PDHA and PEG-PDHA with SiRNA	5	Paclitaxel	Anticancer drug delivery	[55]
PLGA and Eudragit S100	7	Curcumin	Antiinflammatory bowel disease	[56]
PLH-PLGA-TPGS	5	Doxorubicin	Chemotherapy	[57]
Acupa	5.4	BSA and Cytochrome C	Antitumor drug delivery	[58]
RΔF	2	Doxorubicin	Anticancer drug delivery	[59]
Eudragit S100	7.2	Curcumin and Celecoxib	Anti-inflammatory bowel disease	[42]
PS-b-PEG	<6	Tannic acid	Antioxidant activity	[60]
PEG-b-PPLG	5–5.5	Doxorubicin	Anticancer drug delivery	[61]
PEG-Dextran	5.5	Doxorubicin	Anticancer drug delivery	[62]

The CC-loaded polymeric nanoparticles were investigated in different pH for drug-release kinetics that showed the speedy release of CC at neutral pH than acidic pH (>4.5) [56]. Li et al. prepared poly(L-histidine)-poly(lactide-co-glycolide)-tocopheryl polyethylene glycol succinate (PLH-PLGA-TPGS) biocompatible nanoparticles loaded with DOX for chemotherapy application. These DOX-loaded nanocarriers exhibited swelling properties in acidic pH 5 medium that triggers the drug release to the tumor cell microenvironment [57].

Li et al. fabricated biocompatible and cancer targeting 2-[3-[5-amino-1-carboxypentyl]-ureido]-pentanedioic acid (Acupa) pH-responsive chimeric polymersomes (Acupa-CPs) to convey therapeutic proteins to breast cancer cells. Bovine serum albumin (BSA) and cytochrome C have been embedded in the Acupa polymeric nanoparticles. At a mild acidic pH 5.4 environment, the proteins released to the cancer cell and showed an efficient antitumor effect [58]. Singh et al. developed self-assembled pH-responsive nanoparticle from a dipeptide arginine-α, β-dehydrophenylalanine (RΔF). These nanoparticles were loaded with DOX and studied for pH-depended drug release in cancer cells. They reported that RΔF nanoparticles showed sustain and speedy release of DOX at pH 2 (90.27 ± 1.45%, 8 h) and pH 10 (96.39 ± 3.87%, 12 h) in comparison to pH 7 (97.64 ± 4.71%, 36 h). DOX-loaded RΔF nanoparticles exhibited enhanced cellular internalization with high cytotoxic potential [59].

In another study, Gugulothu et al. formulated CC (as antiinflammatory agent) and celecoxib (as an antioxidant)-loaded Eudragit S100 nanoparticles for pH-responsive drug delivery in ulcerative colitis (UC). They investigated the combinatorial therapy with pH-dependent drug-release kinetics in the UC rat model. The S100 nanoparticles showed 70% individual drug-release drugs at pH 7.2 and less than 4% at lower pH 1.2 and 4.5 thus indicating the control release of drugs in the GI tract [42]. Tang et al. developed pH-sensitive, multifunctional nanoparticles to facilitate the intracellular delivery of antioxidants. By polymer-directed self-assembly process, antioxidant tannic acid (TA) was incorporated with block copolymer polystyrene-b-poly(ethylene oxide) (PS-b-PEG). The resulting nanoparticles showed complete unloading of drugs in acidic pH (<6) condition by solubilizing the polymeric nanoparticles [60]. Quadir et al. designed ligand-decorated polypeptide-based nanospheres for pH-regulated drug release of systemic chemotherapy to tumors. Folic acid (FA) residues were immobilized to poly(ethylene glycol)-b-poly(γ-propargyl L-glutamate) (PEG-b-PPLG) block copolymers to synthesize FA-functionalized polymeric nanoparticles. These nanospheres exhibited controlled unloading of DOX from folate-bearing PEG-b-PPLG at lysosomal pH (5–5.5), and in vivo tumor growth suppression [61]. On similar lines, Lian et al. prepared dual pH and reduction-responsive PEG-dextran nanoparticles (CPD NPs) for anticancer drug-delivery application. The cross-linked CPD NPs were loaded with DOX and studied for efficient nanoparticle internalization and DOX release in cancer (A549 and MCF-7) cells. Owing to reducing condition, the DOX-loaded NPs exhibited higher drug release, which showed good cytotoxicity against cancer cells [62].

4.1.2 Dendrimers

Dendrimers are highly branched tree-like three-dimensional globular nanostructures. The discrete branches of dendrimers diverge out symmetrically from a central core region. After several layers of branching, it converts into a spherical shape. The internal branched region is availed with an extensive volume of solvent filled void space that can cargo ample amounts of therapeutic agents [63]. These dendritic molecules show monodisperse properties at the nanoscopic level. The particular architecture of dendrimer is divided into three distinctive parts: (i) core, the epicenter of the dendrimer molecule consisting at least two identical functional groups; (ii) middle, branches radiating from core region consisting of multiple branch junctions; (iii) outer, periphery of the dendrimers with different functional groups subjected to host–guest chemistry (Fig. 2.2B) [64,65]. These unique structural features can also improve by the polymer selection for multiple functional groups on the surface. The functional group may possess positive, negative, or neutral charges, which play a vital role in drug immobilization and delivery. The biocompatibility of dendrimer depends mainly on the surface functional groups. It has been investigated that the terminal amine group of the polymer shows concentration-depended toxicity and hemolysis in several cases [66,67]. Subsequently, this can be overcome by peripheral functional group modification or the addition of negatively charged or neutral groups. The synthesis process of dendrimers includes the repetitive addition of monomeric units to the central core multifunctional molecule. The size of the dendrimers increases upon the addition of monomers to the exposed functional groups [64,68]. Divergent and convergent methods are frequently used for the synthesis of dendrimers. The divergent method activates the surface functional groups to bind the next monomer; whereas in the convergent method, monomers couple iteratively to produce a divergent core and then monomers bind subsequently to yield a complete dendrimer. Polyamidoamine (PAMAM), polyaryl ether, polylysine, 2,2-bis(hydroxymethyl)propionic acid, and polyglycerol are commonly used polymers to synthesize the nanoscale dendrimers [69]. The application of dendrimers increasingly explored the biological system for imaging, drug targeting, and drug delivery (Table 2.3) [70,71].

Ali et al. synthesized pH-responsive dendrimers as MRI contrast agents for the detection of kidney disease and cancer. The gadolinium-based PAMAM dendrimers have been to study the pH-dependent relaxivity. When the pH of the solution decreases from pH 9 to 6, the relaxivity increases from 10.8 to 24 $mm^{-1}s^{-1}$ per Gd^{3+} ion [72]. Tekade et al. successfully encapsulated two antileukemic drugs, that is, methotrexate (MTX) and all-trans retinoic acid (ATRA) inside a PAMAM dendrimer. They have investigated the release kinetics of both drugs under three different pH conditions such as pH 4, 7.4, and 10. The drug-loaded PAMAM dendrimers were found to be more stable, and the efficient release of drugs resulted in high protonation conditions at pH 4 [73]. Lui et al. constructed zwitterionic chitosan (ZWC)-PAMAM dendrimer complex as a drug carrier to inhibit solid tumor growth with acidic microenvironments. The presence of ZWC over the dendrimer prevents the cytotoxic effect of amine-terminated PAMAM dendrimer and protects red blood cells and fibroblast cells from the hemolytic reaction. At a low pH environment, ZWC degrades due to charge conversion and

TABLE 2.3
Dendrimers Employed for pH-Sensitive Drug-Release Applications.

Dendrimers	pH-Sensitive	Drug	Application	References
PAMAM	6	Gadolinium	Detection of kidney disease and cancer	[72]
PAMAM	4	Methotrexate and retinoic acid	Anti-leukemic drug delivery	[73]
PAMAM-TN	5	Doxorubicin	Anticancer drug delivery	[75]
FPD-HZN-Dex	4.5–6	Doxorubicin	Anticancer drug delivery	[76]
MSNs-PAMAM	5	Curcumin	Anticancer drug delivery	[77]

allows PAMAM dendrimers to enter the targeted cells [74]. Similarly, Kumar et al. prepared a dendrimer of PAMAM-Tryptophan-N-acetyl glucosamine (PTN) as a potential drug nanocarrier for anticancer drug delivery application. The use of N-acetyl glucosamine (NAG) was for GLUT receptor targeting which is the comparably more efficient system than the FA receptor. The PTN dendrimers were loaded with DOX and demonstrated low pH drug delivery against breast cancer cells [75].

Ma et al. formulated dextran functionalized amphiphilic fluorinated peptide (FPD-HZN-Dex) dendrons for intracellular drug delivery. The amphiphilic FPD-HZN-Dex particles were encapsulated with DOX at physiological pH conditions. In response to endosomal/lysosomal pH, these DOX-loaded dendrimers disassembled their structure and enabled the control release of encapsulated drugs [76]. Xu et al. prepared pH-sensitive and self-fluorescent mesoporous silica nanoparticles (MSNs) incorporated PAMAM dendrimers for biomedical imaging and drug-delivery system. The fluorescent properties of PAMAM dendrimers add a functional advantage in the diagnosis-mediated drug-delivery approach. The MSNs-PAMAM dendrimers were loaded with CC molecules. At low pH conditions, the synthesized PAMAM dendrimers showed conformational changes by releasing of drugs to the external environment [77].

4.1.3 Core–shell polymeric nanoparticles

Core–shell polymeric nanoparticles are spherical, branched with biphasic structures having core region and surrounding outer shell region. The core and shell architecture is obtained by the utilization of two different types of materials or polymers with two different domains [78,79]. The unique features arise from the core and shell combinations, where the core region protects the cargo and shell region materials improve the reactivity, stability, solubility, and biocompatibility of the nanoparticles (Fig. 2.2C) [80]. The synthesis of core–shell polymeric nanoparticles in the solution phase comprises two steps: (1) synthesis of the core structure and (2) coating the core structure by shell materials or by the polymerization method. Under suitable conditions, the polymer molecules are combined together to form a three-dimensional nanostructure. The polymerization process is further divided according to the polymer's phase behavior such as bulk, emulsion, solution, and suspension [79]. Although these nanoparticles are less cytotoxic, the biocompatibility of the nanoparticles can be enhanced by the conjugation of biomolecules to the shell region. The outer shell materials also govern the hydrophilicity of the nanoparticles. For the hydrophobic cargo materials delivery, the core materials will be hydrophobic and the surrounding shell materials will be hydrophilic. Owing to these interesting/tunable features, the core–shell polymeric nanoparticles are mainly utilized in healthcare applications such as sustained/target-specific drug release, biosensing, and theranostics (Table 2.4) [81–84].

Mukhopadhyay et al. prepared chitosan-alginate (CS/ALG) core–shell polymeric nanoparticles for oral insulin delivery. The authors reported the formation of spherical CS/ALG nanoparticles with an average hydrodynamic size of 100–200 nm and capable of almost 85% insulin encapsulation. At simulated gastric fluid pH 1.2 conditions, the insulin-loaded CS/ALG nanoparticles showed burst release due to the weakening of electrostatic interaction between insulin and surface polyelectrolytes. However, in intestinal (duodenum pH 6.8, ileum pH 7.4) pH condition, the CS/ALG nanoparticles showed sustained and prolonged release of insulin to the external environment [85]. In a similar study, Mukhopadhyay et al. investigated quercetin (QUE) drug-release kinetics from pH-sensitive CS/ALG core–shell–corona nanoparticles for diabetes therapeutics. The QUE molecules were encapsulated

TABLE 2.4
Core–Shell Polymeric Nanoparticles for pH-Sensitive Drug-Release Applications.

Core-shell Polymeric Nanoparticles	pH-Sensitive	Drug	Application	References
CS/ALG	1.2	Insulin	Oral delivery for diabetes treatment	[85]
CS/ALG	6.8 and 7.2	Quercetin	Oral delivery for diabetes treatment	[86]
P(NIPAAm-co-DMAAm-co-UA)	5	Doxorubicin	Anticancer drug delivery	[87]
PEG-PH-PLLA	5	Doxorubicin	Anticancer drug delivery	[88]

in CS/ALG core–shell nanoparticles by the ionic cross-linking method. The resulted core–shell nanoparticles exhibited almost 88%–95% pH-dependent sustainable drug release at pH 6.8 and 7.4 [86].

Soppimath et al. synthesized pH-responsive, self-assembled poly(N-isopropylacrylamide-co-N,N-dimethylacrylamide-co-10-undecenoic acid) (P(NIPAAm-co-DMAAm-co-UA)) core–shell nanoparticles for anticancer drug delivery. The free amine group of polymers was conjugated with FA for targeting overexpressed FA receptor cancer cells. An anticancer drug DOX was internalized in the hydrophobic corona of P(NIPAAm-co-DMAAm-co-UA) nanoparticles. These FA-linked nanosystems demonstrated much faster drug release, which is approximately 64% and 54% at pH 5 and 6.6, respectively [87]. Lui et al. prepared pH-responsive poly(ethylene glycol)-poly(L-histidine)-poly(L-lactide) (PEG-PH-PLLA) core–shell nanoparticles for antitumor drug delivery. The PEG-PH-PLLA nanoparticles were encapsulated with anticancer drug DOX to explore the pH-sensitive release kinetics. The nanoparticles showed a faster drug-release profile at pH 5 than pH 7.4, which gives rise to sustainable antitumor drug delivery in HepG2 cells [88].

4.1.4 Polymer-drug conjugates

Polymer-drug conjugates can be synthesized by covalently conjugating the therapeutic molecules with pH-responsive polymers, or encapsulating them by hydrophobic/electrostatic interactions. The conjugation of drug to a polymer is also called as "polymeric-prodrug" (Fig. 2.2D) [89,90]. The advantages of drug conjugation include (i) enhance the water solubility of drugs, (ii) stability and prevent drugs from degradation, (iii) smart stimuli-responsive delivery, (iv) site-specific drug targeting [89]. Owing to these advantages, the polymer-drug conjugates are deliberately used as a nanocarrier for drug-delivery applications (Table 2.5). Depending upon the working mechanism and polymeric chain structure, different designs have been fabricated; they include homopolymers, polymer blocks, and polymeric nanocages. For pH-responsive drug release, a pH-responsive linker is required for the activation of nanocarrier to release the drugs. Few pH-responsive bonds such as hydrazone, acetal, *cis*-acotinyl, linkers containing Schiff-base, and β-thiopropionate were used for pH-sensitive polymer-drug conjugates [91–94].

Zhang et al. designed DOX and CC-loaded PEG pH-responsive prodrug nanocarrier for combination therapy to treat cancer. By Schiff-base reaction, DOX was conjugated to PEG polymer chain; at pH 7.4, the PEG-DOX and CC were self-assembled to produce a stable nanoparticle with a hydrophobic corona structure. After the internalization of PEG-DOX-CC nanoparticles in a cancer cell, the Schiff-base linker will break under acidic conditions thus releasing the drugs DOX and CC to the cell cytoplasm. The codelivery of two anticancer drugs showed efficient antitumor activity [95]. Similarly, Zhao et al. formulated pH-sensitive polyethylenimine (PEI)-linked PEG prodrug by Schiff-base linkage for tumor drug-delivery applications. The PEGylated-PEI nanoassemblies were loaded with indomethacin (IND) and docetaxel (DTX) through multiple noncovalent interactions. These drug-loaded nanoassemblies exhibited acidic pH-sensitive PEG cleavage to release the drugs in the tumor microenvironment [96].

Du et al. designed dual pH (tumor extracellular and intracellular)-sensitive prodrug to achieve the intracellular drug delivery for cancer therapy. They prepared monomethoxyl poly(ethylene glycol)-b-poly-(allyl ethylene phosphate) (mPEG-b-PAEP) copolymer diblock by the process of ring-opening polymerization. Then, DOX was conjugated to mPEG-b-PAEP polymer through an acid-labile hydrazone bond. The mPEG-b-PAEP-Cya-DOX/PPC-DOX was linked with 2,3-dimethylmaleic anhydride (DMMA) to acquire the pH-responsive charge alteration of PPC-DOX-DA prodrug. The synthesized negatively charged PPC-DOX-DA nanoparticles displayed positive charge at pH 6.8 and remained negatively charged at pH 7.4. So the negatively charged behavior of nanoparticles at the tumor

TABLE 2.5
Applications of Polymer-Drug Conjugate for pH-Sensitive Drug Release.

Polymer-Drug Conjugate	pH-Sensitive	Drug	Application	References
PEG	5	Doxorubicin and Curcumin	Anticancer drug delivery	[95]
PEGylated-PEI	5.7	Indomethacin and Docetaxel	Antitumor drug delivery	[96]
mPEG-b-PAEP	6.8	Doxorubicin	Anticancer drug delivery	[97]
mPEG-b-P(LL-DTPA)	5	Paclitaxel	Antitumor drug delivery	[98]

cell surface rendered the enhanced cellular internalization and due to charge-conversion behavior in acidic pH, nanoparticles exhibited drug release in cellular cytoplasm [97].

Lv et al. synthesized PTX conjugated, redox, and pH-responsive polymer drug carrier for antitumor therapy. The polymer 3,3′-dithiodipropionic acid-functionalized poly(ethylene glycol)-b-poly(L-lysine) (mPEG-b-P(LL)-DTPA) was formulated, and PTX was conjugated through a disulfide bond. The polymer-drug conjugation converted into a micelle formation (P(L-ss-PTX)). Under acidic pH 5, PTX released rapidly from the P(L-ss-PTX) micelles and showed improved cytotoxicity against the diversity of tumor cells [98].

4.1.5 Nanocomposite hydrogels

The incorporation of nanocomposite materials to the three-dimensional hydrogel system exhibits the synergistic properties of hydrogel and nanomaterials. The pH-responsive polymer forms nanocages (hydrophobic pocket) like structures in the hydrogel matrix to trap inside the hydrophobic drugs. The novel combination enables the control release of therapeutic agents from the hydrogels by the stimuli-responsive activation. Commonly, the nanohydrogels are prepared by self-assembly, chemical cross-linking, and polymerization method with stimuli-sensitive polymers for target-specific drug delivery (Table 2.6) [99,100]. Natural polymers such as proteins (gelatin, collagen, fibrin, lysozyme, silk) and carbohydrates (chitosan, dextran, cellulose, agarose, hyaluronic acid, carrageenan) and biodegradable synthetic polymers such as polyvinyl alcohol (PVA), poly(vinyl pyrrolidone) (PVP), poly(methacrylic acid) (PMAA), and acrylamide (AAm), have been used to prepare stable nanohydrogels. Smart stimuli-responsive nanohydrogels respond to external environmental stimuli like pH, temperature, and electrolyte compounds. The polymers with ionizable (acidic/basic) functional groups can be used for pH-sensitive nanocomposite hydrogels [99,101]. At a certain pH, the therapeutic compound gets entrap in nanohydrogels and when the external pH change, the pore size of the hydrogels increases and releases the drugs to the microenvironment (Fig. 2.3A) [102].

Jaiswal et al. developed both pH and temperature-responsive nanohydrogels for the combined effect of thermochemotherapeutics. They have studied anticancer drug DOX release from the nanohydrogels under acidic pH ~4.6 condition and hyperthermia temperature ~42°C condition [103]. Jin et al. fabricated a multiple (pH/protease/redox) stimuli-sensitive biodegradable nanohydrogels for cancer treatment. Methacrylic acid (MAA) monomer was cross-linked with N, N-bis(acryloyl)-cystamine (BACy) through the reflux precipitation polymerization method. These nanohydrogels were decorated with FA and PEG for active targeting and longer in vivo circulation lifetime, and they were loaded with two anticancer hydrophobic drugs such as DOX and PTX. At pH 5 and reduced microenvironment [glutathione (GSH)] in cancer cells, the amido bonds of nanohydrogels broke and released the hydrophobic drugs into the targeted site [104]. Similarly, Rao et al. synthesized pH-responsive polymeric nanogels composed of gelatin proteins and poly(acrylamidoglycolic acid) for cancer therapy applications. They reported that CC encapsulated nanogels swelled at higher pH ~7.2 to release the drug to the surrounding microenvironment [105].

TABLE 2.6
Summary of Nanocomposite Hydrogels for pH-Sensitive Drug-Release Applications.

Nanocomposite Hydrogels	pH-Sensitive	Drug	Application	References
PNIPA and Chitosan	~4.6	Doxorubicin	Anticancer drug delivery	[103]
MAA and BACy (-FA-PEG)	5	Paclitaxel and Doxorubicin	Anticancer drug delivery	[104]
Gelatin and PAGA	7.2	Curcumin	Anticancer drug delivery	[105]
MPEG-b-PNLG	~6.4	Doxorubicin	Tumor growth inhibition	[106]
FA-PEG	5	Doxorubicin	Tumor growth inhibition	[107]
PMPB	5	Doxorubicin	Anticancer drug delivery	[108]
TG and GMA	7	Quercetin	Oral drug delivery	[109]
NIPA-co-VMDZ	5.5	Metformin, terfenadine and 5-fluorouracil	Endosomal/lysosomal delivery	[110]

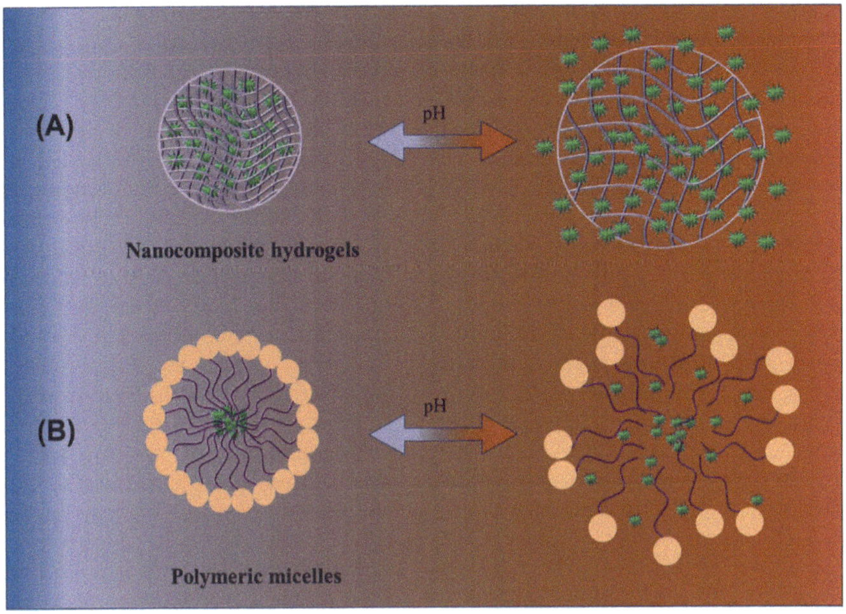

FIG. 2.3 Schematic representation of pH-dependent drug release: **(A)** nanocomposite hydrogels showing swelling activity and **(B)** polymeric micelles showing dissociation of drugs and polymer micelle complex in different pH conditions.

Li et al. prepared pH-sensitive DOX-loaded polypeptide-based methoxy poly-(ethylene glycol)-b-poly[N-[N-(2-aminoethyl)-2-aminoethyl]-L-glutamate] (MPEG-b-PNLG) nanogels for potential drug-delivery applications. At tumoral low pH conditions ~6.4, the imine bonds were cleaved and MPEG-b-PNLG nanogels released the drugs at the targeted site [106]. Zhang et al. synthesized DOX-loaded FA-PEG-conjugated zinc (II)-cross-linked polymeric nanohydrogels (FPZCLNs-15) for tumoral growth inhibition. These dual (pH and glutathione) stimuli respond FPZCLNs-15 nanogels showed controlled drug release and antitumor effect when exposed to low pH 5 and GSH-rich environment [107].

In a similar study, Zhou et al. synthesized dual pH and reduction stimuli-responsive monodisperse biodegradable poly[methacrylic acid-co-poly(ethylene glycol) methyl ether methacrylate-co-N,N-bis(acryloyl) cystamine] (PMPB) nanohydrogels for controlled anticancer drug delivery. PMPB nanohydrogels were encapsulated with DOX and showed high drug encapsulation efficiency (more than 96%) in basal pH conditions with good stability. At lower pH 5 and in the presence of GSH, DOX released almost more than 85% (within 30 h) from the nanohydrogels [108]. Hemmati et al. prepared quercetin-loaded pH-responsive tragacanth gum (TG) polysaccharide and glycidyl methacrylate (GMA) nanohydrogels for the oral drug-delivery system. The TG-GMA nanohydrogel particles were tested at pH 2 and 7 for drug-release kinetics. They found the particle size increased from 80 to 140 nm with an increase of pH from 2 to 7, and this might be due to the ionization of the carboxylic group that generated electrostatic repulsion on the particles' surface. The swelling effect of nanohydrogels formed a larger pore size in the gel structure to release the drug at basal pH conditions [109]. Bautista et al. synthesized pH and temperature-sensitive nanohydrogels by polymerizing N-isopropylacrilamide and 1-vinyl imidazole. The synthesized NIPA-co-VMDZ hydrogel nanoparticles showed swelling properties by decreasing the pH. They loaded with three different types of drugs such as metformin, terfenadine, and 5-fluorouracil to investigate the release kinetics. At endosomal/lysosomal pH 5.5, the NIPA-co-VMDZ hydrogel nanoparticles showed a faster release of all three drugs with a complete endocytosis process [110].

4.1.6 Micelles

Polymeric micelles are spherical shaped and self-assembled structures, composed of amphiphilic polymers. In an aqueous medium, the hydrophobic domains of the polymers interact with each other to form a nanosize micelle-like structure. The specific number of amphiphilic polymer blocks and their hydrophobic/ionic interactions play an essential role in the formation of micelles (Fig. 2.3B). These are water-soluble, biodegradable/biocompatible nanocarriers with drug-loading pockets [8,111]. As a potential drug nanocarrier system, micelles can easily internalize the drugs into the inner hydrophilic core region. Owing to these properties, micelles have been explored as a latent drug nanocarrier arrangement with enhanced permeability and retention effect (EPR). Furthermore, micelles have been improved by the incorporation of other stimuli (pH, temperature, and electrolytes)-responsive agents to their structures [112]. The site-specific drug delivery has been improved by the functionalization of ligands, monoclonal antibodies, and fluorescent magnetic nanoparticles [113−116]. The pH-sensitive micelles have been used to deliver the drug in lower physiological pH conditions such as tumor targeting and chemotherapy (Table 2.7) [16,22,117].

Wu et al. developed PTX-loaded poly(histidine)-polyethylene glycol (PEG) and 1,2-distearoyl-sn-glycero-3-phosphoethanolamine (DSPE)-PEG micelles systems for pH-sensitive cytosolic drug delivery. The hybrid polymeric micelles exhibited excellent release of the drugs in response to the acidic pH conditions of endosomes in the cell [118]. Wang et al. designed PTX-loaded pH-sensitive histidine (His)-modified auricularia auricular polysaccharide (AAP) micelles for antitumor drug delivery. The PTX-loaded His-AAP micelles showed increased drug release at pH 5. They also tested PTX-loaded His-AAP micelles in tumor-bearing mice, which showed significant inhibition of tumor growth after drug delivery [119]. Yang et al. formulated a pH-sensitive micellar system with poly(-ethylene glycol)-benzoicimine-poly(γ-benzyl-L-aspartate)-b-poly(1-vinylimidazole) block copolymer (PPBV) for codelivery of PTX and CC to eliminate breast cancer stem cells (bCSC) as well as non-bCSC. These drug-loaded micelles have intelligent properties to switch its surface charge from neutral to positive in response to the acidic pH condition. These, micelles deshield their surface covered PEG layer at acidic pH, thus reducing their size for deep tumor penetration and cellular uptake. The result showed the combinatorial therapy of PTX and CC inhibited tumor growth and effectively killed bCSCs in vivo [120].

Shi et al. synthesized pH-sensitive star-like micelles with amphiphilic diblock copolymers poly(2-(diisopropylamino) ethylmethacrylate)-b-poly[(ethylene glycol) methyl ether methacrylate] (PDPA-b-POEGMA) grafted from β-cyclodextrin for effective antitumor drug delivery. The produced unimolecular micelles showed excellent stability in aqueous medium at pH 7.4. These micelles displayed effective drug-loading capacity with DOX and at acidic pH 4, it released drug to tumor cells (HeLa and MCF-7), which implied its tumor-targeting efficacy [121]. Yan et al. prepared galactose-grafted pH-sensitive poly(ortho ester diamide) graft lactobionic acid (POEAd-g-LA) micelles for DOX delivery to liver cancer cells. These galactose-POEAd-g-LA micelles interact with the glucose receptor of liver cells and released DOX at extracellular tumoral pH conditions. The results showed significant drug accumulation and inhibition in tumor growth [122]. Teo et al. synthesized pH and GSH-sensitive micelles

TABLE 2.7
Applications of Polymeric Micelles for pH-Sensitive Drug-Release Applications.

Polymeric Micelles	pH-Sensitive	Drug	Application	References
Poly histidine-PEG/DSPE-PEG	5.5	Paclitaxel	Cytosolic drug delivery	[118]
HIS-APP	5	Paclitaxel	Antitumor drug delivery	[119]
PPBV	5	Paclitaxel and Curcumin	Anticancer drug delivery	[120]
PDPA-b-POEGMA and β-cyclodextrin	4	Doxorubicin	Anticancer drug delivery	[121]
POEAd-g-LA and galactose	5.5	Doxorubicin	Anticancer drug delivery	[122]
PEG	5	Doxorubicin	Anticancer drug delivery	[123]
HA	5	Iridium (III)	Anticancer drug delivery	[124]

by copolymerization and functionalization of PEG. The PEG micelles were loaded with DOX through ionic interactions. When these DOX-encapsulated micelles were exposed to an endosomal pH of 5, the drug released rapidly and exhibited greater cytotoxicity against cancer cells [123].

Cai et al. synthesized pH and reduction responsive, anticancer drug iridium (III) (Ir(III))-conjugated nanomicelles for cancer therapy. For more effectively CD44-targeted drug delivery, Ir(III) drugs were conjugated with amphiphilic hyaluronan (HA) polymers through disulfide bonds containing cystamin-pyrenyl (Py) complex. Ir(III)-loaded HA-Py nanomicelles revealed excellent stability in physiological condition and released drug in tumor cell environment (high GSH and low pH). It demonstrated a high tumor inhibition effect in vivo in comparison with free Ir(III) anticancer drugs [124].

4.2 Liposomes

Liposomes are spherical shaped self-assembled vesicles. These can be constituted with single or multilayers of concentric lipid bilayers by generating a hollow core structure. The outer layer of liposomes interacts with external aqueous solution whereas the inner layer of liposome surrounded by hydrophobic lipid heads (Fig. 2.2E) [8,125]. As hydrophilic drugs cannot pass the lipid bilayer, this unique nanocarrier structure can only encapsulate the hydrophobic drugs in its core void region. The surface and site-specific targeting properties can be altered by the addition of external ligands [126]. Due to the lipid bilayer structure, the liposomes are comparatively more stable than other nanocarrier systems (Table 2.8). In addition, in comparison to micelles structures, liposomes are more bio-/cyto-compatible and biodegradable. However, low drug encapsulation and high drug-release rate are two limitations for liposome nanocarrier in drug-delivery applications. The liposome structures are characterized by the dimension, number of lamellae, lipid composition, and polymers/ligands [127]. For pH-sensitive liposome synthesis, acidic/basic functional group comprising polymers has been incorporated in the lipid bilayer structure. At different physiological pH conditions, the functional groups of the polymers undergo the process of protonation/deprotonation. Thus, the charge differentiation of functional groups produces an unstable condition to disrupt the lipid bilayer system and release the drugs to the external microenvironment [128].

Xu et al. synthesized poly(2-ethyl-2-oxazoline)-cholesterol hemisuccinate (PEtOz-CHEMS) pH-reactive liposomes for a fluorescent drug DOX release. These CHEM-conjugated PEtOzylated liposomes showed acidic pH-responsive swelling. In vitro study demonstrated an efficient cellular uptake of nanoliposomes and a significant increase of fluorescent intensity at pH 6.4 in comparison with biological pH 7.4. So, the pH-responsive PEtOz liposomes confirmed a promising material for pH-sensitive drug delivery [129]. Yuba et al. prepared a dextran derivative of 3-methylglutarylated residues (MGlu-Dex) pH-responsive liposomes as a potent antigen carrier system for cancer immunotherapy. The surface modification of MGlu-Dex liposomes was done by the integration of egg yolk phosphatidylcholine (PTCh) to produce highly pH-responsive liposomes, which were unchanged at physiological pH and highly unbalanced at weakly acidic pH condition. The PTCh-modified MGlu-Dex liposomes (P-MGlu-Dex) were internalized efficiently by dendritic cells and released the drug ovalbumin (OVA) into the cytoplasm. The OVA-loaded P-MGlu-Dex showed excellent tumor suppression effect in E.G7-OVA tumor-bearing mice [130].

TABLE 2.8
Various Liposomes Used for pH-Sensitive Drug-Release Applications.

Liposomes	pH-Sensitive	Drug	Application	References
PEtOz-CHEMS	6.4	Doxorubicin	Anticancer drug delivery	[129]
P-MGlu-Dex	<5	Ovalbumin	Cancer immunotherapy	[130]
PG-(MGlu)-HA/PG-(CHex)-HA	<5	Doxorubicin	Cancer chemotherapy	[131]
MGlu-HPG	5	TRX and CpG-DNA	Cancer immunotherapy	[132]
IM-Chol-PC	5	Curcumin	Targeted drug delivery	[133]
RGD-PE	5	Docetaxel/Coumarin-6	Anticancer drug delivery	[134]

Miyazaki et al. fabricated modified HA-based pH-sensitive and target-specific liposomes for cancer chemotherapy. HA polymers were subjected to target CD 44 expressing cell, which is a known cancer cell surface marker. To introduce the pH sensitivity, HA polymers were modified with carboxyl groups. Such a process resulted in two charged derivatives such as 3-methyl glutarylated (MGlu)-HA and 2-carboxycyclohexane-1-carboxylated (CHex)-HA. These two derivatives of HA were conjugated with poly(glycidol)s liposomes encapsulated with anticancer drug DOX. Both modified HA containing liposomes showed content release in lower pH environment but CHex-HA-modified liposomes exhibited more efficient drug delivery than (MGlu)-HA-modified liposomes [131]. Similarly, Yoshizaki et al. used 3-methyl glutarylated hyperbranched poly(glycidol) (MGlu-HPG) liposomes as an antigen delivery system for cancer immunotherapy. There are two important functions required for the efficient vaccine carrier in cancer immunotherapy: (1) antigen delivery to dendritic cells (DCs) and (2) an adjuvant activator for DCs. Yoshizaki et al. incorporated a cationic lipid 5-Didodecyloxybenzamidine hydrochloride (TRX) and a ligand CpG-DNA to the MGlu-HPG liposome to activate Toll-like receptor 9 (TLR9) expressing over the endosomes of DCs. These antigen-loaded modified liposomes promoted cytokines production, thus exhibiting a strong antitumor effect [132].

Ju et al. prepared unique pH-sensitive liposomes by using N-(3-Aminopropyl) imidazole-cholesterol (IM-Chol) and phosphatidylcholine for targeted drug delivery. These IM-chol liposomes assembled uniformly at pH 7. However, in acidic conditions, at pH 5, these liposomes lost its structural integrity by exhibiting pH stimuli-responsive property. CC-loaded IM-chol liposomes were later investigated for in vitro drug-release study and showed enhanced cytotoxic effects in EC109 cancer cells [133]. Chang et al. developed arginine–glycine–aspartic acid (RGD)-modified phosphatidylethanolamine pH-sensitive liposomes for anticancer drug delivery into tumor cells. The synthesized liposomes were loaded with anticancer drugs docetaxel and coumarin-6 and showed faster release from liposomes at acidic pH conditions [134].

4.3 Inorganic–Organic Nanocomposites

A wide variety of inorganic nanocomposites has been used in pH-sensitive drug-delivery applications. They include silica, silver, gold, zinc oxide, and iron oxide (magnetic) nanoparticles. Owing to their rigid surface, they exhibit highly stable and controlled drug-delivery applications [135,136]. Various organic materials, as well as inorganic materials, can modify the surface of inorganic nanoparticles easily. The other advantages of inorganic nanoparticles include thermal stability, good biocompatibility, and high amount of drug encapsulations [137,138]. These inorganic materials are mostly synthesized by the chemical synthesis method. Generally, drug molecules are loaded onto these inorganic nanoparticles with/without linker by chemical bonds such as covalent bonds, hydrogen bonds, and electrostatic interactions. The pH-sensing action of inorganic nanoparticles is guided by incorporated ionizable compounds or polymers [139]. For pH-sensitive drug-delivery applications, two types of inorganic–organic hybrid nanocomposites [mesoporous silica nanoparticles (MSNs), and metal nanocomposites] are used (Table 2.9).

4.3.1 Mesoporous silica nanoparticles

Mesoporous silica nanoparticles are porous and spherical shaped nanomaterials possessing some unique properties such as facile surface functionalization, mesoporosity, and excellent biocompatibility (Fig. 2.4A) [140]. Owing to these unique structural properties, MSNs offer an opportunity to load a high amount of drugs within the nanocarrier. MSNs are synthesized using sol–gel methods (Stöber synthesis) using a template polymer like tetraethyl orthosilicate (TEOS) and different surfactants like cetyltrimethylammonium bromide (CTAB), or spray drying method [141,142]. Furthermore, the surface of MSNs contains ample silanol (Si–O–H) functional groups, which facilitate the surface coating and drug loading. In comparison with conventional MSNs, hollow MSNs have high drug-loading space due to their core hollow structures. Thus, in chemotherapy, hollow MSNs are used for high therapeutic efficiency. Considering all these unique characteristics, MSNs are continuously explored for different biomedical applications, that is, drug delivery, biosensing, and diagnosis [143,144].

Xu et al. developed pH-responsive hollow MSNs (HMSNs) for anticancer drug-delivery applications. HMSNs were synthesized using tetradecyltrimethylammonium bromide (TTAB) as the porogen with the assistance of triethanolamine (TEA). HMSNs were loaded with DOX and showed sustained drug release in a low pH environment around 50 wt.% DOX at pH 5 and 65 wt.% DOX at pH 3 [145]. Llinàs et al. prepared a pH-sensitive MSN-based nanovehicle for dual drug DOX and camptothecin (CPT) delivery. The drug CPT was internalized through pores of the MSNs, while DOX was covalently conjugated with an aldehyde-

TABLE 2.9
Summary of Inorganic–Organic Nanocomposites for pH-Sensitive Drug-Release Applications.

Inorganic–Organic Nanocomposites	pH-Sensitive	Drug	Application	References
MESOPOROUS SILICA NANOPARTICLES				
Hollow MSNs	3	Doxorubicin	Anticancer drug delivery	[145]
MSNs-diHz-PEG	<4	Doxorubicin/ Camptothecin	Cancer chemotherapy	[146]
MSNs-P4VP	4	Methotroxate	Anticancer therapeutics	[147]
MSNs-DMAEA-Azobenzene	5	Camptothecin	Anticancer drug delivery	[148]
MSNs-PAA-ConA	<5.5	Doxorubicin	Bone cancer treatment	[149]
Au-MSNs-ATPMS -TGA	6	Doxorubicin	Tumor growth inhibition	[150]
Pd–Ag-MSNs	5	Doxorubicin	Cancer phototherapy and chemotherapy	[151]
MSN-ANA-α-CD and MSN-MBI-β-CD	4	Moxifloxacin	*F. tularensis* infection treatment	[152]
METAL NANOCOMPOSITES				
AuNPs-PEG-PAMAM	5	Doxorubicin	Anticancer drug delivery	[160]
CMC-AuNPs	<5.3	Doxorubicin	Anticancer drug delivery	[161]
MBA/SMART ligand-AuNPs	~5.5	–	Cancer photothermal therapy	[162]
DNA-GNR	~5	Doxorubicin	Anticancer drug delivery	[163]
P(HEO2MA-co-MACPT)-AgNPs	5	Camptothecin	Anticancer drug delivery	[164]
ZnO NPs	5	Doxorubicin	Anticancer drug delivery	[165]
ZnO QDs- diCPEG-HA	5	Doxorubicin	Anticancer drug delivery	[166]
Fe_3O_4/PDMAEMA -NPs	3	Phenolphthalein	Targeted drug delivery	[167]
Fe_3O_4-mSiO_2-P2	5.8	Doxorubicin	Anticancer drug delivery	[168]
MnFe-LDH	<6	Methotrexate	Anticancer drug delivery	[169]

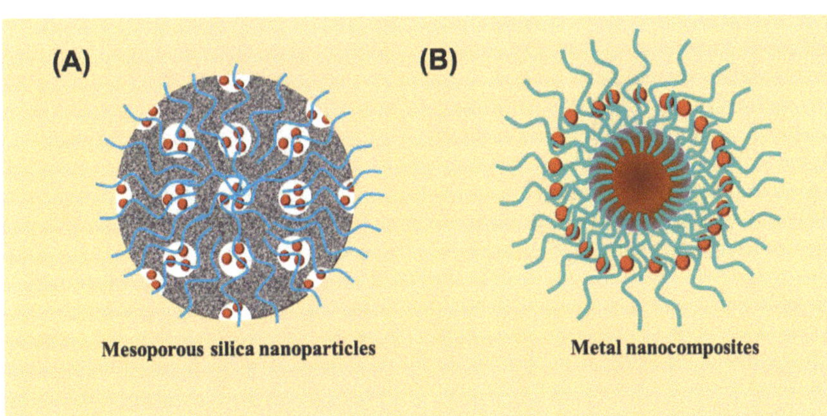

FIG. 2.4 Schematic diagram of inorganic–organic nanocomposites: **(A)** mesoporous silica nanoparticles; **(B)** metal nanoparticles.

functionalized MSNs through a dihydrazide-polyethylene glycol chain. In this formulation, DOX and the linker played a pH-responsive surface coating for drug delivery. At acidic pH condition, the drug-loaded MSNs rupture the DOX surface coat and sustain released CPT that resulted in an acidity-activated synergistic chemotherapy [146]. Rafi et al. synthesized pH-sensitive methotrexate (MTX)-loaded MSNs (MCM-41) for continuous drug release for anticancer therapeutics. The pores of the MCM-41 were polymerized with poly(4-vinylpyridine) as a pH-responsive gatekeeper for drug unloading. MTX-loaded MCM-41 exhibited excellent drug release about 93% at pH 4, and 51% at pH 5.8 in 3-h time frame [147].

Tian et al. fabricated surface-modified MSNs for both light- and pH-sensitive drug-delivery systems. So, the MSNs were modified with amino- and acrylic groups followed by the conjugation of azobenzene and polymerized DMAEA, respectively. The modified MSNs were filled with a model drug rhodamine B (RhB) to understand the drug-unloading efficiency of this novel nanocarrier. The release profile of RhB showed high drug-unloading efficiency of 95% at pH 5 under light irradiation at 413 nm, compared to only 55% in pH 7.4 in dark suggesting that the anticancer drug CPT-loaded MSNs killed the cancer cells efficiently with the IC_{50} value of 0.02 μg mL^{-1} upon exposure of 413 nm light [148]. Carmona et al. developed a multifunctional DOX-loaded MSNs for bone cancer treatment. This nanoassembly constituted of MSN with a capping layer polyacrylic acid (PAA) and a targeting ligand plant lectin concanavalin A (ConA), which can recognize overexpressed sialic acid (SA) in the bone tumor cell. This PAA-ConA-MSNs exhibited a high degree of internalization and a 100% cytotoxic effect in human osteosarcoma cells [149]. Xu et al. synthesized gold (Au) particle-conjugated MSNs (Au-MSNs) as pH and GSH-sensitive nanocarrier for intracellular anticancer drug delivery. In this study, MSNs were functionalized with 3-aminopropyl trimethoxysilane (ATPMS) followed by thioglycolic acid to acquire MSNs–SH. In a further study, the DOX-loaded Au-MSNs showed a faster release of almost 45.6% within 11 h at pH 6 with 10 mM GSH and in vitro release showed an inhibitory effect on tumor growth of Tca8113 cells with controlled drug release [150].

Fang et al. prepared mesoporous silica-coated Pd–Ag nanoparticles for a dual photo- and pH-sensitive drug release for cancer chemo-/phototherapy. The near-infrared (NIR) light was absorbed by Pd–Ag nanoparticles and converted into heat. By this action of radiation, DOX-loaded Pd–Ag-MSNs nanocarrier released drugs at the endosomal pH of cancer cells and resulted in higher killing efficiency with the synergistic effect of phototherapy and chemotherapy [151]. Li et al. developed pH-sensitive nanovalves of modified MSNs loaded with moxifloxacin (MXF) for treating *Francisella tularensis* infections. They synthesized two differently modified nanosystems for the comparative analysis of drug-unloading efficiency. First nanosystem, MCM-41 MSNs framework was covalently attached with stalk anilinoalkane (ANA) and a cap α-cyclodextrin (CD); second nanosystem, MCM-41 framework was attached with stalk 1-methyl-1H-benzimidazole (MBI) and a cap β-CD. Both nanosystems were closed at pH 7.4 and open to release the drug at pH 6 or lower. Then, the MXF-loaded MSN-ANA and MSN-MBI demonstrated efficient drug unloading *F. tularensis* infected macrophages and 2.7-fold more efficient than free MXF administration [152].

4.3.2 Metal nanocomposites

Metallic nanoparticles (MNPs) can be functionalized with pH-sensitive polymers that can be natural or synthetic (Fig. 2.4B). Owing to several unique properties such as rapid and easy synthesis process, facile surface functionalization, optical, electrical, magnetic, and surface plasmon behavior, MNPs are being explored more for potential drug nanocarrier [153,154]. In general, MNPs can be synthesized and stabilized by two methods such as chemical (photo-/electro-/chemical reduction) and physical (mechanical milling, sputtering, and physical deposition) [155]. For drug-delivery applications, chemical reduction methods are probable approaches for the surface functionalization of pH-sensitive polymers and drug molecules. Depending on the metals, MNPs show additional characteristics such as fluorescent, magnetic, and photoactive in drug-delivery applications. Metals such as silver and gold nanoparticles show higher surface plasmon resonance. Iron nanoparticles show supermagnetic properties, and zinc quantum dots show fluorescence properties for imaging applications [156–159].

Khutale et al. designed a gold (Au) nanoparticle-based pH-sensitive drug-delivery system for cancer chemotherapy. The Au nanoparticles were surface coated and functionalized with PEG polymer and PAMAM G4 dendrimers, respectively. Anticancer drug DOX was then conjugated to Au-PEG-PAMAM nanoparticles via the amide bond of PAMAM dendrimer. The resulted DOX-loaded Au-PEG-PAMAM exhibited pH-responsive drug release in the acidic/endosomal pH 5 environment and unveiled greater cytotoxicity than free DOX in A549 cells [160]. On the contrary,

Madhusudhan et al. synthesized carboxymethyl chitosan (CMC)-capped Au nanoparticles for anticancer drug delivery. DOX drug was incorporated to surface chitosan by the noncovalent interaction of an amino group of DOX to the carboxyl group of CMCs. DOX-loaded CMC-Au nanoparticles showed effective internalization by cervical cancer cells and released drugs by the induction of nigericin, which is an ionophore that causes intracellular acidification [161].

Jung et al. reported Au nanoparticle-based pH-sensitive theranostic approach through Raman imaging and photothermal therapy. To fabricate this agent, spherical Au nanoparticles (AuNPs) were decorated with a pH-responsive molecule (2-(6,8-dimercaptooctanamido)ethylamino)-3-methyl-4-oxobut-2-enoic acid ("smart" ligand) and Raman reporter surface molecule 4-mercaptobenzoic acid (4-MBA) to produce MBA/SMART-AuNP. At a mild acidic condition of pH ~5.5, these MBA/SMARTAuNPs aggregated rapidly and this aggregation concurrently provides an active spot for surface-enhanced Raman spectroscopy (SERS) with shifts in absorption to the far-red and NIR. Such spectral shifts were effectively used during photothermal therapy for the detection of B16 F10 mouse melanoma cells [162].

Zhang et al. developed pH-sensitive DNA-conjugated gold nanorod (GNR) as a smart nanomaterial for dual pH- and near-infrared (NIR) radiation stimuli-responsive drug delivery. The DNA-GNR was further factionalized with cancer cell targeting ligand such as thiolated poly(ethylene glycol)-biotin, which specifically binds to the overexpressed biotin receptors on the cancer cells. DOX was then loaded to DNA-GNR by ionic interacting with pH-responsive double-stranded DNA. Under an acidic pH condition pH ~5 and/or an 808 nm NIR laser radiation, the DNA-GNR-DOX nanocarrier released its 89.5% DOX payload within 30 min to the external environment, which is a much higher and quick release. Further, this DNA-GNR-DOX nanocomposite exhibited enhanced the cytotoxicity (~67 fold) than free DOX in MCF-7/ADR cancer cells [163]. Qui et al. formulated pH-sensitive prodrug-linked silver nanoparticles (AgNPs) for anticancer drug delivery. The AgNP was encapsulated with a fluorescent anticancer drug CPT conjugated polymer, poly(2-(2-hydroxyethoxy) ethyl methacrylate-co-methacryloyloxy-3-thiahexanoyl-CPT) (P(HEO2MA-co-MACPT). Thus, the resulted P(HEO2MA-co-MACPT)-AgNPs complex at pH 7.4 experienced the fluorescence quenching of the CPT molecule due to nanoparticle surface energy transfer (NSET). However, under low pH conditions (pH 5), no such behavior is resulted, as the CPT molecules were gradually released from P(HEO2MA-co-MACPT)-AgNPs nanocarrier complex due to the cleavage of acid-liable bonds. Thus, the acidic pH-sensitive drug-delivery property of this nanocarrier complex exhibited in vitro drug release in HeLa cells by showing stronger blue fluorescence of CPT after 8–12 h of incubation [164].

Liu et al. developed pH-responsive zinc oxide nanoparticles (ZnONPs) for intracellular drug-delivery applications. DOX-loaded ZnONPs were prepared by the chemical synthesis method. In DOX-ZnONPs, DOX does not show fluorescent properties at pH 7.5 because of the "π– π stacking" of a rigid planar structure. However, under acidic pH conditions, DOX is released from the nanoparticles and can be directly visualized by fluorescent signal recovery. Owing to this pH responsive, DOX-ZnO NPs effectively delivered DOX, which significantly inhibited the growth of MCF-7S and MCF-7R drug resistant cancer cells [165]. Similarly, Cai et al. formulated a pH-responsive ZnO quantum dots (ZnO QDs) for intracellular drug delivery applications. The ZnO QDs were surface factionalized with dicarboxyl-terminated PEG polymer that is stable in physiological pH conditions. PEG-coated ZnO QDs were then decorated with a CD44-targeting ligand, HA and anticancer drug DOX. In acidic/endosomal/lysosomal pH conditions, the fully functionalized DOX-ZnO QDs-HA nanocomplex dissociated metal-drug complex and exhibited controlled drug release. DOX-ZnO QDs-HA nanocomplex showed antitumor synergistic by the addition of the antitumor effect of Zn^{2+} ions in combination with anticancer drug [166].

Zhou et al. synthesized pH-sensitive magnetic nanoparticles as a potent drug nanocarrier for target-specific drug delivery. They prepared N-bromoisobutyric acid functionalized magnetite nanoparticles followed by the coating of pH-responsive poly((2-dimethylamino) ethyl methacrylate) (PDMAEMA) polymer to produce pH-sensitive Fe_3O_4/PDMAEMA core–shell nanoparticles. This process of the surface coating was done by atom transfer radical polymerization (ATRP) method. pH sensitivity and magnetic properties of Fe_3O_4/PDMAEMA nanoparticles were characterized by dynamic light scattering spectroscopy (DLS) and vibrating sample magnetometry (VSM). These Fe_3O_4/PDMAEMA nanoparticles exhibited an increase in their hydrodynamic size by decreasing the pH of the solution and superparamagnetism with very low coercivity and no obvious hysteresis. To investigate the pH-dependent drug-releasing properties, a model drug phenolphthalein was loaded onto

Fe_3O_4/PDMAEMA nanoparticles, which resulted in a controlled drug release of almost 56.4% at pH 3 condition in 48 h [167].

Yang et al. prepared polymer-modified core–shell Fe_3O_4@$mSiO_2$ magnetic nanoparticles for anticancer drug-delivery application. The Fe_3O_4@$mSiO_2$ nanoparticles were prepared by chemical synthesis and were further grafted with pH-responsive β-thiopropionate-PEG (P2) polymer as a blocking layer to inhibit the premature release of drugs. The DOX-loaded Fe_3O_4@$mSiO_2$-P2 nanoparticles showed more than 90% drug release at pH 5.8 after 24 h of incubation. Furthermore, the DOX-loaded Fe_3O_4@$mSiO_2$-P2 nanoparticles were quickly internalized by cancer cells and showed selectively drug-release properties with the magnetic targeted site [168].

Huang et al. designed manganese–iron-layered double hydroxide (MnFe-LDH) as a pH-responsive nanocarrier for sustained drug release along with magnetic resonance imaging (MRI) contrast enhancement agent. The MnFe-LDH nano-platform structure responded to the acidic microenvironment and released Mn^{2+} and Fe^{3+} paramagnetic ions as T_1 MRI contract agents. The MnFe-LDL nanocarrier was loaded with chemotherapeutic drug MTX by in situ coprecipitation process and studied its drug-release profile in HeLa cells. T_1-weighted images of HeLa cells showed enhanced bright MRI signals with the elapsed time due to Mn^{2+} and Fe^{3+} ions in endosomes/lysosomes microenvironments. MnFe-LDH showed an excellent drug release in an acidic environment that was approximately 90%, 85%, and 80% at pH 4, 5, and 6, respectively. In addition, MnFe-LDH exhibited excellent in vivo therapeutic performance with good MRI contrast and inhibitory effect against S180 tumor-bearing BALB/c nude mice models [169].

5 ADVANTAGES AND DISADVANTAGES

Designing and development of a pH-sensitive nanocarrier have been demonstrated to be a potent and efficient strategy for the smart release of drugs. The pH-sensitive nanosystem emerged as a potential vehicle as both active and passive targets to overcome biological barriers. The employment of pH-responsive polymers to nanomaterials has increased the biocompatibility as well as its pH-sensing properties. The pH-responsive nanocarriers have also provided a platform for hydrophobic or low solubility drug delivery by forming a hydrophobic corona inside. In addition, the external environment of the hydrophobic corona covered with hydrophilic pH-sensitive polymers to make contact with aqueous surroundings. The presence of hydrophobic corona and hydrophilic external layers makes the nanocarrier more stable and provides prolonged circulation time in the bloodstream. Several pH-sensitive nanocarriers were functionalized with cell-specific ligands allowed the ligands to bind to their complementary receptors on targeted cells. Furthermore, drug-loaded pH-sensitive nanocarriers were also engineered with different bioimaging and diagnostic agents to have a synergistic impact on theranostics. With the release of drugs, nanocarriers also release bioimaging agents simultaneously to help in monitoring the drug release as well as to locate the target site. Many diseases and complications with pH variant microenvironment provide a great opportunity to pH-sensitive nanocarrier for targeted drug delivery with negligible side effects.

Despite numerous advantages, pH-sensitive nanomaterials still face many problems related to smart drug release. In physiological pH conditions, it is difficult to differentiate between the targeted site and healthy surroundings. In addition, the biodegradability or excretion of nanocarrier and unused/remained/uncharacterized drugs brings new challenges to researchers for pH-sensitive drug delivery and development. The incompatible renal excretion of remained hydrophobic drugs may cause accumulation and aggregation in different organs such as liver and kidney [170,171]. Hence, there is a need to optimize the drug concentrations and monitor all parameters associated with drug delivery to improve drug delivery, biocompatibility, and biodegradability.

6 CONCLUDING REMARKS

In the context of the smart drug-delivery system, pH-responsive smart drug release has gained rising interest to overcome the challenges and limitations in therapeutics as well as diagnosis. This chapter reviewed different pH-sensitive materials with their significant improved mechanism for smart drug release in specific targets. The pH-sensitive nanocarrier has been improved by the combination of different biopolymers and other stimuli-responsive agents, that is, redox potential, thermal, photo, and magnetic. The applications of pH-sensitive drug delivery have considerably increased in the drug-delivery processes in chemotherapeutics. Several target-specific ligands have been incorporated within nanocarriers to overcome the undesirable side effects. It is worth noting that the maximum studies reviewed here are just proof of concept and experiments have been optimized at maximum in vitro and/or at the cell-based level. Although many new concepts have emerged for the smart drug delivery, it is still considered at its infancy as further efforts are needed to completely elucidate the bioavailability, biodegradability, and excretion of nanocarriers after drug release using animal models.

ACKNOWLEDGMENTS

KMP acknowledges the receipt of DBT-IYBA-BT/07/IYBA/2013-19 and SERB-SB/YS/LS-380/2013 grants from the Government of India. MKM acknowledges the MHRD fellowship from IIT-Roorkee for pursuing PhD.

REFERENCES

[1] Zhang L, Gu F, Chan J, Wang A, Langer R, Farokhzad O. Nanoparticles in medicine: therapeutic applications and developments. Clinical Pharmacology and Therapeutics 2008;83(5):761−9.

[2] Sounderya N, Zhang Y. Use of core/shell structured nanoparticles for biomedical applications. Recent Patents on Biomedical Engineering 2008;1(1):34−42.

[3] Li Z, Barnes JC, Bosoy A, Stoddart JF, Zink JI. Mesoporous silica nanoparticles in biomedical applications. Chemical Society Reviews 2012;41(7):2590−605.

[4] Elsabahy M, Wooley KL. Design of polymeric nanoparticles for biomedical delivery applications. Chemical Society Reviews 2012;41(7):2545−61.

[5] Grabrucker AM, Ruozi B, Belletti D, et al. Nanoparticle transport across the blood brain barrier. Tissue Barriers 2016;4(1):e1153568.

[6] Saraiva C, Praça C, Ferreira R, Santos T, Ferreira L, Bernardino L. Nanoparticle-mediated brain drug delivery: overcoming blood−brain barrier to treat neurodegenerative diseases. Journal of Controlled Release 2016;235:34−47.

[7] Liu J, Huang Y, Kumar A, et al. pH-sensitive nano-systems for drug delivery in cancer therapy. Biotechnology Advances 2014;32(4):693−710.

[8] Karimi M, Eslami M, Sahandi-Zangabad P, et al. pH-Sensitive stimulus-responsive nanocarriers for targeted delivery of therapeutic agents. Wiley Interdisciplinary Reviews: Nanomedicine Nanobiotechnology 2016; 8(5):696−716.

[9] Meng F, Zhong Z, Feijen J. Stimuli-responsive polymersomes for programmed drug delivery. Biomacromolecules 2009;10(2):197−209.

[10] Mura S, Nicolas J, Couvreur P. Stimuli-responsive nanocarriers for drug delivery. Nature Materials 2013;12(11): 991.

[11] Davis ME, Chen Z, Shin DM. Nanoparticle therapeutics: an emerging treatment modality for cancer. Nanoscience and Technology: A Collection of Reviews from Nature Journals 2010:239−50. World Scientific.

[12] Zhang J, Rana S, Srivastava R, Misra R. On the chemical synthesis and drug delivery response of folate receptor-activated, polyethylene glycol-functionalized magnetite nanoparticles. Acta Biomaterialia 2008;4(1):40−8.

[13] Liu L, Yao W, Rao Y, Lu X, Gao J. pH-Responsive carriers for oral drug delivery: challenges and opportunities of current platforms. Drug Delivery 2017;24(1):569−81.

[14] Balamuralidhara V, Pramodkumar T, Srujana N, et al. pH sensitive drug delivery systems: a review. American Journal of Drug Discovery and Development 2011; 1(1):25.

[15] Webb BA, Chimenti M, Jacobson MP, Barber DL. Dysregulated pH: a perfect storm for cancer progression. Nature Reviews Cancer 2011;11(9):671.

[16] Ganta S, Devalapally H, Shahiwala A, Amiji M. A review of stimuli-responsive nanocarriers for drug and gene delivery. Journal of Controlled Release 2008;126(3): 187−204.

[17] Dai S, Ravi P, Tam KC. pH-Responsive polymers: synthesis, properties and applications. Soft Matter 2008;4(3): 435−49.

[18] Gil ES, Hudson SM. Stimuli-reponsive polymers and their bioconjugates. Progress in Polymer Science 2004; 29(12):1173−222.

[19] Zhu L, Smith PP, Boyes SG. pH-responsive polymers for imaging acidic biological environments in tumors. Journal of Polymer Science Part B: Polymer Physics 2013; 51(14):1062−7.

[20] Bazban-Shotorbani S, Hasani-Sadrabadi MM, Karkhaneh A, et al. Revisiting structure-property relationship of pH-responsive polymers for drug delivery applications. Journal of Controlled Release 2017;253: 46−63.

[21] Zheng Y, Wang L, Lu L, Wang Q, Benicewicz BC. pH and thermal dual-responsive nanoparticles for controlled drug delivery with high loading content. ACS Omega 2017;2(7):3399−405.

[22] Nasongkla N, Bey E, Ren J, et al. Multifunctional polymeric micelles as cancer-targeted, MRI-ultrasensitive drug delivery systems. Nano Letters 2006;6(11):2427−30.

[23] Lim EK, Huh YM, Yang J, Lee K, Suh JS, Haam S. pH-triggered drug-releasing magnetic nanoparticles for cancer therapy guided by molecular imaging by MRI. Advanced Materials 2011;23(21):2436−42.

[24] Syamchand SS, Sony G. Multifunctional hydroxyapatite nanoparticles for drug delivery and multimodal molecular imaging. Microchimica Acta 2015;182(9−10): 1567−89.

[25] Boelsma E, Van De Vijver LP, Goldbohm RA, Klöpping-Ketelaars IA, Hendriks HF, Roza L. Human skin condition and its associations with nutrient concentrations in serum and diet. American Journal of Clinical Nutrition 2003;77(2):348−55.

[26] Felber AE, Dufresne M-H, Leroux J-C. pH-sensitive vesicles, polymeric micelles, and nanospheres prepared with polycarboxylates. Advanced Drug Delivery Reviews 2012;64(11):979−92.

[27] Boron WF. Regulation of intracellular pH. Advances in Physiology Education 2004;28(4):160−79.

[28] Ferris DG, Francis SL, Dickman ED, Miler-Miles K, Waller JL, McClendon N. Variability of vaginal pH determination by patients and clinicians. The Journal of the American Board of Family Medicine 2006;19(4):368–73.
[29] Linhares IM, Summers PR, Larsen B, Giraldo PC, Witkin SS. Contemporary perspectives on vaginal pH and lactobacilli. American Journal of Obstetrics Gynecology 2011;204(2). 120.e121–120.e125.
[30] Schwalfenberg GK. The alkaline diet: is there evidence that an alkaline pH diet benefits health? Journal of Environmental Public Health 2012;2012:727630.
[31] Mohammed HM, Abdelatief DA. Easy blood gas analysis: implications for nursing. Egyptian Journal of Chest Diseases and Tuberculosis 2016;65(1):369–76.
[32] Cardone RA, Casavola V, Reshkin SJ. The role of disturbed pH dynamics and the Na+/H+ exchanger in metastasis. Nature Reviews Cancer 2004;5(10):786.
[33] Shirmanova MV, Druzhkova IN, Lukina MM, et al. Intracellular pH imaging in cancer cells in vitro and tumors in vivo using the new genetically encoded sensor SypHer2. Biochimica et Biophysica Acta (BBA) - General Subjects 2015;1850(9):1905–11.
[34] Jiang B. Aerobic glycolysis and high level of lactate in cancer metabolism and microenvironment. Genes Diseases 2017;4(1):25–7.
[35] Pu H-L, Chiang W-L, Maiti B, et al. Nanoparticles with dual responses to oxidative stress and reduced pH for drug release and anti-inflammatory applications. ACS Nano 2014;8(2):1213–21.
[36] Steen KH, Steen AE, Reeh PW. A dominant role of acid pH in inflammatory excitation and sensitization of nociceptors in rat skin, in vitro. Journal of Neuroscience 1995;15(5):3982–9.
[37] Machida T. A study of intragastric pH in patients with peptic ulcer-with special reference to the clinical significance of basal pH value. Gastroenterologia Japonica 1981;16(5):447–58.
[38] Alhouayek M, Lambert DM, Delzenne NM, Cani PD, Muccioli GG. Increasing endogenous 2-arachidonoylglycerol levels counteracts colitis and related systemic inflammation. The FASEB Journal 2011;25(8):2711–21.
[39] Markman JL, Rekechenetskiy A, Holler E, Ljubimova JY. Nanomedicine therapeutic approaches to overcome cancer drug resistance. Advanced Drug Delivery Reviews 2013;65(13–14):1866–79.
[40] Peer D, Karp JM, Hong S, Farokhzad OC, Margalit R, Langer R. Nanocarriers as an emerging platform for cancer therapy. Nature Nanotechnology 2007;2(12):751.
[41] Jabr-Milane LS, van Vlerken LE, Yadav S, Amiji MM. Multi-functional nanocarriers to overcome tumor drug resistance. Cancer Treatment Reviews 2008;34(7): 592–602.
[42] Gugulothu D, Kulkarni A, Patravale V, Dandekar P. pH-sensitive nanoparticles of curcumin–celecoxib combination: evaluating drug synergy in ulcerative colitis model. Journal of Pharmaceutical Sciences 2014; 103(2):687–96.
[43] Wang X-Q, Zhang Q. pH-sensitive polymeric nanoparticles to improve oral bioavailability of peptide/protein drugs and poorly water-soluble drugs. European Journal of Pharmaceutics and Biopharmaceutics 2012;82(2): 219–29.
[44] Huh KM, Kang HC, Lee YJ, Bae YH. pH-sensitive polymers for drug delivery. Macromolecular Research 2012; 20(3):224–33.
[45] Cerletti A, Drewe J, Fricker G, Eberle A, Huwyler J. Endocytosis and transcytosis of an immunoliposome-based brain drug delivery system. Journal of Drug Targeting 2000;8(6):435–46.
[46] Bareford LM, Swaan PW. Endocytic mechanisms for targeted drug delivery. Advanced Drug Delivery Reviews 2007;59(8):748–58.
[47] Liu Y, Wang W, Yang J, Zhou C, Sun J. pH-sensitive polymeric micelles triggered drug release for extracellular and intracellular drug targeting delivery. Asian Journal of Pharmaceutical Sciences 2013;8(3):159–67.
[48] Meng F, Zhong Y, Cheng R, Deng C, Zhong Z. pH-sensitive polymeric nanoparticles for tumor-targeting doxorubicin delivery: concept and recent advances. Nanomedicine 2014;9(3):487–99.
[49] Fleige E, Quadır MA, Haag R. Stimuli-responsive polymeric nanocarriers for the controlled transport of active compounds: concepts and applications. Advanced Drug Delivery Reviews 2012;64(9):866–84.
[50] Lungu II, Radulescu M, Mogosanu GD, Grumezescu AM. pH sensitive core-shell magnetic nanoparticles for targeted drug delivery in cancer therapy. Romanian Journal of Morphology and Embryology 2016;57(1):23–32.
[51] Kocak G, Tuncer C, Bütün V. pH-Responsive polymers. Polymer Chemistry 2017;8(1):144–76.
[52] Schmaljohann D. Thermo-and pH-responsive polymers in drug delivery. Advanced Drug Delivery Reviews 2006; 58(15):1655–70.
[53] Alvarez-Lorenzo C, Blanco-Fernandez B, Puga AM, Concheiro A. Crosslinked ionic polysaccharides for stimuli-sensitive drug delivery. Advanced Drug Delivery Reviews 2013;65(9):1148–71.
[54] Singh A, Garg G, Sharma P. Nanospheres: a novel approach for targeted drug delivery system. International Journal of Pharmaceutical Sciences Review and Research 2010;5(3):84–8.
[55] Tang S, Yin Q, Su J, et al. Inhibition of metastasis and growth of breast cancer by pH-sensitive poly (β-amino ester) nanoparticles co-delivering two siRNA and paclitaxel 2015;48:1–15.
[56] Beloqui A, Coco R, Memvanga PB, Ucakar B, des Rieux A, Préat V. pH-sensitive nanoparticles for colonic delivery of curcumin in inflammatory bowel disease. International Journal of Pharmaceutics 2014;473(1–2): 203–12.
[57] Li Z, Qiu L, Chen Q, et al. pH-sensitive nanoparticles of poly (L-histidine)–poly (lactide-co-glycolide)–tocopheryl polyethylene glycol succinate for anti-tumor drug delivery. Acta Biomaterialia 2015;11: 137–50.

[58] Li X, Yang W, Zou Y, Meng F, Deng C, Zhong Z. Efficacious delivery of protein drugs to prostate cancer cells by PSMA-targeted pH-responsive chimaeric polymersomes. Journal of Controlled Release 2015; 220:704–14.
[59] Singh PK, Chibh S, Dube T, Chauhan VS, Panda JJ. Arginine-α, β-dehydrophenylalanine dipeptide nanoparticles for pH-responsive drug delivery. Pharmaceutical Research 2018;35(2):35.
[60] Tang C, Amin D, Messersmith PB, Anthony JE, Prud'homme RK. Polymer directed self-assembly of pH-responsive antioxidant nanoparticles. Langmuir 2015;31(12):3612–20.
[61] Quadir MA, Morton SW, Mensah LB, et al. Ligand-decorated click polypeptide derived nanoparticles for targeted drug delivery applications. Nanomedicine: Nanotechnology, Biology and Medicine 2017;13(5): 1797–808.
[62] Lian H, Du Y, Chen X, et al. Core cross-linked poly (ethylene glycol)-graft-Dextran nanoparticles for reduction and pH dual responsive intracellular drug delivery. Journal of Colloid Interface Science 2017; 496:201–10.
[63] Menjoge AR, Kannan RM, Tomalia DA. Dendrimer-based drug and imaging conjugates: design considerations for nanomedical applications. Drug Discovery Today 2010;15(5–6):171–85.
[64] Kesharwani P, Jain K, Jain NK. Dendrimer as nanocarrier for drug delivery. Progress in Polymer Science 2014; 39(2):268–307.
[65] Frechet JM. Functional polymers and dendrimers: reactivity, molecular architecture, and interfacial energy. Science 1994;263(5154):1710–5.
[66] Chen H-T, Neerman MF, Parrish AR, Simanek EE. Cytotoxicity, hemolysis, and acute in vivo toxicity of dendrimers based on melamine, candidate vehicles for drug delivery. Journal of the American Chemical Society 2004;126(32):10044–8.
[67] Malik N, Wiwattanapatapee R, Klopsch R, et al. Dendrimers: relationship between structure and biocompatibility in vitro, and preliminary studies on the biodistribution of 125I-labelled polyamidoamine dendrimers in vivo. Journal of Controlled Release 2000; 65(1–2):133–48.
[68] Jevprasesphant R, Penny J, Jalal R, Attwood D, McKeown N, D'emanuele A. The influence of surface modification on the cytotoxicity of PAMAM dendrimers. International Journal of Pharmaceutics 2003;252(1–2):263–6.
[69] Gillies ER, Frechet JM. Dendrimers and dendritic polymers in drug delivery. Drug Discovery Today 2005; 10(1):35–43.
[70] Luong D, Kesharwani P, Deshmukh R, et al. PEGylated PAMAM dendrimers: enhancing efficacy and mitigating toxicity for effective anticancer drug and gene delivery. Acta Biomaterialia 2016;43:14–29.
[71] Sharma AK, Gothwal A, Kesharwani P, Alsaab H, Iyer AK, Gupta U. Dendrimer nanoarchitectures for cancer diagnosis and anticancer drug delivery. Drug Discovery Today 2017;22(2):314–26.
[72] Ali MM, Woods M, Caravan P, et al. Synthesis and relaxometric studies of a dendrimer-based pH-responsive MRI contrast agent. Chemistry–A European Journal 2008;14(24):7250–8.
[73] Tekade RK, Dutta T, Gajbhiye V, Jain NK. Exploring dendrimer towards dual drug delivery: pH responsive simultaneous drug-release kinetics. Journal of Microencapsulation 2009;26(4):287–96.
[74] Liu KC, Yeo Y. Zwitterionic chitosan–polyamidoamine dendrimer complex nanoparticles as a pH-sensitive drug carrier. Molecular Pharmaceutics 2013;10(5): 1695–704.
[75] Kumar P, Paknikar K, Gajbhiye V. A robust pH-sensitive unimolecular dendritic nanocarrier that enables targeted anti-cancer drug delivery via GLUT transporters. Colloids and Surfaces B: Biointerfaces 2018;171:437–44.
[76] Ma S, Zhou J, Wali ARM, et al. Self-assembly of pH-sensitive fluorinated peptide dendron functionalized dextran nanoparticles for on-demand intracellular drug delivery. Journal of Materials Science: Materials in Medicine 2015;26(8):219.
[77] Xu X, Lü S, Gao C, et al. Facile preparation of pH-sensitive and self-fluorescent mesoporous silica nanoparticles modified with PAMAM dendrimers for label-free imaging and drug delivery. Chemical Engineering Journal 2015;266:171–8.
[78] Nomoev AV, Bardakhanov SP, Schreiber M, et al. Structure and mechanism of the formation of core–shell nanoparticles obtained through a one-step gas-phase synthesis by electron beam evaporation. Beilstein Journal of Nanotechnology 2015;6:874.
[79] Ghosh Chaudhuri R, Paria S. Core/shell nanoparticles: classes, properties, synthesis mechanisms, characterization, and applications. Chemical Reviews 2011;112(4): 2373–433.
[80] Chatterjee K, Sarkar S, Rao KJ, Paria S. Core/shell nanoparticles in biomedical applications. Advances in Colloid Interface Science 2014;209:8–39.
[81] Surnar B, Sharma K, Jayakannan M. Core–shell polymer nanoparticles for prevention of GSH drug detoxification and cisplatin delivery to breast cancer cells. Nanoscale 2015;7(42):17964–79.
[82] Yang YY, Wang Y, Powell R, Chan P. Polymeric core-shell nanoparticles for therapeutics. Clinical and Experimental Pharmacology and Physiology 2006;33(5-6): 557–62.
[83] Kim I-S, Jeong Y-I, Cho C-S, Kim S-H. Core-shell type polymeric nanoparticles composed of poly (L-lactic acid) and poly (N-isopropylacrylamide). International Journal of Pharmaceutics 2000;211(1–2):1–8.
[84] Cao Y, Wang B, Wang Y, Lou D. Polymer-controlled core–shell nanoparticles: a novel strategy for sequential drug release. RSC Advances 2014;4(57):30430–9.
[85] Mukhopadhyay P, Chakraborty S, Bhattacharya S, Mishra R, Kundu P. pH-sensitive chitosan/alginate core-shell nanoparticles for efficient and safe oral insulin

delivery. International Journal of Biological Macromolecules 2015;72:640–8.
[86] Mukhopadhyay P, Maity S, Mandal S, Chakraborti AS, Prajapati A, Kundu P. Preparation, characterization and in vivo evaluation of pH sensitive, safe quercetin-succinylated chitosan-alginate core-shell-corona nanoparticle for diabetes treatment. Carbohydrate Polymers 2018;182:42–51.
[87] Soppimath KS, Liu LH, Seow WY, et al. Multifunctional core/shell nanoparticles self-assembled from pH-induced thermosensitive polymers for targeted intracellular anticancer drug delivery. Advanced Functional Materials 2007;17(3):355–62.
[88] Liu R, Li D, He B, et al. Anti-tumor drug delivery of pH-sensitive poly (ethylene glycol)-poly (L-histidine-)-poly (L-lactide) nanoparticles. Journal of Controlled Release 2011;152(1):49–56.
[89] Khandare J, Minko T. Polymer–drug conjugates: progress in polymeric prodrugs. Progress in Polymer Science 2006;31(4):359–97.
[90] Larson N, Ghandehari H. Polymeric conjugates for drug delivery. Chemistry of Materials 2012;24(5):840–53.
[91] Gu Y, Zhong Y, Meng F, Cheng R, Deng C, Zhong Z. Acetal-linked paclitaxel prodrug micellar nanoparticles as a versatile and potent platform for cancer therapy. Biomacromolecules 2013;14(8):2772–80.
[92] Zou J, Jafr G, Themistou E, et al. pH-Sensitive brush polymer-drug conjugates by ring-opening metathesis copolymerization. Chemical Communications 2011; 47(15):4493–5.
[93] Hu F-Q, Liu L-N, Du Y-Z, Yuan H. Synthesis and antitumor activity of doxorubicin conjugated stearic acid-g-chitosan oligosaccharide polymeric micelles. Biomaterials 2009;30(36):6955–63.
[94] Zou J, Zhang F, Zhang S, et al. Poly (ethylene oxide)-block-polyphosphoester-graft-paclitaxel conjugates with acid-labile linkages as a pH-sensitive and functional nanoscopic platform for paclitaxel delivery. Advanced Healthcare Materials 2014;3(3):441–8.
[95] Zhang Y, Yang C, Wang W, et al. Co-delivery of doxorubicin and curcumin by pH-sensitive prodrug nanoparticle for combination therapy of cancer. Scientific Reports 2016;6:21225.
[96] Zhao G, Long L, Zhang L, et al. Smart pH-sensitive nanoassemblies with cleavable PEGylation for tumor targeted drug delivery. Scientific Reports 2017;7(1):3383.
[97] Du J-Z, Du X-J, Mao C-Q, Wang J. Tailor-made dual pH-sensitive polymer–doxorubicin nanoparticles for efficient anticancer drug delivery. Journal of the American Chemical Society 2011;133(44):17560–3.
[98] Lv S, Tang Z, Zhang D, et al. Well-defined polymer-drug conjugate engineered with redox and pH-sensitive release mechanism for efficient delivery of paclitaxel. Journal of Controlled Release 2014;194:220–7.
[99] Dalwadi C, Patel G. Application of nanohydrogels in drug delivery systems: recent patents review. Recent Patents on Nanotechnology 2015;9(1):17–25.
[100] Qian Z-Y, Fu S-Z, Feng S-S. Nanohydrogels as a prospective member of the nanomedicine family. Nanomedicine 2013;8(2):161–4.
[101] Akram M, Hussain R. Nanohydrogels: history, development, and applications in drug delivery. Nanocellulose Nanohydrogel Matrices: Biotechnological Biomedical Applications 2017:297–330.
[102] Swain SK, Prusty K. Biomedical applications of acrylic-based nanohydrogels. Journal of Materials Science 2018;53(4):2303–25.
[103] Jaiswal MK, Pradhan A, Banerjee R, Bahadur D. Dual pH and temperature stimuli-responsive magnetic nanohydrogels for thermo-chemotherapy. Journal of Nanoscience and Nanotechnology 2014;14(6):4082–9.
[104] Jin S, Wan J, Meng L, et al. Biodegradation and toxicity of protease/redox/pH stimuli-responsive PEGlated PMAA nanohydrogels for targeting drug delivery. ACS Applied Materials and Interfaces 2015;7(35):19843–52.
[105] Rao KM, Rao KK, Ramanjaneyulu G, Ha C-S. Curcumin encapsulated pH sensitive gelatin based interpenetrating polymeric network nanogels for anti cancer drug delivery. International Journal of Pharmaceutics 2015; 478(2):788–95.
[106] Li Y, Bui QN, Duy LTM, Yang HY, Lee DS. One-step preparation of pH-responsive polymeric nanogels as intelligent drug delivery systems for tumor therapy. Biomacromolecules 2018;19(6):2062–70.
[107] Zhang Z, Wan J, Sun L, Li Y, Guo J, Wang C. Zinc finger-inspired nanohydrogels with glutathione/pH triggered degradation based on coordination substitution for highly efficient delivery of anti-cancer drugs. Journal of Controlled Release 2016;225:96–108.
[108] Zhou T, Zhao X, Liu L, Liu P. Preparation of biodegradable PEGylated pH/reduction dual-stimuli responsive nanohydrogels for controlled release of an anti-cancer drug. Nanoscale 2015;7(28):12051–60.
[109] Hemmati K, Masoumi A, Ghaemy M. Synthesis and characterization of pH-responsive nanohydrogels as biocompatible drug carriers based on chemically modified tragacanth gum polysaccharide. RSC Advances 2015;5(104):85310–8.
[110] Álvarez-Bautista A, Duarte C, Mendizábal E, Katime I. Controlled delivery of drugs through smart pH-sensitive nanohydrogels for anti-cancer therapies: synthesis, drug release and cellular studies. Designed Monomers Polymers 2016;19(4):319–29.
[111] Letchford K, Burt H. A review of the formation and classification of amphiphilic block copolymer nanoparticulate structures: micelles, nanospheres, nanocapsules and polymersomes. European Journal of Pharmaceutics Biopharmaceutics 2007;65(3):259–69.
[112] Lv Y, Huang H, Yang B, Liu H, Li Y, Wang J. A robust pH-sensitive drug carrier: aqueous micelles mineralized by calcium phosphate based on chitosan. Carbohydrate Polymers 2014;111:101–7.
[113] Zhong Y, Yang W, Sun H, et al. Ligand-directed reduction-sensitive shell-sheddable biodegradable micelles actively

deliver doxorubicin into the nuclei of target cancer cells. Biomacromolecules 2013;14(10):3723–30.

[114] Chan DP, Owen SC, Shoichet MS. Double click: dual functionalized polymeric micelles with antibodies and peptides. Bioconjugate Chemistry 2013;24(1):105–13.

[115] Zhang L, Lin Y, Zhang Y, et al. Fluorescent micelles based on star amphiphilic copolymer with a porphyrin core for bioimaging and drug delivery. Macromolecular Bioscience 2012;12(1):83–92.

[116] Hickey RJ, Haynes AS, Kikkawa JM, Park S-J. Controlling the self-assembly structure of magnetic nanoparticles and amphiphilic block-copolymers: from micelles to vesicles. Journal of the American Chemical Society 2011;133(5):1517–25.

[117] Kataoka K, Harada A, Nagasaki Y. Block copolymer micelles for drug delivery: design, characterization and biological significance. Advanced Drug Delivery Reviews 2001;47(1):113–31.

[118] Wu H, Zhu L, Torchilin VP. pH-sensitive poly (histidine)-PEG/DSPE-PEG co-polymer micelles for cytosolic drug delivery. Biomaterials 2013;34(4):1213–22.

[119] Wang Y, Li P, Chen F, et al. A novel pH-sensitive carrier for the delivery of antitumor drugs: histidine-modified auricularia auricular polysaccharide nano-micelles. Scientific Reports 2017;7(1):4751.

[120] Yang Z, Sun N, Cheng R, et al. pH multistage responsive micellar system with charge-switch and PEG layer detachment for co-delivery of paclitaxel and curcumin to synergistically eliminate breast cancer stem cells. Biomaterials 2017;147:53–67.

[121] Shi X, Ma X, Hou M, et al. pH-Responsive unimolecular micelles based on amphiphilic star-like copolymers with high drug loading for effective drug delivery and cellular imaging. Journal of Materials Chemistry B 2017;5(33):6847–59.

[122] Yan G, Wang J, Hu L, et al. Stepwise targeted drug delivery to liver cancer cells for enhanced therapeutic efficacy by galactose-grafted, ultra-pH-sensitive micelles. Acta Biomaterialia 2017;51:363–73.

[123] Teo JY, Chin W, Ke X, et al. pH and redox dual-responsive biodegradable polymeric micelles with high drug loading for effective anticancer drug delivery. Nanomedicine: Nanotechnology, Biology and Medicine 2017;13(2):431–42.

[124] Cai Z, Zhang H, Wei Y, Wei Y, Xie Y, Cong F. Reduction- and pH-sensitive hyaluronan nanoparticles for delivery of iridium (III) anticancer drugs. Biomacromolecules 2017;18(7):2102–17.

[125] Gregoriadis G, Florence A. Liposomes in drug delivery. Drugs 1993;45(1):15–28.

[126] Huang W, Chen L, Kang L, et al. Nanomedicine-based combination anticancer therapy between nucleic acids and small-molecular drugs. Advanced Drug Delivery Reviews 2017;115:82–97.

[127] Sercombe L, Veerati T, Moheimani F, Wu SY, Sood AK, Hua S. Advances and challenges of liposome assisted drug delivery. Frontiers in Pharmacology 2015;6:286.

[128] Paliwal SR, Paliwal R, Vyas SP. A review of mechanistic insight and application of pH-sensitive liposomes in drug delivery. Drug Delivery 2015;22(3):231–42.

[129] Xu H, Hu M, Yu X, et al. Design and evaluation of pH-sensitive liposomes constructed by poly (2-ethyl-2-oxazoline)-cholesterol hemisuccinate for doxorubicin delivery. European Journal of Pharmaceutics and Biopharmaceutics 2015;91:66–74.

[130] Yuba E, Tajima N, Yoshizaki Y, Harada A, Hayashi H, Kono K. Dextran derivative-based pH-sensitive liposomes for cancer immunotherapy. Biomaterials 2014;35(9):3091–101.

[131] Miyazaki M, Yuba E, Hayashi H, Harada A, Kono K. Hyaluronic acid-based pH-sensitive polymer-modified liposomes for cell-specific intracellular drug delivery systems. Bioconjugate Chemistry 2017;29(1):44–55.

[132] Yoshizaki Y, Yuba E, Sakaguchi N, Koiwai K, Harada A, Kono K. pH-sensitive polymer-modified liposome-based immunity-inducing system: effects of inclusion of cationic lipid and CpG-DNA. Biomaterials 2017;141:272–83.

[133] Ju L, Cailin F, Wenlan W, Pinghua Y, Jiayu G, Junbo L. Preparation and properties evaluation of a novel pH-sensitive liposomes based on imidazole-modified cholesterol derivatives. International Journal of Pharmaceutics 2017;518(1–2):213–9.

[134] Chang M, Lu S, Zhang F, et al. RGD-modified pH-sensitive liposomes for docetaxel tumor targeting. Colloids and Surfaces B: Biointerfaces 2015;129:175–82.

[135] Gao W, Chan JM, Farokhzad OC. pH-responsive nanoparticles for drug delivery. Journal of Molecular Pharmaceutics 2010;7(6):1913–20.

[136] Giner-Casares JJ, Henriksen-Lacey M, Coronado-Puchau M, Liz-Marzan LM. Inorganic nanoparticles for biomedicine: where materials scientists meet medical research. Materials Today 2016;19(1):19–28.

[137] Cao Y, Xie Y, Liu L, et al. Influence of phytochemicals on the biocompatibility of inorganic nanoparticles: a state-of-the-art review. Phytochemistry Reviews 2017;16(3):555–63.

[138] Jiao M, Zhang P, Meng J, et al. Recent advancements in biocompatible inorganic nanoparticles towards biomedical applications. Biomaterials Science 2018;6(4):726–45.

[139] Lim E-K, Chung BH, Chung SJ. Recent advances in ph-sensitive polymeric nanoparticles for smart drug delivery in cancer therapy. Current Drug Targets 2018;19(4):300–17.

[140] Bharti C, Nagaich U, Pal AK, Gulati N. Mesoporous silica nanoparticles in target drug delivery system: a review. International Journal of Pharmaceutical Investigation 2015;5(3):124.

[141] Vazquez NI, Gonzalez Z, Ferrari B, Castro Y. Synthesis of mesoporous silica nanoparticles by sol–gel as nanocontainer for future drug delivery applications. Boletín de la Sociedad Española de Cerámica y Vidrio 2017;56(3):139–45.

[142] Watermann A, Brieger J. Mesoporous silica nanoparticles as drug delivery vehicles in cancer. Nanomaterials 2017;7(7):189.
[143] Wen J, Yang K, Liu F, Li H, Xu Y, Sun S. Diverse gatekeepers for mesoporous silica nanoparticle based drug delivery systems. Chemical Society Reviews 2017;46(19):6024–45.
[144] Florek J, Caillard R, Kleitz F. Evaluation of mesoporous silica nanoparticles for oral drug delivery–current status and perspective of MSNs drug carriers. Nanoscale 2017;9(40):15252–77.
[145] Xu H, Zhang H, Wang D, Wu L, Liu X, Jiao Z. A facile route for rapid synthesis of hollow mesoporous silica nanoparticles as pH-responsive delivery carrier. Journal of Colloid Interface Science 2015;451:101–7.
[146] Llinàs MC, Martínez-Edo G, Cascante A, Porcar I, Borrós S, Sánchez-García D. Preparation of a mesoporous silica-based nano-vehicle for dual DOX/CPT pH-triggered delivery. Drug Delivery 2018;25(1):1137–46.
[147] Rafi AA, Mahkam M, Davaran S, Hamishehkar H. A Smart pH-responsive Nano-Carrier as a Drug Delivery System: a hybrid system comprised of mesoporous nanosilica MCM-41 (as a nano-container) & a pH-sensitive polymer (as smart reversible gatekeepers): preparation, characterization and in vitro release studies of an anti-cancer drug. European Journal of Pharmaceutical Sciences 2016;93:64–73.
[148] Tian Y, Kong Y, Li X, Wu J, Ko AC-T, Xing M. Light-and pH-activated intracellular drug release from polymeric mesoporous silica nanoparticles. Colloids and Surfaces B: Biointerfaces 2015;134:147–55.
[149] Martínez-Carmona M, Lozano D, Colilla M, Vallet-Regí M. Lectin-conjugated pH-responsive mesoporous silica nanoparticles for targeted bone cancer treatment. Acta Biomaterialia 2018;65:393–404.
[150] Xu S, Li Y, Chen Z, et al. Mesoporous silica nanoparticles combining Au particles as glutathione and pH dual-sensitive nanocarriers for doxorubicin. Materials Science and Engineering: C 2016;59:258–64.
[151] Fang W, Yang J, Gong J, Zheng N. Photo-and pH-triggered release of anticancer drugs from mesoporous silica-coated Pd@ Ag nanoparticles. Advanced Functional Materials 2012;22(4):842–8.
[152] Li Z, Clemens DL, Lee B-Y, Dillon BJ, Horwitz MA, Zink JI. Mesoporous silica nanoparticles with pH-sensitive nanovalves for delivery of moxifloxacin provide improved treatment of lethal pneumonic tularemia. ACS Nano 2015;9(11):10778–89.
[153] Adeyemi OS, Sulaiman FA. Evaluation of metal nanoparticles for drug delivery systems. Journal of Biomedical Research 2015;29(2):145.
[154] Ahmad MZ, Akhter S, Jain GK, et al. Metallic nanoparticles: technology overview & drug delivery applications in oncology. Expert Opinion on Drug Delivery 2010;7(8):927–42.
[155] Ealia SAM, Saravanakumar M. A review on the classification, characterisation, synthesis of nanoparticles and their application. IOP Conference Series: Materials Science and Engineering 2017;263(3):032019.
[156] Brown SD, Nativo P, Smith J-A, et al. Gold nanoparticles for the improved anticancer drug delivery of the active component of oxaliplatin. Journal of the American Chemical Society 2010;132(13):4678–84.
[157] Cobley CM, Au L, Chen J, Xia Y. Targeting gold nanocages to cancer cells for photothermal destruction and drug delivery. Expert Opinion on Drug Delivery 2010;7(5):577–87.
[158] Liong M, Lu J, Kovochich M, et al. Multifunctional inorganic nanoparticles for imaging, targeting, and drug delivery. ACS Nano 2008;2(5):889–96.
[159] Mody VV, Cox A, Shah S, Singh A, Bevins W, Parihar H. Magnetic nanoparticle drug delivery systems for targeting tumor. Applied Nanoscience 2014;4(4):385–92.
[160] Khutale GV, Casey A. Synthesis and characterization of a multifunctional gold-doxorubicin nanoparticle system for pH triggered intracellular anticancer drug release. European Journal of Pharmaceutics and Biopharmaceutics 2017;119:372–80.
[161] Madhusudhan A, Reddy GB, Venkatesham M, et al. Efficient pH dependent drug delivery to target cancer cells by gold nanoparticles capped with carboxymethyl chitosan. International Journal of Molecular Sciences 2014;15(5):8216–34.
[162] Jung S, Nam J, Hwang S, et al. Theragnostic pH-sensitive gold nanoparticles for the selective surface enhanced Raman scattering and photothermal cancer therapy. Analytical Chemistry 2013;85(16):7674–81.
[163] Zhang W, Wang F, Wang Y, et al. pH and near-infrared light dual-stimuli responsive drug delivery using DNA-conjugated gold nanorods for effective treatment of multidrug resistant cancer cells. Journal of Controlled Release 2016;232:9–19.
[164] Qiu L, Li J-W, Hong C-Y, Pan C-Y. Silver nanoparticles covered with pH-sensitive camptothecin-loaded polymer prodrugs: switchable fluorescence "off" or "on" and drug delivery dynamics in living cells. ACS Applied Materials and Interfaces 2017;9(46):40887–97.
[165] Liu J, Ma X, Jin S, et al. Zinc oxide nanoparticles as adjuvant to facilitate doxorubicin intracellular accumulation and visualize pH-responsive release for overcoming drug resistance. Molecular Pharmaceutics 2016;13(5):1723–30.
[166] Cai X, Luo Y, Zhang W, Du D, Lin Y. pH-Sensitive ZnO quantum dots–doxorubicin nanoparticles for lung cancer targeted drug delivery. ACS Applied Materials and Interfaces 2016;8(34):22442–50.
[167] Zhou L, Yuan J, Yuan W, et al. Synthesis, characterization, and controllable drug release of pH-sensitive hybrid magnetic nanoparticles. Journal of Magnetism and Magnetic Materials 2009;321(18):2799–804.
[168] Yang C, Guo W, Cui L, et al. pH-responsive magnetic core–shell nanocomposites for drug delivery. Langmuir 2014;30(32):9819–27.
[169] Huang G, Zhang K-L, Chen S, et al. Manganese-iron layered double hydroxide: a theranostic nanoplatform

with pH-responsive MRI contrast enhancement and drug release. Journal of Materials Chemistry B 2017; 5(20):3629–33.

[170] Danhier F, Feron O, Préat V. To exploit the tumor microenvironment: passive and active tumor targeting of nanocarriers for anti-cancer drug delivery. Journal of Controlled Release 2010;148(2):135–46.

[171] Hrubý M, Koňák Č, Ulbrich K. Polymeric micellar pH-sensitive drug delivery system for doxorubicin. Journal of Controlled Release 2005;103(1):137–48.

CHAPTER 3

Smart Nanomaterials for Tumor Targeted Hyperthermia

JAIDIP M. JAGTAP, PHD • ABDUL K. PARCHUR, PHD • GAYATRI SHARMA, PHD

1 INTRODUCTION

Cancer, an uncontrolled and unwanted cells growth, is still the most common cause of death globally even after the progress in early diagnosis and treatment, though recent studies looking promising for tumor regression. The changes in lifestyle and advancement in the fields of radiotherapy, chemotherapy, immunotherapy and molecular imaging are helping to understand the dose calibration and its importance in personalized medicine and monitoring. The American Cancer Society reported that in 2019, US cancer death rate is dropped to 27% in the last 25 years and projected the new cancer cases to be 1,762,450 and 606,880 cancer deaths to occur in the United States [1]. It has been now understood that not all the patients respond same way to the therapy and needs personalized calibrated dose for therapy. Hyperthermia is one of the non-invasive methods which works by increasing tumor temperature more than 42–43°C to cause death of cancer cells. Research has shown that heat can damage or kill cancer cells by destructing proteins and structure within cells thereby shrink tumors [2,3]. Hyperthermia can make damage to the tumor cells either prone to radiation or which are not damaged by radiation, and thus hyperthermia can provide applications in wide field of cancers types [2,4,5]. After heat shock, all the cell types exhibit thermal resistance for 24–48 h. The biomolecular mechanisms of hyperthermia are under investigation for cell membranes, DNA repair, the cytoskeleton and various cell organelles. The heat-shock proteins (HSP) expressing genes may get upregulated or downregulated by heat and may also be influencing the immunogenicity of some cancer cells [6–8]. Increased temperature may regulate molecular functions like apoptosis and DNA repairs. Some research studies have also reported the immunological reactions of human leukocytes on heat-treatment which could impact on natural killer cells and cytokine depletion [9–15].

Numerous side effects of conventional hyperthermia treatments had been seen such as collateral damage to healthy cells and non-homogeneous heating inside tumor make possibility of relapse of tumor after treatment. By adding hyperthermia to other techniques such as radiation therapy, ultrasound, magnetic and light therapy has made advancement in the efficacy of tumor treatment and reducing damage to collateral healthy tissue. In current practice, radiation therapy and chemotherapy are in use for various cancer treatment, while the major concern of healthy tissue damages due to nonspecific, nontargeted nanomaterials and toxicity of these materials still in need of improvement [16–19]. The non-targeted treatment is also impacting the patient's lifestyle, treatment frequencies and length of durations which also put uncertainty and may lack the patient's willingness and motivation in pursuing nanomedicine treatment for clinical practice. The hyperthermia therapy also needs to be calibrated for individual patient responses resulting from tumor heterogeneity [20]. In ideal treatment method, the disease needs to be noninvasively treated by targeting selected area with limiting side effects of drug and heat damage. The non-specific targeted therapy deposits only about 1% of drugs in tumors and majority of injected drugs remain in other tissue or gets excreted [21]. Thus, treatment efficacy is majorly affected due to not enough drug at the tumor site or excess remaining at undesired organs. Nanoparticles has potential as drug carrier and can be tuned for size, shape for long circulation in body, resulting for efficient thermal ablation for tumor treatment and limiting collateral damage to healthy tissue via less invasive surgical procedures [22,23].

Recently, many clinical experiments of radiotherapy or chemotherapy assisted with hyperthermia conducted on head and neck, breast, brain, bladder, bone and cervix tumors has shown significantly improved tumor control and patient survival rates

Intelligent Nanomaterials for Drug Delivery Applications. https://doi.org/10.1016/B978-0-12-817830-0.00003-5

[24–29]. Researchers also explored potential of combining magnetic and photothermal hyperthermia with the aim to produce a cumulative and complementarity heating for a synergic effect at the targeted site [30–34]. With magnetic hyperthermia, the solid tumors at depth can treated and MRI helps in tumor imaging whereas photothermal ablation can also produce heat for damaging cancer cells. Novel Pt-Janus designed nanomaterial possess the nanozyme activity for synergistic catalyst-enhanced sonodynamic and photothermal cancer therapy [35]. Hyperthermia with chemotherapy is also an interested approach where minimal tissue heating increase blood flow and this enhances the concentration of chemotherapeutic agents [18,36]. For superficial lesions, nanophotothermal therapy is most adequate because the optical energy get efficiently absorbed and produces thermal injury. Our recent study presented interventional photothermal therapy has a new approach to overcome depth penetration by using minimally invasive procedure for catheter use to deliver nanoparticle and light therapy in colorectal cancer treatment [37]. These nanotheranostic particle can be used as agent for optical and magnetic imaging and also possess photothermal ablation property for hyperthermia.

Nanomaterials assisted hyperthermia have great potential to revolutionize conventional hyperthermia methods and further understanding for its risk's free benefits. In this chapter, recent promising nanomaterials utilized in different hyperthermia modalities are broadly summarized into four groups as; (1) Photothermal Hyperthermia Nanomaterials (Palladium, Copper, Carbon, and Au etc., nanomaterials), (2) Radiofrequency Hyperthermia Nanomaterials (Silicon, Carbon, Quantum dots, Cobalt, Au, Iron Oxide, and Platinum etc., nanomaterials), (3) Magnetic Hyperthermia nanomaterials Iron Oxide, Iron Platinum, and Iron Cobalt etc., nanomaterials, and (4) Ultrasound Hyperthermia Nanomaterials (Iron Oxide, Silicon, and Au etc., nanomaterials). Nanoparticle assisted photothermal therapy and magnetic hyperthermia are currently most exciting approaches for targeted treatment. All four methods have been discussed with their progress as individual as well as possibilities for combined multimodal method with novel nanomaterials.

2 INTERVENTIONAL IMAGE-GUIDED NIR LIGHT HYPERTHERMIA THERAPY

Nanoscientist are studying to treat the solid tumors using hyperthermia by delivering near-infrared light (using a laser) and the theranostic nanoparticles to the solid tumors. Theranostic nanoparticles greatly enhance the treatment response. Nanoparticles absorbed near-infrared laser light and transfer it to heat energy thereby creating a hyperthermia around tumor [38–41]. However, deep solid tumors undergo additional challenges in terms of its treatment. Interventional imaging-guided procedures using theranostic contrast agents for an effective cancer therapy has a great advantage in clinical intervention where no surgical resection possible, inaccessible tumor location, circumferential resection margins, volume of liver remain after post-surgery and bilobar distribution of their hepatic metastases [37]. Parchur et al., used an interventional image-guided procedure in treating colorectal liver metastasis (CRLM) tumor beating rats using magneto-plasmonic sub-100 nm core/shell nanostructure having Au nanorods as core and Gd(III) as shell (Fig. 3.1A and B). These theranostic magneto-plasmonic nanoparticles are focused to effectively deliver to CRLM tumor sites using interventional image-guided procedures followed by NIR light delivery to cook tumor. Wilhelm et al. [21,42,43], recently analyzed research publications published between 2005 and 2015, the delivery efficiency of nanoparticles to the solid tumors is 0.7% of the injected dose, whereas providing active targeting molecular strategies the uptake efficacy increase by around 0.3%. Other 99% of off-target nanoparticles results uncomfortable implications for the clinical translation due to its high toxicity concerns [21,44]. By delivering the theranostic magneto-plasmonic nanomaterials via hepatic portal vein instead systemic injection overcome uptake efficacy concerns. A schematic illustration interventional image-guided nanoparticles delivery followed by photothermal therapy procedure is depicted in Fig. 3.1A, where same catheter is used to deliver both the nanoparticles and NIR light to the CRLM tumors. The Au core of the nanoparticles (Gold nanorod) with surface plasmon resonance properties at NIR region (around 808 nm) act as photothermal therapeutic and X-ray contrast agent, and Gd(III) shell as T1 contrast magnetic resonance agent [45,46]. Parchur et al., have highlighted both imaging and therapeutic functionalities by using same nanoconstruct [37]. A 3D reconstruction of DynaCT image CRLM tumor bearing rat having a catheter placed in the liver via a portal vein is shown in Fig. 3.1C and X-ray contrast of CRLM tumor evaluated for (i) pre-TNPs injection, (ii) post-TNPs injection, and (iii) post-PTT. Furthermore, dark-field image of CRLM tumor tissue confirms the uptake nanoparticles (as depicted inset of Fig. 3.1C). Furthermore, authors have tracked the uptake of nanoparticle delivery using

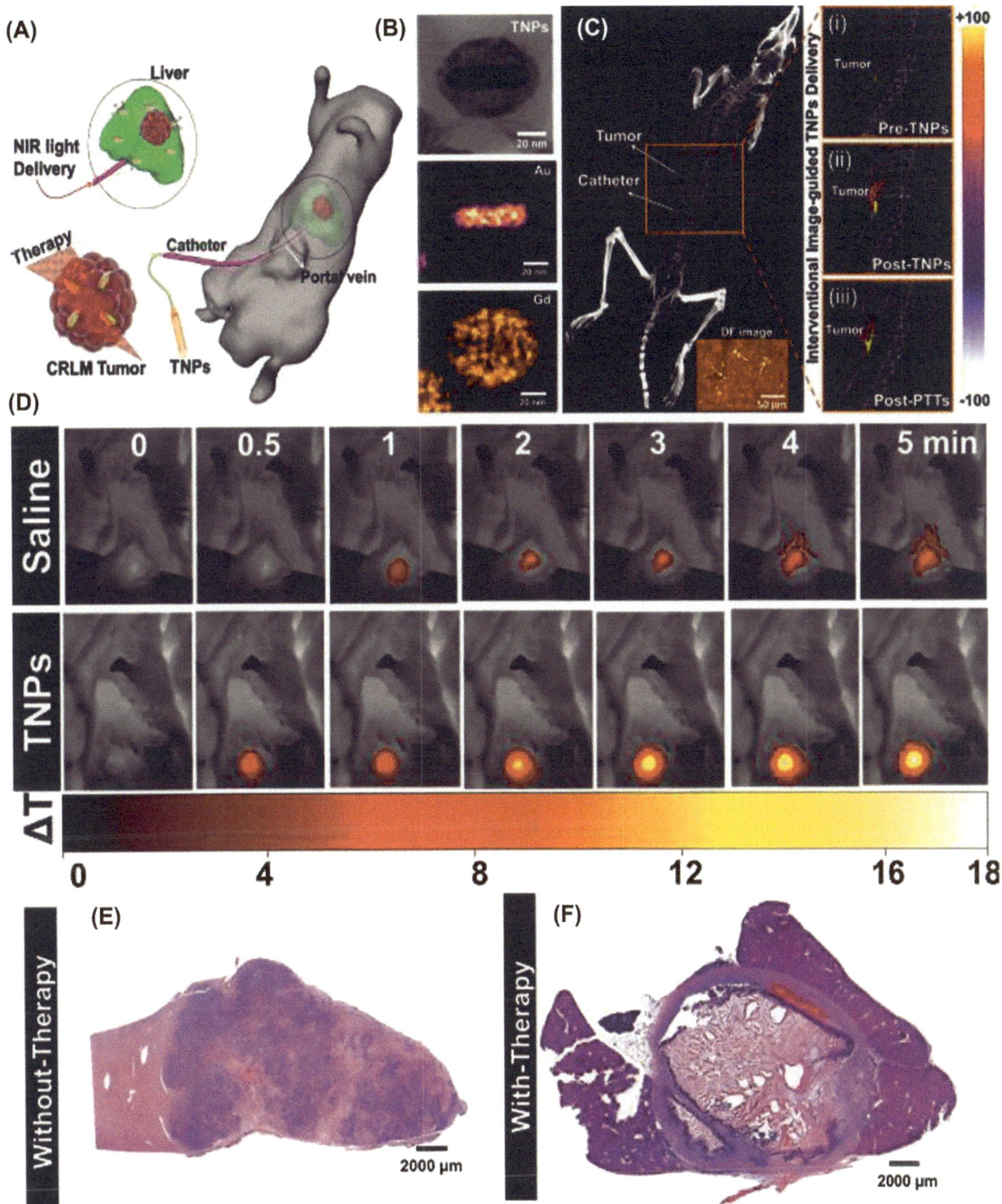

FIG. 3.1 **(A)** Schematic illustration interventional image-guided NPs delivery followed by PTT procedure. **(B)** Scanning transmission electron microscope (STEM) image of TNPs and elemental maps of Au in core and Gd in shell. **(C)** A 3D reconstruction of DynaCT image of a rat having colorectal liver metastasis (CRLM), where a catheter has been placed in the liver via portal vein. Rats were injected with TNPs followed by saline. DynaCT image of rats (i) pre-TNPs injection, (ii) post-TNPs injection, and (iii) post-PTT. Inset of **(C)** shows the dark-field

T1-weighted MR imaging in both portal vein (5 min post-injection) and systemic (4 h post-injection) delivery roots. Tumor to background signal ratio enhancement through hepatic portal vein delivery found to be 111% higher as compared to systemic TNPs delivery. Results effective therapeutic contrast loading into the targeted tumor site and clear identification of tumor margins [37]. Dynamic FLIR thermal imaging has been performed after injecting saline and theranostic nanoparticles in CRLM tumor bearing rat using 808-nm, 0.55 W/cm^2 NIR light for 5 min. FLIR images at different time points (0−5 min) for both saline and theranostic nanoparticles treated groups is depicted in Fig. 3.1D. Higher raise in temperature of CRLM tumors in theranostic nanoparticles group also confirms effective killing of tumor cells. Furthermore, representative H&E stained images of CRLM tumors tissue in untreated and theranostic nanoparticles treated rats clearly depicts the thermal damage in tumor tissue [37]. This procedure shows excellent thermal damage to tumor tissues leaving almost no damage to healthy tissue in the body.

3 IMAGE-GUIDED MAGNETIC HYPERTHERMIA THERAPY

Magnetic nanoparticles (MNPs) are vastly used in clinical application for MR imaging, magnetic hyperthermia and drug delivery. MNPs can image the malignancies in MRI and generates hyperthermia in high frequency alternating magnetic field which can also be used as nanocarrier and release under magnetic field. Inhomogeneity in heat distribution is major problem in conventional hyperthermia where MNPs can distribute in small regions to help in enhancing the temperature inside tumor [16,47−49]. Currently, a numerous MNPs are in early clinical trials for as an imaging contrast agent (summarized in Table 3.1). Lubbe et al. in 1996 first time MNPs were used in animal to test the epirubicin-loaded nanoparticles for targeting pancreatic cancer [50]. The MNPs biocompatibility and physical properties helps in efficient drug delivery in targeted cancer therapy. Several trials have been ongoing with magnetic nanoparticles assisted treatments on lung and prostate cancer [51,52]. Maier et al. have reported the prolonged survival in glioblastoma patients [53] by SPION-coated aminosilane treatment. Superparamagnetic iron oxide nanoparticles (SPIONs, diameter of 5−50 nm) has shown potential as an imaging contrast agent for MRI [54]. Iron oxide possess size-dependent magnetic properties, biocompatibility, conjugation possibility with inorganic nonmetallic and metallic materials, less toxicity and high stability in tumor tissue [16,53,55].

In therapeutic technique, MNPs injected in target and exposed to alternating magnetic field rages from kHz up to 10 MHz to generate heat >45°C. Hysteresis loss and relaxation phenomena are responsible for heat generation. Recently Tay et al. has demonstrated magnetic particle imaging (MPI) method, noninvasive 3D tomographic imaging with SPIONs for quantitative image-guidance in treatment planning and spatial localization of magnetic hyperthermia to tumor (MDA-MB-231) while minimizing collateral damage to the adjacent organs, liver (10−20 mm distant) [56]. Magnetic hyperthermia has potential to image real time deep tissues without invasive catheters [57,58]. Fig. 3.2A demonstrate the schematic of MPI and the image guided magnetic hyperthermia scanner. Once tumor or region of interest localized with MPI scanner, field-free-line was held to gather and focus over the spot for generating higher frequency (354 kHz) and thereby locally heat only the region of interest. Fig. 3.2B exhibits the MPI imaging and localization of magnetic fields for hyperthermia to heat tumor while sparing the liver. SPIONs were injected to see tumor and tail vein for liver clearance and observed negligible temperature increase with 20 kHz frequency. Second and third subfigures demonstrate the mouse excited by 354 kHz having SPIONs without MPI gradient and not received SPIONs with MPI gradient respectively and observed temperature increase to 43°C in 12 min. Last subfigures explain the impact of MPI gradient on selected tumor in dual tumor bearing mouse having SPIONs injected in tumors. Temperature curves confirm the heating from selected area with MPI gradient hyperthermia. Fig. 3.2C demonstrate an algorithm developed for MPI to predict the spatial distribution of heating.

Further, luciferase activity experiment carried out on dual tumor bearing mice and found out

image of liver tissue conforming the uptake of TNPs; a pointed white color arrows indicate scattering from the TNPs. **(D)** FLIR thermal images of CRLM tumor-bearing rats on site-selective delivery of TNPs and saline, irradiated with an 808 nm NIR laser at 0.55 W/cm^2 at 0, 0.5, 1, 2, 3, 4, and 5 min time points. Representative H&E stained images of tumors in **(E)** untreated animals and **(F)** TNPs and laser treatment rats. (Adapted from Parchur AK, et al. Vascular interventional radiology-guided photothermal therapy of colorectal cancer liver metastasis with theranostic gold nanorods. ACS Nano 2018;12(7):6597−11, with permission.)

TABLE 3.1
Magnetic Nanoparticles Commercially Available at Market as Imaging Contrast Agent.

MNPs Role	Commercial Name	MR Target	Status
Imaging	Feridex/ferumoxides, AMAG pharma; Feridex I.V., Berlex laboratories; Endorem, Guerbet	Liver	Approved
Imaging	Ferucarbotran, Schering AG; Resovist, Bayer healthcare	Liver	Approved
Imaging	Ferumoxsil/Lumirem, Guerbet; GastroMARK, AMAG pharma	Bowel	Approved
Imaging	Abdoscan, Nycomed imaging	Bowel	Approved
Imaging	Feraheme (ferumoxytol), AMAG pharma	Vasculature	Phase II
Imaging	Ferumoxtran-10/Combidex, AMAG pharma; Sinerem, Guerbet	Lymph node metastases	Phase III
Imaging	Clariscan, Nycomed imaging (part of GE healthcare)	Tumor microvasculature	Discontinued owing to safety concern

Adapted from Singh A, Sahoo SK. Magnetic nanoparticles: a novel platform for cancer theranostics. Drug Discovery Today 2014;19(4):474–481, with the permission.

bioluminescence disappeared in MPI gradient treated tumor. Fig. 3.3D describes the increase in temperature and less bioluminescence signal from MPI gradient treated tumor which was against in spared tumor. Thus, the magnetic hyperthermia has the potential to become as an imaging and targeted therapeutic tool. Domenech et al. has investigated the lysosomal membrane permeabilization in cancer cells by iron oxide assisted magnetic hyperthermia. The tumor cell viability was decreased post treatment and observed an increased cytosolic activity of the lysosomal protease cathepsin B through confocal microscopy [60].

Apart from iron oxide, dextran, starch, porphyrin, dimercaptosuccinic acid and aminosilane have been tested in *in vivo* studies [61,62]. Table 3.2 has summarized the clinical trials with MNPs as therapeutic and diagnostic agent. Wang et al. studied ultra-small sized magnetic iron oxide (USMIO) for lymph node metastasis in cervical cancer of rabbit pyriform sinus carcinoma [61] and reported apoptosis rate of 100% with USMIO assisted hyperthermia compared to 20% in healthy. Kossatz et al. [62] demonstrated treatment in breast and pancreas xenograft model and observed significant tumor suppression at 4 week. Johannsen et al. presented first clinical application with MNPs at 100 kHz frequency for prostate cancer [63]. Another clinical trial by Maier-Hauff et al., reported the MNPs assisted hyperthermia potential on the survival of glioblastoma patients [53]. Hua et al. had used DOX coated magnetic nanoparticles conjugated to COOH to treat tumor in presence of AC magnetic field [64]. In another technique, MNPs used as drug carrier [65–67]. Hayashi et al. investigated smart hybrid NPs (Fe3O4/DOX/PPy-PEG-FA) to release DOX under magnetic field to treat subcutaneous xenograft tumor in mice [68]. Baeza et al. had used hybrid magnetic mesoporous SiO_2 nanoparticles by coating a temperature sensitive polymeric shell for controlled release of drug in presence of magnetic field [69]. There are still some limiting factors for magnetic hyperthermia like patient discomfort in strong field of magnets and need further study on heat distribution in tumor.

4 IMAGE-GUIDED RADIOFREQUENCY HYPERTHERMIA THERAPY

Radiofrequency (RF) is non-ionizing thermal therapy method to enhance the impact of radiotherapy and chemotherapy effect on tumor. RF thermal ablation have advantage over NIR superficial thermal ablation to reach in deep for cancer therapy of hepatocellular carcinoma, colorectal metastases [70,71]. In radiofrequency hyperthermia method, cancer/tumor tissue is exposed with alternating current (AC) magnetic field to high temperatures (up to 50°C) to damage proteins, structures and kill cancer cells. Hyperthermia results in shrinking tumors. In radiofrequency hyperthermia, thermal fluctuations produced by AC magnetic field in

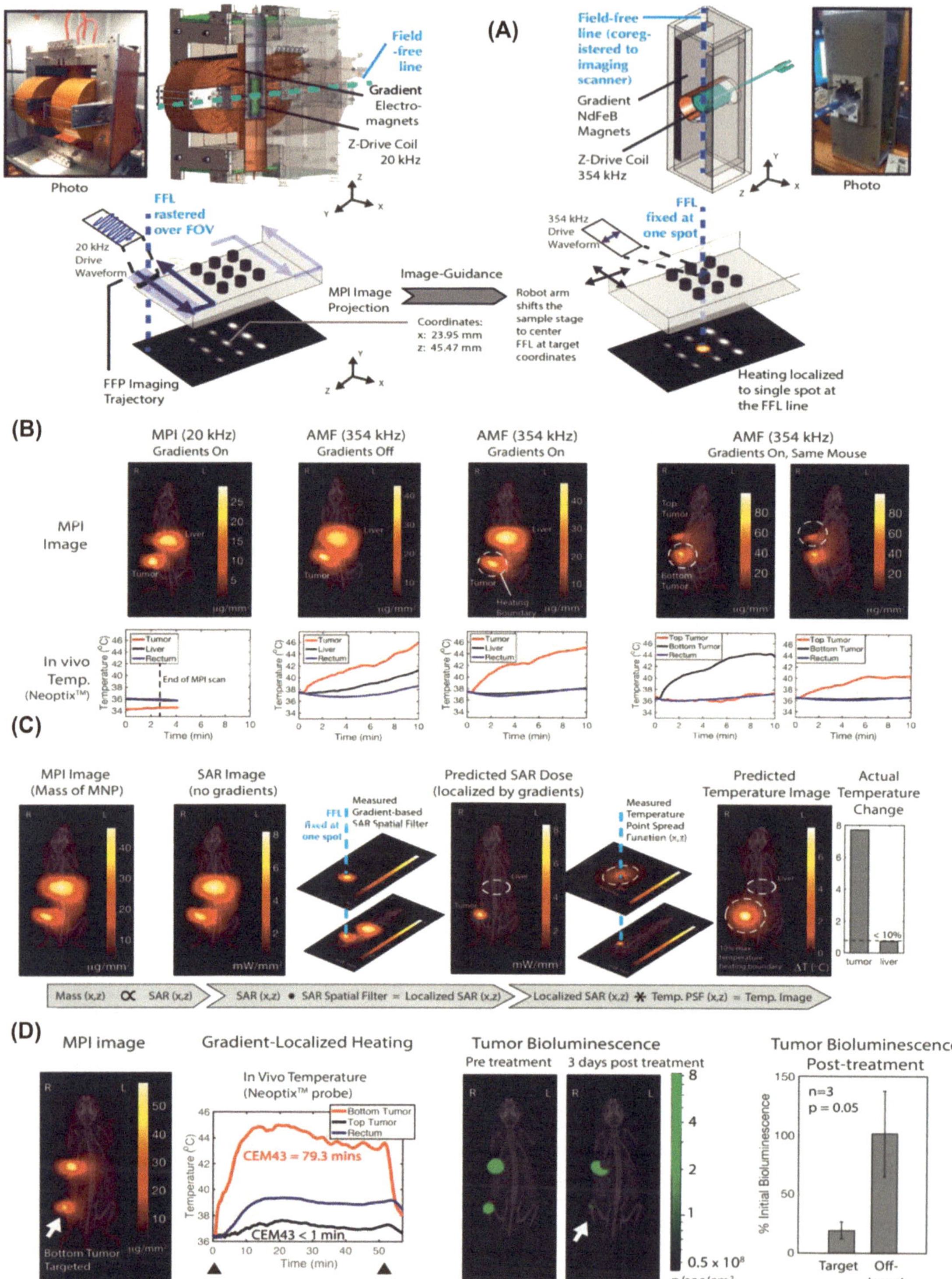

FIG. 3.2 **(A)** The schematic of the image guided magnetic hyperthermia and MPI scanner. **(B)** The MPI imaging and localization of magnetic fields for hyperthermia treatment. The last subfigures with MPI gradient

magnetic nanoparticles leads to magnetic energy transforms into heat. The heat generated may destroy tumor cells either through thermal ablation or indirectly through a synergic reinforcement of radiation/chemo efficacy. Typically, an AC magnetic fields of 1−15 kA/m, frequency of 40−100 kHz at 1−25 mT used for clinical applications. It's very important to maintain animal body temperature at 37°C and monitor coil heating. Heating depends on properties of magnetic nanoparticles such as type, size, stabilization, functionalization, and also the AC magnetic field strength and frequency. For the efficacy of treatment, controlled drug release with external stimuli at particular location is important. Herein, the systems triggered by AC magnetic fields by surface-functionalizing nanoparticles with a thermally labile molecule, linking them to a therapeutic drug. Nano-RF hyperthermia has advantage over conventional RF hyperthermia such as improved specificity for tumor heating, thermo-chemotherapy, diagnostic capability through nano-theranostics, however it still needs to overcome the inhomogeneous distribution of nanoparticles and difficulties in predicting temperature distribution. Joule heating phenomena explains RF heat generated via nanomaterials by $W=I^2R$ (Where W is power dissipated in heat, current, I, and R is resistance). Further heat generation depends on conductivity, cross-sectional area, electrical field strength and dielectric property of nanomaterials.

Ashokan et al. [59], reported single phase calcium phosphate biomineral nanoparticle (nCP) having dual-mode M1-M2 MR contrast and RF mediated thermal response shows potential for assisting the MR image guided RF ablation of tumor. MR imaging helps in directing the intratumoral injection in subcutaneous rat glioma. The orthotopic liver tumor models gives clear visualization of nanomaterials in MR imaging which further aids in quantifying nanomaterial distribution inside the tumor. Fig. 3.3 illustrates the schematic of interventional image-guided PTT procedure. The schematic lattice arrangement of nCP:Fe (as shown in Fig. 3.3A), where $Fe3+$ doping in place of some of $Ca2+$ changed color to brown from white (B). TEM images illustrated in Fig. 3.3C shows the size distribution of nCP. After optimizing the properties, the testing of nCP nanomaterial performed under RF ablation as shown in schematic of non-invasive RF in Fig. 3.3D. Based on temperature increment and cell viability test (Fig. 3.3E−F), significant death observed in nCP:Fe treated cells at 300 ug/mL operated at RF power 100W for 5 min. RF treatment with nCP:Fe noted 26% increased cell death than undoped nCP treated cells. (G) T2 MR contrast images of rat liver model shows significant differences before and 1 h after the injection of nCP:Fe (10 mg/kg), respectively. The temperature measure and monitoring performed with infrared camera (IR). Fig. 3.3H−H′ photograph and IR image after nCP:fe injected, (I−I′) at liver tumor model with RF treatment and RF treatment at healthy liver (J−J′). A drastic increase in temperature to ~81°C in nCP:fe injected liver tumor model compared to ~51°C in healthy liver. Thus, by utilizing RF mediated thermal response, nano-biomineral theranostic agent shows promising approach in the treatment of radiofrequency ablation of solid tumors.

Similar study was performed by Merkle et al. with iron oxide nanoparticles for MR contrast guided RF ablation of liver, however did not observe any major increase in coagulation diameter as found in the in vivo studies [72]. Other nanostructures (e.g., Au [73], graphene [74] and carbon nanotubes [75]) were reported for improving the effectiveness of RF ablation but due non-biodegradable property of material they could have the long-term circulation leading toxicity concern. Glazer et al. has recently reported the study considering nano-radio-frequency ablation (NaRFA) on human pancreatic carcinoma xenograft model (Capan-1 and Panc-1) developed monoclonal antibody-conjugated gold nanoparticles for targeted tumor by RF triggering. The hyperthermia treatment has shown great potential for targeted intracellular cytotoxicity and reported less damage to adjacent healthy tissues [73].

Tamarov at al. have also demonstrated the Si nanoparticles assisted RF wave (27 MHz frequency) triggered treatment for mice bearing lung carcinoma tumors. Nanoparticle assisted RF hyperthermia treatment shown declined in tumor growth with 20%−25% prolonged lifetime in studied animals [70]. Table 3.3 demonstrated the nanoparticles assisted radio frequency hyperthermia for comparison in *in vivo* studies [70,73,75−79]. Table 3.3 also briefly explains nanoparticles size and dose dependent effect with RF exposure on tumor growth.

Au nanoparticles assisted techniques were prominently used in RF hyperthermia due their thermo-

on SPIONs injected dual tumor mouse confirm increase in temperature from 354 kHz frequency gradient field targeted tumor. **(C)** An algorithm demonstrated for MPI to predict the spatial distribution of heating. **(D)** Bioluminescence decreased from MNPs assisted magnetic hyperthermia treatment received tumor. (Adapted from Tay ZW, et al. Magnetic particle imaging-guided heating in vivo using gradient fields for arbitrary localization of magnetic hyperthermia therapy. ACS Nano 2018;12(4):3699−3713, with permission.)

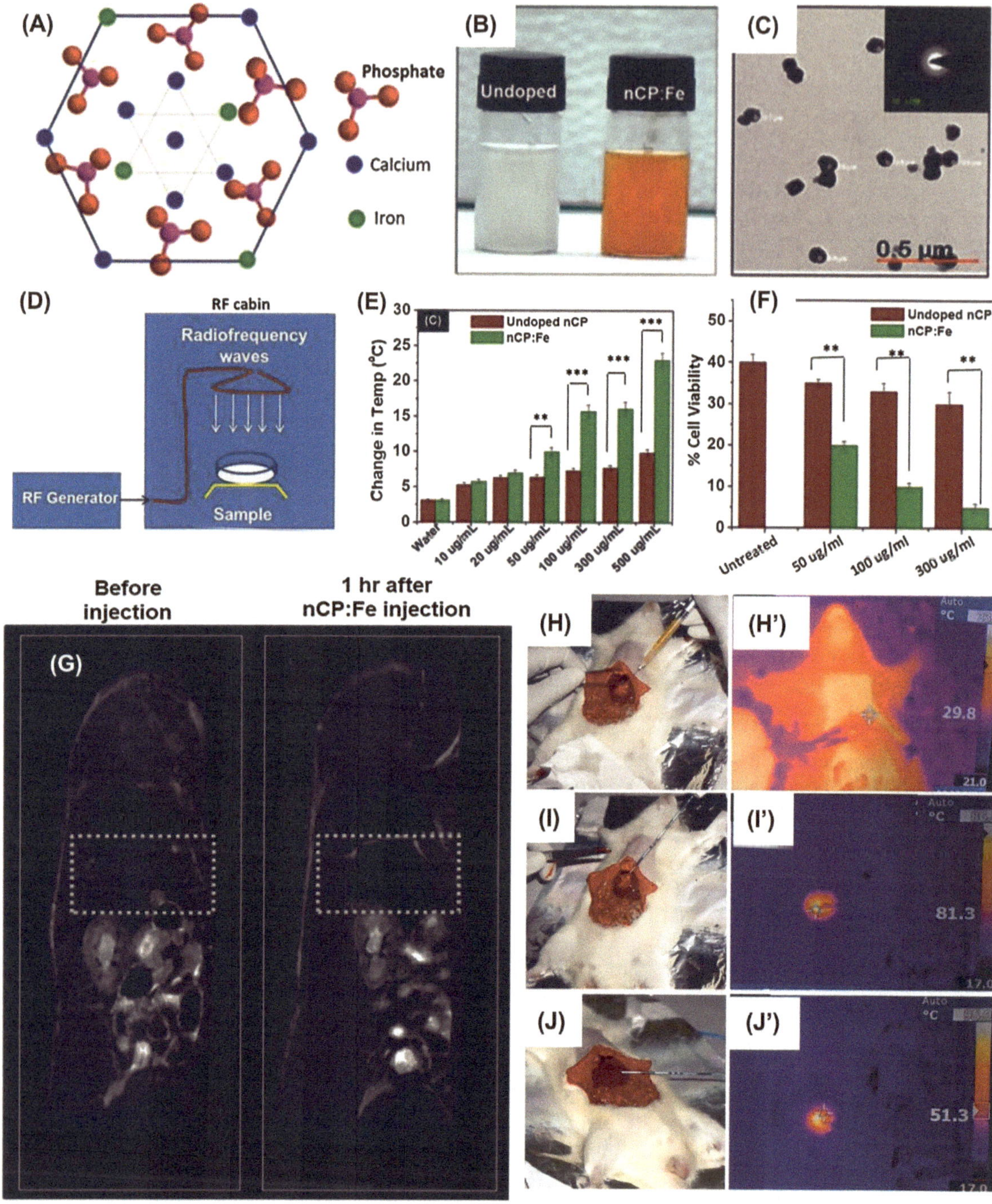

FIG. 3.3 Schematic of interventional image-guided nano-RF hyperthermia. **(A)** Lattice arrangement of nCP:Fe where few Ca2+ were replaced by Fe3+ atoms **(B)** Solution of undoped nCP and nCP:Fe at concentration 10 mg/mL. Fe doped nCP turns **(C)** TEM image for size distribution of nCP:Fe. **(D)** Schematic of RF system **(E)** Temperature variation from various nanoparticles concentrations under RF exposure at 100W for 1 min. **(F)** Cells viability measured at 4 h after a treatment of 5 min RF exposure **(G)** T2 weighted MR contrast images (coronal section) of rat recorded at pre and post 1 h of nCP:Fe injection. Liver shown by *dotted box*.

response in heat production from alternating electrical field helping in the cancer treatment [73,80–83]. The Au nanoparticles have advantage in biocompatibility, higher thermal conductivity, RF heating ability which could be beneficial with existing oncology therapies. Further, by conjugating Au nanoparticles with antibodies has promised targeted RF hyperthermia to destroy the cancer cells [81]. In particular, pancreatic cancer treatment with cetuximab conjugated Au had found more promising approach in targeting tumor and RF based hyperthermia [73,84,85]. Au core with drug loaded polymer shell releasing on RF exposure could be dual mode effective therapy.

Iron oxide is FDA approved contrast agent [86–88] and utilized in RF hyperthermia due to highly biocompatible, and easiness to conjugate [16,89]. The magnetic moment helps in generating heat unlike in electric moment in Au nanoparticles. Bae et al. [79] used Chitosan Oligosaccharide-Stabilized Ferrimagnetic Iron Oxide nanocubes to investigate by monitoring lung tumor growth (A549 cells) in mice. With the intratumorally injected dose of 375 μg (Fe/kg body weight) and 20 min AC magnetic field, the author has noted 70% decrease in the growth of tumor volume. For the success of this therapy, the critical points are alternating current for magnetic field strength, the frequency, and the size and concentration of nanoparticles.

Antibody conjugated with RF responsive quantum dots (QDs) can be targeted to tumor cells treatment. EGFR is known for the cell growth, survival and proliferation and reported to be overexpressing in cancer cells. Glazer et al. [90] conjugated EGFR targeting cetuximab (C225) monoclonal antibody with quantum dots (CdSe and InGaP) and measured RF heating as concentration varies in the in vitro study. Results demonstrated that healthy cells didn't exhibit any toxicity, however 58.2% of Pan-1 viability with InGaP and 52.5% of Cama-1 viability with CdSe has been achieved with RF assisted QDs treatment. The major setback for these QDs is the leaching property of Cd and the retaining long time for in vivo applications. Apart from nanomaterials based study, Krzykawska et al. [91] recently used abraxane and gemcitabine to report an ultrasound imaging using doppler mode to estimate the effect of combined RF and chemotherapy in a murine 4T1 breast tumors. This study emphasizes the tumor perfusion and area of impact which are important parameters in determining the response to hyperthermia.

Local hyperthermia is important for targeted tumor treatment and RF hyperthermia have potential in effective treatment with controlled drug release in external stimuli. Thus, targeted Nano-RF hyperthermia has advantage over conventional RF hyperthermia, however there is a need to engineer suitable candidate for the nontoxic, short lifetime exhibiting nanoparticles for advancement of this technique.

5 IMAGE-GUIDED ULTRASOUND HYPERTHERMIA THERAPY

Ultrasound is widely used in diagnosis and therapy of various tumors as this modality is non-invasive, safe and cost less [92]. In conventional ultrasound, absorption of sound generates heat in the tissue. However, to achieve effective temperature to destroy tumor tissues high intensity focused ultrasound waves (HIFU) ($1-100 W/cm^2$) are needed. This method has been approved for treatment of uterine fibroids [93], prostate cancer [94] and for palliative care in bone metastasis patients. The preclinical and clinical studies are undergoing for other malignancies such as brain [95], liver [96] and kidney malignancies [97]. The major limitation of HIFU is prolonged treatment for large tumors which may harm the surrounding normal tissues and need specialized instrument. These drawbacks can be overcome by using nano sensitizers. As a result, the combination of ultrasound with recently emerging field of nanotechnology has attracted interest in treatment of cancer with hyperthermia and delivery of chemotherapy [98]. Heat generated by ultrasound can be remotely controlled and high energy waves can be targeted to focused point at several depths [99]. The interaction of this energy source with nanoparticles results in higher damage to tumor region with least effect on the adjoining normal tissue. Inorganic nanoparticles with high heating efficiency are commonly used for ultrasound mediated hyperthermia. These include gold, graphene oxide, superparamagnetic iron oxide nanoparticles (SPIONs). Now, several non-invasive, temperature imaging techniques have been developed to monitor real time change in temperature of the targeted region which provides better control over the treatment.

Photograph and IR images measuring body temperature while, **(H–H′)** nCP:Fe injected in orthotopic liver tumor at body temp ~ 29.8°C, **(I–I′)** RF treatment in nCP:Fe injected liver tumor with temperature reaching ~81°C, **(J–J′)** RF treatment in control liver (no tumor) without nCP:Fe where temperature reaching ~51°C. (Adapted from Ashokan A, et al. Biomineral nano-theranostic agent for magnetic resonance image guided, augmented radiofrequency ablation of liver tumor. Scientific Reports 2017;7(1):14481, with permission.)

TABLE 3.2
Clinical Trials of Magnetic Nanoparticles as a Diagnostic and Therapeutic Agents.

Study	Start Date	Condition	Drug/Device
Detection of lymphoblasts by a novel magnetic needle and nanoparticles in patients with leukemia	August 2010	Leukemia	Device: MagProbe
Ferumoxytol-enhanced MRI in adult/pedi sarcomas	September 2012	Soft tissue sarcoma	Drug: Ferumoxytol
Pre-operative staging of pancreatic cancer using superparamagnetic iron oxide magnetic resonance imaging (SPIO-MRI)	July 2009	Pancreatic cancer	Drug: Superparamagnetic iron oxide MR imaging
Assessing dynamic MRI in patients with recurrent high-grade glioma receiving chemotherapy	October 2008	Brain neoplasms	Drug: Ferumoxytol

Adapted from Singh A, Sahoo SK. Magnetic nanoparticles: a novel platform for cancer theranostics. Drug Discovery Today 2014;19(4):474–481, with the permission.

TABLE 3.3
Summarizes Some of Highly Impacted Research From Recent Progress and Study Performed With Nano-RF Hyperthermia.

Reference	Tumor	Nanomaterials	Injection [mg/Kg], Route	Tumor Declination [%]
Gannon et al. [76]	Liver, Rat	SWNT	0.18, Intratumorally	>90
Gannon et al. [75]	Hepatic VX2 Rabbit	SWNT	500, Intratumorally	–
Glazer et al. [73]	Pancreas, Murine	Au-cetuximab, Au-PAM4	10, Intraperitoneal	>85 ~40
Sadeghi et al. [77]	Colon Mice	Au–Au-sulfide	1, Intratumorally	>65
Tamarov et al. [70]	Lung, Rat	Porous Si, Laser ablated-Si	~15, ~2.5 Intratumorally	25, 40
Raoof et al. [78]	Liver, Mice	Au-gemcitabine Au-cetuximab	10, Intraperitoneal	>85
Bae et al. [79]	Lung, Mice	Chito-FIONs	~0.38, Intratumorally	>70

Some of these real time imaging techniques include MR [100], ultrasound [101] and photoacoustic [102], but among these MR based imaging technique referred as Magnetic resonance–guided focused ultrasound (MRgFUS) is clinically-applied [100]. The strategy to ablate large area of tumor by ultrasound involves either to cover desired area in circular pattern or grid pattern of ablation foci across the desired volume of interest. The temperature rises rapidly in either way and provide maximum toxicity to the targeted region with minimal damage to surrounding normal tissue.

Ultrasound and nanoparticles mediated hyperthermia destroys malignant cells directly. This approach increases the cytotoxic effect of chemotherapeutic drugs (thermo-chemo sensitization) and even triggered drug release, delivers sufficient amount of drug at the target site [103–106]. Doxorubicin loaded in temperature sensitive liposomes triggered with ultrasound (US)-induced hyperthermia, have been demonstrated to result in tumor regression [107]. Recently, Fite et al. [108], demonstrated the effect of temperature-sensitive liposomal formulations of doxorubicin on mammary tumor using a syngeneic mice model. They observed that temperature-sensitive liposomal formulations of doxorubicin in combination with ultrasound resulted in tumor regression. The analysis of histological sections of treated tumors indicated red blood cell extravasation within tumors treated with temperature-sensitive doxorubicin and ultrasound hyperthermia as compared to untreated or only ultrasound treated contralateral tumors. This observation led them to hypothesize that the hyperthermic release of doxorubicin alters the vascular transport rate constants and high local concentration of drug have a direct tumoricidal effects. They further tested this hypothesis by using Dynamic contrast enhanced MRI (DCE-MRI) and ultrasound imaging with microbubble. They demonstrated that blood flow did not significantly altered after 30 min of the treatment, but later retention and accumulation of drug was more in tumors treated with temperature-sensitive liposomal doxorubicin and ultrasound hyperthermia. Recently, Ferrara et al. [109], group demonstrated the effect of ablation on immunotherapy. They demonstrate the combined effect on syngeneic model of mammary cancer, treatment with TLR9 agonist and checkpoint blockade followed by MRgFUS reduced macrophages and myeloid-derived suppressor cells and increased CD8+ T cells, the M1 macrophages, and PD-L1 expression on CD45 + cells and tumors showed complete response. The immunotherapy treatment alone or combined treatment of immunotherapy with ablation (primed ablation), resulted in a complete response in the majority of mice but primed MRgFUS expanded CD8+ T cells as compared with all control groups. They further demonstrated that increase in tumor burden with further implantation of tumors, followed by immunotherapy and two primed ablations resulted in increased survival time of mice as compared to mice treated with immunotherapy alone. In summary, it appeared that ablation after initial immunotherapy can be curative and is more effective in the presence of a high tumor burden. Various clinical studies have demonstrated efficacy of combination treatment of ultrasound mediated hyperthermia and ionizing radiation [110–113]. It has been effective in the treatment for head and neck cancer [114], breast cancer [115–117], prostate cancer [118–120], cancer in the pelvis [121–123]. Tumor ablation by hyperthermia helps to de bulks tumor, then reduces interstitial pressure which enhances nanoparticle accumulation and release of drug at the target [124], overall tumor and tumor microenvironment is modified by ultrasound mediated hyperthermia. Furthermore, mechanical injury caused by hyperthermia activates the immune system and increases migration of immune cells toward target region. The continued advancement of the technology and further understanding of mechanism will help to push this new ultrasound HT technology toward common clinical application.

6 CHALLENGES ASSOCIATED WITH HYPERTHERMIA TREATMENT

Radiation therapy is given to more than 50% for cancer patients in the clinic. However, some cancer cells are unresponsive to radiation treatment where hyperthermia may provide solution. Conventional non targeted nanoparticle-based hyperthermia technique lacks homogeneous heating which leads to relapse of tumor, collateral damage to healthy tissue around tumor area, irregular response to radiation therapy by heterogeneous cancer type and non-targeted uncontrolled drug delivery and drug toxicity are major concern for success of hyperthermia. Table 3.4 has demonstrated the comparison of four hyperthermia methods with their principle, benefits and disadvantages at current knowledge [125,126]. With immense research ongoing at laboratory and preclinical level, researchers have come a long way to understand the challenges in hyperthermia for clinical use and trying to solve by designing multifunctional nanomaterials for combined hyperthermia treatment.

Engineered magnetic particles need to be extensively looked for the toxicity at cellular and molecular level [20], and limitations due to not enough strength of external magnetic field to trigger the MNP–drug complexes at targeted location. Iron oxide is the most popular agent along with CuS and few other magnetic nanomaterials [51–53]. Inhomogeneous heating in tumors can be overcome by considering strong magnetic core facility externally triggering for controlled drug release. In photothermal therapy, light does not penetrate tissue enough for deep cancer treatment. Our recent study with trimodal nanoparticle shown potential in interventional photothermal therapy, via catheter to deliver nanoparticle for heat treatment and optical imaging, which could be one of the ways to treat tumors at depth level [37]. Nanoparticle assisted RF ablation gives advantage over NIR superficial photothermal therapy to achieve therapy for cancers of hepatocellular

TABLE 3.4
Comparison of Nanoparticle Assisted Hyperthermia Methods.

	Photothermal	Magnetic	Radio Frequency	Ultrasound
Heat source	Laser	Oscillating magnetic field	Radio frequency	Ultrasound
Source strength	400–1100 nm	50–500 kHz	300 kHz–300 MHz	0.5–5.0 MHz
Principle	Surface plasmon resonance	Relaxation and hysteresis losses	Joule heating	Mechanical and thermal effects
Benefits	Tumors at depth can be assessed for treatment; safe	No surface heating; multiple treatment with single dose of nanoparticles is possible; safe.	Non-invasive; deep penetration; High rate of specific absorption, vast scope to explore.	Non-invasive; focusing achieves deep penetration; doesn't overheat fat.
Disadvantages	Deep tumors treatment achievable by invasive path via interstitial fiber, less penetration, small scope;	Hot spots at bone interfaces, require higher dose of nanoparticles, low rate of specific absorption, inhomogeneous heating;	Inhomogeneous heating, excessive heating of fat.	Restrained by air cavity, not enough literature, hot spots at bone interfaces.
Action domains	Surface tumor, lung cancer, head and neck cancers.	Deep tumors such as brain, prostate, liver, esophagus and cervix tumors	Deep tumors for pelvis and abdomen, kidney, lung, liver, pancreas, breast and bone tumors.	Both deep and surface tumors can be treated.

Partially adapted from Beik J, et al. Nanotechnology in hyperthermia cancer therapy: from fundamental principles to advanced applications. Journal of Controlled Release 2016;235:205–221, with permission.

carcinoma and colorectal metastases located at depth. Iron oxide and Au are mostly used nanomaterials in this category [72,90]. Recently, hybrid nanoparticles are being used to conjugate with antibodies and release at tumor site [70,79,84]. Small sized targeted nanoparticle remains longer in circulation which could benefit for more nanoparticles accumulation in tumor thereby enhancing treatment effect. Recent developments in ultrasound combination with nanotechnology for hyperthermia treatment could help in the delivery of chemotherapy [98], remotely controlled and focused high energy can go at longer depths [99,108]. There are obstacles in this method success due to air cavity presence. There is not much published data available in ultrasound hyperthermia therapy and further research could help in understanding the benefits and disadvantages of this method.

Though progress to take hyperthermia methods in clinical use is time taking, nanoparticle assisted hyperthermia have great potential due to its theranostic capability for all cancer types. Further understanding can also achieve the goal of personalized nanomedicines. The progress in hyperthermia gives hopes toward future translation of nanoparticle-based hyperthermia into the clinic.

7 CONCLUSION

Nanotechnology applications in nanomedicine are developing rapidly to improve the delivery of drugs, as a contrast agent in multimodal optical imaging and tumor for targeted therapy. The property of heat generation upon external triggering can play an important role for controlled drug release at site and thereby controlling the collateral damage to healthy cells near the tumor region. There is still significant work needed on designing personalized medicine and calibrate the dose of drug since the heterogeneous treatment response has been observed from various patients for same therapy. For efficient hyperthermia treatment, nanomaterials need to be tuned for long circulation, maximizing drug loading capacity, non-toxicity and possibility of multifunctional modality for imaging and therapy. Multifunctional nanoparticle assisted hyperthermia with optical image guidance (NIR fluorescence/MRI/photoacoustic/X-ray) can completely treat the tumor by minimal collateral damage to healthy cells near the tumor. Hybrid nanoparticle assist combined hyperthermia methods may further explore nanotechnology for therapeutic nanomedicine at wide range of various cancers in clinics.

ACKNOWLEDGMENTS

The authors would like to thank the Department of Biomedical engineering, Medical college of Wisconsin, Milwaukee for providing the resources for the study.

REFERENCES

[1] Siegel RL, Miller KD, Jemal A. Cancer statistics, 2019. CA: A Cancer Journal for Clinicians 2019;69(1):7–34.

[2] van der Zee J, et al. Comparison of radiotherapy alone with radiotherapy plus hyperthermia in locally advanced pelvic tumours: a prospective, randomised, multicentre trial. Lancet 2000;355(9210):1119–25.

[3] Wust P, et al. Hyperthermia in combined treatment of cancer. The Lancet Oncology 2002;3(8):487–97.

[4] Carroll NM, Alexander Jr HR. Isolation perfusion of the liver. The Cancer Journal 2002;8(2):181–93.

[5] Kapp DS, et al. Two or six hyperthermia treatments as an adjunct to radiation therapy yield similar tumor responses: results of a randomized trial. International Journal of Radiation Oncology, Biology, Physics 1990; 19(6):1481–95.

[6] Saleh A, et al. Negative regulation of the Apaf-1 apoptosome by Hsp70. Nature Cell Biology 2000;2(8):476–83.

[7] Li GC, Mivechi NF, Weitzel G. Heat shock proteins, thermotolerance, and their relevance to clinical hyperthermia. International Journal of Hyperthermia 1995;11(4):459–88.

[8] Suto R, Srivastava PK. A mechanism for the specific immunogenicity of heat shock protein-chaperoned peptides. Science 1995;269(5230):1585–8.

[9] Bouchama A, Knochel JP. Heat stroke. New England Journal of Medicine 2002;346(25):1978–88.

[10] Leon LR, Bouchama A. Heat stroke. Comparative Physiology 2015;5(2):611–47.

[11] Baronzio G, Gramaglia A, Fiorentini G. Hyperthermia and immunity. A brief overview. In Vivo 2006;20(6A):689–95.

[12] Atanackovic D, et al. 41.8 degrees C whole body hyperthermia as an adjunct to chemotherapy induces prolonged T cell activation in patients with various malignant diseases. Cancer Immunology Immunotherapy 2002;51(11–12):603–13.

[13] Multhoff G, et al. Heat shock protein 72 on tumor cells: a recognition structure for natural killer cells. The Journal of Immunology 1997;158(9):4341–50.

[14] Robins HI, et al. Cytokine induction by 41.8 degrees C whole body hyperthermia. Cancer Letters 1995;97(2): 195–201.

[15] Shen RN, et al. Influence of elevated temperature on natural killer cell activity, lymphokine-activated killer cell activity and lectin-dependent cytotoxicity of human umbilical cord blood and adult blood cells. International Journal of Radiation Oncology, Biology, Physics 1994; 29(4):821–6.

[16] Banobre-Lopez M, Teijeiro A, Rivas J. Magnetic nanoparticle-based hyperthermia for cancer treatment. Reports of Practical Oncology and Radiotherapy 2013; 18(6):397–400.

[17] Chatterjee DK, Diagaradjane P, Krishnan S. Nanoparticle-mediated hyperthermia in cancer therapy. Therapeutic Delivery 2011;2(8):1001–14.

[18] Hildebrandt B, et al. The cellular and molecular basis of hyperthermia. Critical Reviews in Oncology 2002;43(1): 33–56.

[19] Roti Roti JL. Introduction: radiosensitization by hyperthermia. International Journal of Hyperthermia 2004;20(2):109–14.

[20] Seoane J. CANCER Division hierarchy leads to cell heterogeneity. Nature 2017;549(7671):164–6.

[21] Wilhelm S, et al. Analysis of nanoparticle delivery to tumours. Nature Reviews Materials 2016;1:16014.

[22] Pelaz B, et al. Diverse applications of nanomedicine. ACS Nano 2017;11(3):2313–81.

[23] Fiocchi C. Tailoring treatment to the individual patient—will inflammatory bowel disease medicine be personalized? Digestive Diseases 2015;33:82–9.

[24] Vernon CC, et al. Radiotherapy with or without hyperthermia in the treatment of superficial localized breast cancer: results from five randomized controlled trials. International Journal of Radiation Oncology, Biology, Physics 1996;35(4):731–44.

[25] Colombo R, et al. Multicentric study comparing intravesical chemotherapy alone and with local microwave hyperthermia for prophylaxis of recurrence of superficial transitional cell carcinoma. Journal of Clinical Oncology 2003;21(23):4270–6.

[26] Harima Y, Sawada S. Hyperthermia classic commentary: 'A randomized clinical trial of radiation therapy versus thermoradiotherapy in stage IIIB cervical carcinoma' by Yoko Harima, International Journal of Hyperthermia 2001;17:97-105. International Journal of Hyperthermia 2009;25(5):344–6.

[27] Huilgol N, Gupta S, Sridhar C. Hyperthermia with radiation in the treatment of locally advanced head and neck cancer: a report of randomized trial. Journal of Cancer Research and Therapeutics 2010;6(4):492–6.

[28] Hua YH. Author's reply to Paulides' commentary on developing effective hyperthermia application for nasopharyngeal cancer. International Journal of Hyperthermia 2011;27(5):526.

[29] Ariyafar T, et al. Evaluating the effectiveness of combined radiotherapy and hyperthermia for the treatment

response of patients with painful bony metastases: a phase 2 clinical trial. Journal of Thermal Biology 2019; 84:129–35.
[30] Kim J, et al. Designed fabrication of multifunctional magnetic gold nanoshells and their application to magnetic resonance imaging and photothermal therapy. Angewandte Chemie International Edition 2006; 45(46):7754–8.
[31] Yan H, et al. "All-in-One" nanoparticles for trimodality imaging-guided intracellular photo-magnetic hyperthermia therapy under intravenous administration. Advanced Functional Materials 2018;28(9):1705710.
[32] Espinosa A, et al. Can magneto-plasmonic nanohybrids efficiently combine photothermia with magnetic hyperthermia? Nanoscale 2015;7(45):18872–7.
[33] Roper DK, Ahn W, Hoepfner M. Microscale heat transfer transduced by surface plasmon resonant gold nanoparticles. The Journal of Physical Chemistry C, Nanomaterials and Interfaces 2007;111(9):3636–41.
[34] Liu MX, et al. Room-temperature synthesis of covellite nanoplatelets with broadly tunable localized surface plasmon resonance. Chemistry of Materials 2015; 27(7):2584–90.
[35] Liang S, et al. Intelligent hollow Pt-CuS janus architecture for synergistic catalysis-enhanced sonodynamic and photothermal cancer therapy. Nano Letters 2019; 19(6):4134–45.
[36] Curley SA, et al. Radiofrequency ablation of hepatocellular cancer in 110 patients with cirrhosis. Annals of Surgery 2000;232(3):381–91.
[37] Parchur AK, et al. Vascular interventional radiology-guided photothermal therapy of colorectal cancer liver metastasis with theranostic gold nanorods. ACS Nano 2018;12(7):6597–611.
[38] Parchur AK, Li Q, Zhou A. Near-infrared photothermal therapy of Prussian-blue-functionalized lanthanide-ion-doped inorganic/plasmonic multifunctional nanostructures for the selective targeting of HER2-expressing breast cancer cells. Biomaterials Science 2016;4(12): 1781–91.
[39] O'Neal DP, et al. Photo-thermal tumor ablation in mice using near infrared-absorbing nanoparticles. Cancer Letters 2004;209(2):171–6.
[40] Li Q, Parchur AK, Zhou A. In vitro biomechanical properties, fluorescence imaging, surface-enhanced Raman spectroscopy, and photothermal therapy evaluation of luminescent functionalized CaMoO4:Eu@Au hybrid nanorods on human lung adenocarcinoma epithelial cells. Science and Technology of Advanced Materials 2016;17(1):346–60.
[41] Kim HS, Lee DY. Near-infrared-responsive cancer photothermal and photodynamic therapy using gold nanoparticles. Polymers 2018;10(9).
[42] Degrauwe N, et al. Theranostics in interventional oncology: versatile carriers for diagnosis and targeted image-guided minimally invasive procedures. Frontiers in Pharmacology 2019;10:450.
[43] Schoenberg SO, et al. Developing a roadmap for interventional oncology. The Oncologist 2018;23(10): 1162–70.
[44] Bennie LA, McCarthy HO, Coulter JA. Enhanced nanoparticle delivery exploiting tumour-responsive formulations. Cancer Nanotechnology 2018;9(1):10.
[45] Henderson L, et al. Routes to potentially safer T1 magnetic resonance imaging contrast in a compact plasmonic nanoparticle with enhanced fluorescence. ACS Nano 2018;12(8):8214–23.
[46] Qu A, et al. Quantitative zeptomolar imaging of miRNA cancer markers with nanoparticle assemblies. Proceedings of the National Academy of Sciences of the United States of America 2019;116(9):3391–400.
[47] Tian C, et al. Poly(acrylic acid) bridged gadolinium metal-organic framework-gold nanoparticle composites as contrast agents for computed tomography and magnetic resonance bimodal imaging. ACS Applied Materials and Interfaces 2015;7(32):17765–75.
[48] Moffat BA, et al. A novel polyacrylamide magnetic nanoparticle contrast agent for molecular imaging using MRI. Molecular Imaging 2003;2(4):324–32.
[49] Singh A, Sahoo SK. Magnetic nanoparticles: a novel platform for cancer theranostics. Drug Discovery Today 2014;19(4):474–81.
[50] Lubbe A. Preclinical experiences with magnetic drug targeting: tolerance and efficacy and clinical experiences with magnetic drug targeting: a phase I study with 4'-epi-doxorubicin in 14 patients with advanced solid tumors-reply. Cancer Research 1997;57(14):3064–5.
[51] Johannsen M, et al. Thermotherapy of prostate cancer using magnetic nanoparticles: feasibility, imaging, and three-dimensional temperature distribution. European Urology 2007;52(6):1653–62.
[52] Fourmy D, Carrey J, Gigoux V. Targeted nanoscale magnetic hyperthermia: challenges and potentials of peptide-based targeting. Nanomedicine 2015;10(6):893–6.
[53] Maier-Hauff K, et al. Efficacy and safety of intratumoral thermotherapy using magnetic iron-oxide nanoparticles combined with external beam radiotherapy on patients with recurrent glioblastoma multiforme. Journal of Neuro-Oncology 2011;103(2):317–24.
[54] Koenig SH, Kellar KE. Theory of 1/T1 and 1/T2 NMRD profiles of solutions of magnetic nanoparticles. Magnetic Resonance in Medicine 1995;34(2):227–33.
[55] Silva AC, et al. Application of hyperthermia induced by superparamagnetic iron oxide nanoparticles in glioma treatment. International Journal of Nanomedicine 2011;6:591–603.
[56] Tay ZW, et al. Magnetic particle imaging-guided heating in vivo using gradient fields for arbitrary localization of magnetic hyperthermia therapy. ACS Nano 2018;12(4): 3699–713.
[57] Hensley D, et al. Combining magnetic particle imaging and magnetic fluid hyperthermia in a theranostic platform. Physics in Medicine and Biology 2017;62(9): 3483–500.

[58] Rodrigues HF, et al. Precise determination of the heat delivery during in vivo magnetic nanoparticle hyperthermia with infrared thermography. Physics in Medicine and Biology 2017;62(10):4062–82.
[59] Ashokan A, et al. Biomineral nano-theranostic agent for magnetic resonance image guided, augmented radiofrequency ablation of liver tumor. Scientific Reports 2017; 7(1):14481.
[60] Domenech M, et al. Lysosomal membrane permeabilization by targeted magnetic nanoparticles in alternating magnetic fields. ACS Nano 2013;7(6):5091–101.
[61] Wang P, et al. Ultra-small superparamagnetic iron oxide mediated magnetic hyperthermia in treatment of neck lymph node metastasis in rabbit pyriform sinus VX2 carcinoma. Tumour Biology 2015;36(10):8035–40.
[62] Kossatz S, et al. High therapeutic efficiency of magnetic hyperthermia in xenograft models achieved with moderate temperature dosages in the tumor area. Pharmaceutical Research 2014;31(12):3274–88.
[63] Johannsen M, et al. Clinical hyperthermia of prostate cancer using magnetic nanoparticles: presentation of a new interstitial technique. International Journal of Hyperthermia 2005;21(7):637–47.
[64] Hua X, et al. Magnetically triggered drug release from nanoparticles and its applications in anti-tumor treatment. Drug Delivery 2017;24(1):511–8.
[65] Cole AJ, Yang VC, David AE. Cancer theranostics: the rise of targeted magnetic nanoparticles. Trends in Biotechnology 2011;29(7):323–32.
[66] Colombo M, et al. Biological applications of magnetic nanoparticles. Chemical Society Reviews 2012;41(11): 4306–34.
[67] Guardia P, et al. Water-soluble iron oxide nanocubes with high values of specific absorption rate for cancer cell hyperthermia treatment. ACS Nano 2012;6(4): 3080–91.
[68] Hayashi K, et al. Magnetically responsive smart nanoparticles for cancer treatment with a combination of magnetic hyperthermia and remote-control drug release. Theranostics 2014;4(8):834–44.
[69] Baeza A, et al. Magnetically triggered multidrug release by hybrid mesoporous silica nanoparticles. Chemistry of Materials 2012;24(3):517–24.
[70] Tamarov KP, et al. Radio frequency radiation-induced hyperthermia using Si nanoparticle-based sensitizers for mild cancer therapy. Scientific Reports 2014;4:7034.
[71] Xu Y, et al. Cobalt nanoparticles coated with graphitic shells as localized radio frequency absorbers for cancer therapy. Nanotechnology 2008;19(43):435102.
[72] Merkle EM, et al. Effects of superparamagnetic iron oxide on radio-frequency-induced temperature distribution: in vitro measurements in polyacrylamide phantoms and in vivo results in a rabbit liver model. Radiology 1999;212(2):459–66.
[73] Glazer ES, et al. Noninvasive radiofrequency field destruction of pancreatic adenocarcinoma xenografts treated with targeted gold nanoparticles. Clinical Cancer Research 2010;16(23):5712–21.
[74] Sasidharan A, et al. Radiofrequency ablation of drug-resistant cancer cells using molecularly targeted carboxyl-functionalized biodegradable graphene. Advanced Healthcare Materials 2015;4(5):679–84.
[75] Gannon CJ, et al. Carbon nanotube-enhanced thermal destruction of cancer cells in a noninvasive radiofrequency field. Cancer 2007;110(12):2654–65.
[76] Gannon CJ, et al. Intracellular gold nanoparticles enhance non-invasive radiofrequency thermal destruction of human gastrointestinal cancer cells. Journal of Nanobiotechnology 2008;6:2.
[77] Sadeghi HR, et al. Gold-gold sulfide nanoshell as a novel intensifier for anti-tumor effects of radiofrequency fields. Iranian Journal of Basic Medical Sciences 2014; 17(7):516–21.
[78] Raoof M, et al. Gold nanoparticles and radiofrequency in experimental models for hepatocellular carcinoma. Nanomedicine 2014;10(6):1121–30.
[79] Bae KH, et al. Chitosan oligosaccharide-stabilized ferrimagnetic iron oxide nanocubes for magnetically modulated cancer hyperthermia. ACS Nano 2012;6(6): 5266–73.
[80] Corr SJ, et al. Protocols for assessing radiofrequency interactions with gold nanoparticles and biological systems for non-invasive hyperthermia cancer therapy. Journal of Visualized Experiments 2013;78.
[81] Corr SJ, et al. Citrate-capped gold nanoparticle electrophoretic heat production in response to a time-varying radiofrequency electric-field. The Journal of Physical Chemistry C, Nanomaterials and Interfaces 2012; 116(45):24380–9.
[82] Kumari P, Rickard H, Majewski P. Deposition of silver and gold nanoparticles on surface engineered silica particles and their potential applications. Journal of Nanoscience and Nanotechnology 2012;12(10):8001–7.
[83] Cardinal J, et al. Noninvasive radiofrequency ablation of cancer targeted by gold nanoparticles. Surgery 2008; 144(2):125–32.
[84] Raoof M, et al. Luciferase-based protein denaturation assay for quantification of radiofrequency field-induced targeted hyperthermia: developing an intracellular thermometer. International Journal of Hyperthermia 2012;28(3):202–9.
[85] Curley SA, et al. Noninvasive radiofrequency field-induced hyperthermic cytotoxicity in human cancer cells using cetuximab-targeted gold nanoparticles. Journal of Experimental Therapeutics and Oncology 2008;7(4): 313–26.
[86] Spadea MF, et al. Contrast-enhanced proton radiography for patient set-up by using x-ray CT prior knowledge. International Journal of Radiation Oncology, Biology, Physics 2014;90(3):628–36.
[87] Fasang AE, Raab M. Beyond transmission: intergenerational patterns of family formation among middle-class American families. Demography 2014;51(5): 1703–28.
[88] Kowallick JT, et al. Real-time phase-contrast flow MRI of the ascending aorta and superior vena cava as a function

of intrathoracic pressure (Valsalva manoeuvre). British Journal of Radiology 2014;87(1042):20140401.
[89] Carenza E, et al. In vitro angiogenic performance and in vivo brain targeting of magnetized endothelial progenitor cells for neurorepair therapies. Nanomedicine 2014;10(1):225–34.
[90] Glazer ES, Curley SA. Radiofrequency field-induced thermal cytotoxicity in cancer cells treated with fluorescent nanoparticles. Cancer 2010;116(13):3285–93.
[91] Krzykawska-Serda M, et al. Ultrasound Doppler as an imaging modality for selection of murine 4T1 breast tumors for combination radiofrequency hyperthermia and chemotherapy. Translational Oncology 2018;11(4):864–72.
[92] Mehrpour M, et al. Effects of low-intensity continuous ultrasound on hematological parameters of rats. Journal of Biomedical Physics and Engineering 2016;6(3):195–200.
[93] Kong CY, et al. MRI-guided focused ultrasound surgery for uterine fibroid treatment: a cost-effectiveness analysis. American Journal of Roentgenology 2014;203(2):361–71.
[94] Lukka H, et al. High-intensity focused ultrasound for prostate cancer: a systematic review. Clinical Oncology 2011;23(2):117–27.
[95] Yang HW, et al. Non-invasive synergistic treatment of brain tumors by targeted chemotherapeutic delivery and amplified focused ultrasound-hyperthermia using magnetic nanographene oxide. Advanced Materials 2013;25(26):3605–11.
[96] Misra SK, et al. Trimodal therapy: combining hyperthermia with repurposed bexarotene and ultrasound for treating liver cancer. ACS Nano 2015;9(11):10695–718.
[97] Rapoport N, Gao Z, Kennedy A. Multifunctional nanoparticles for combining ultrasonic tumor imaging and targeted chemotherapy. Journal of the National Cancer Institute 2007;99(14):1095–106.
[98] Beik J, Mehdizadeh AR, Shakeri-Zadeh A. Ultrasound in cancer treatment through nanotechnology. Journal of Biomedical Physics and Engineering 2016;6(3):123–6.
[99] Abed Z, et al. Effects of ultrasound irradiation on the release profile of 5-fluorouracil from magnetic polylactic co-glycolic acid nanocapsules. Journal of Biomedical Physics and Engineering 2016;6(3):183–94.
[100] Rieke V, Butts Pauly K. MR thermometry. Journal of Magnetic Resonance Imaging 2008;27(2):376–90.
[101] Lewis MA, Staruch RM, Chopra R. Thermometry and ablation monitoring with ultrasound. International Journal of Hyperthermia 2015;31(2):163–81.
[102] Yao J, et al. Absolute photoacoustic thermometry in deep tissue. Optics Letters 2013;38(24):5228–31.
[103] Sirsi SR, Borden MA. State-of-the-art materials for ultrasound-triggered drug delivery. Advanced Drug Delivery Reviews 2014;72:3–14.
[104] Issels RD. Hyperthermia adds to chemotherapy. European Journal of Cancer 2008;44(17):2546–54.
[105] Aryal M, et al. Ultrasound-mediated blood-brain barrier disruption for targeted drug delivery in the central nervous system. Advanced Drug Delivery Reviews 2014;72:94–109.
[106] Kiessling F, et al. Recent advances in molecular, multimodal and theranostic ultrasound imaging. Advanced Drug Delivery Reviews 2014;72:15–27.
[107] Kheirolomoom A, et al. Complete regression of local cancer using temperature-sensitive liposomes combined with ultrasound-mediated hyperthermia. Journal of Controlled Release 2013;172(1):266–73.
[108] Fite BZ, et al. Dynamic contrast enhanced MRI detects changes in vascular transport rate constants following treatment with thermally-sensitive liposomal doxorubicin. Journal of Controlled Release 2017;256:203–13.
[109] Silvestrini MT, et al. Priming is key to effective incorporation of image-guided thermal ablation into immunotherapy protocols. JCI Insight 2017;2(6):e90521.
[110] Horsman MR, Overgaard J. Hyperthermia: a potent enhancer of radiotherapy. Clinical Oncology 2007;19(6):418–26.
[111] Mallory M, et al. Therapeutic hyperthermia: the old, the new, and the upcoming. Critical Reviews in Oncology 2016;97:56–64.
[112] Rao W, Deng ZS, Liu J. A review of hyperthermia combined with radiotherapy/chemotherapy on malignant tumors. Critical Reviews in Biomedical Engineering 2010;38(1):101–16.
[113] Datta NR, et al. Hyperthermia and radiotherapy in the management of head and neck cancers: a systematic review and meta-analysis. International Journal of Hyperthermia 2016;32(1):31–40.
[114] Xia T, et al. Relationship between thermal parameters and tumor response in hyperthermia combined with radiation therapy. International Journal of Clinical Oncology 2001;6(3):138–42.
[115] Zagar TM, et al. Hyperthermia combined with radiation therapy for superficial breast cancer and chest wall recurrence: a review of the randomised data. International Journal of Hyperthermia 2010;26(7):612–7.
[116] Jones EL, et al. Randomized trial of hyperthermia and radiation for superficial tumors. Journal of Clinical Oncology 2005;23(13):3079–85.
[117] Varma S, et al. Simultaneous radiotherapy and superficial hyperthermia for high-risk breast carcinoma: a randomised comparison of treatment sequelae in heated versus non-heated sectors of the chest wall hyperthermia. International Journal of Hyperthermia 2012;28(7):583–90.
[118] Hurwitz MD, et al. Hyperthermia combined with radiation for the treatment of locally advanced prostate cancer: long-term results from Dana-Farber Cancer Institute study 94-153. Cancer 2011;117(3):510–6.
[119] Hurwitz MD, et al. Hyperthermia combined with radiation in treatment of locally advanced prostate cancer is associated with a favourable toxicity profile. International Journal of Hyperthermia 2005;21(7):649–56.
[120] Algan O, et al. External beam radiotherapy and hyperthermia in the treatment of patients with locally

advanced prostate carcinoma. Cancer 2000;89(2): 399–403.

[121] Diederich CJ, et al. Catheter-based ultrasound hyperthermia with HDR brachytherapy for treatment of locally advanced cancer of the prostate and cervix. Proceedings of SPIE-The International Society for Optical Engineering 2011;7901:79010O.

[122] Wootton JH, Hsu IC, Diederich CJ. Endocervical ultrasound applicator for integrated hyperthermia and HDR brachytherapy in the treatment of locally advanced cervical carcinoma. Medical Physics 2011;38(2): 598–611.

[123] Wootton JH, et al. Implant strategies for endocervical and interstitial ultrasound hyperthermia adjunct to HDR brachytherapy for the treatment of cervical cancer. Physics in Medicine and Biology 2011;56(13): 3967–84.

[124] Watson KD, et al. Ultrasound increases nanoparticle delivery by reducing intratumoral pressure and increasing transport in epithelial and epithelial-mesenchymal transition tumors. Cancer Research 2012;72(6):1485–93.

[125] Beik J, et al. Nanotechnology in hyperthermia cancer therapy: from fundamental principles to advanced applications. Journal of Controlled Release 2016;235: 205–21.

[126] Tilly W, et al. Temperature data and specific absorption rates in pelvic tumours: predictive factors and correlations. International Journal of Hyperthermia 2001;17(2):172–88.

CHAPTER 4

Novel Paradigms of Nanomediated Targeted Drug Delivery in Gastrointestinal Disorders

RAKHI CHAUDHARY, MTECH • BARKHA SINGHAL, PHD

1 INTRODUCTION

The phenomenal growth in the field of nanotechnology has been tremendously recognized nowadays as a promising platform for changing the landscape of the medical and pharmaceutical fields. The technology has presented revolutionary approaches for improving the diagnosis, treatment, and therapeutics of several devastating diseases and paved the way toward personalized medicine in the foreseeable future. One of the most critical endeavors of the pharmaceutical field has been envisaged in the field of drug delivery and targeting systems [1]. The burgeoning quest for finding novel drug-delivery systems and their mechanism of action requires an immersive amalgamation of multidisciplinary scientific approaches for maximizing the bioavailability of drugs and their therapeutic benefits. In this context, the remarkable benefits have been manifested by nanotechnological interventions in the form of differentially modified or engineered nanoparticles/nanostructures in target-oriented delivery of drugs [2]. In more border sense, targeted drug-delivery systems are defined as the innovative use of smart technology for controlled delivery of drug/gene to the target body parts with respect to accurate time and dose. This smart drug-delivery system (SDDS) based on nanotechnology revives the delivery of feeble water-soluble drugs, facilitates the delivery of macromolecular drugs intracellularly, mediates codelivery of combinatorial drugs, enables absorption of drugs against tight epithelial and endothelial barriers, leading to less cumbersome plasma fluctuations and minimize the therapeutic side effects, and also enables the real-time monitoring for effective drug delivery and assessing therapeutic efficacy of drugs in in vivo conditions [3].

Currently, the research from the last decade emphasized that 55% of the global population is suffering from gastrointestinal (GI) disorders. The insurmountable efforts have been made toward the development of novel therapeutic, diagnostic modalities for alleviating these disorders and improving the quality of life of the patients [4]. As each location of the gastrointestinal tract comprising different biochemical and physiological features that may either favor or hinder the absorption of drugs across the gastrointestinal membrane, there is a dire need for targeting the drugs to specific locations along the gastrointestinal tract. This targeted drug delivery enhances the efficiency of drug and reduces the amount of drug as well as possess the ability to treat local diseases, by eliminating systemic side effects of drugs [5].

The stupendous strides have been envisaged in the field of nanotechnology for developing the novel drug-delivery regimes for GI disorders. It is currently being realized that incorporating nanoparticles in drug-delivery vehicles will develop more potent therapy with less systemic side effects and adding life to drugs having low therapeutic indices or premature digestion as well as not effectively absorbed by the GI tract. Apart from that, nanotechnological approaches enhanced the detection of various disease biomarkers as well as mediated the more sensitive binding of these biomarkers that improved the imaging and therapeutic modalities in the GI tract [6]. The significant improvement in drug delivery in the GI tract has been proven with increased cell surface interaction and deposition into specific tissue over conventional pH-sensitive, microbe-assisted, and receptor-dependent models [7]. The amenability of chemical modification of nanoparticles leads to careful and precise consideration of various parameters such as surface chemistry, size, shape, and other physicochemical characteristics for the development of smart drug-delivery systems. An

Intelligent Nanomaterials for Drug Delivery Applications. https://doi.org/10.1016/B978-0-12-817830-0.00004-7

overview of SDDS in the GI tract is represented in Fig. 4.1. The recent understanding of the persuasive role of gut microbiota and their interaction with various nanosystems has been a niche area of investigation for exploring the mechanisms assessing the efficiency of nanotechnology-based drug delivery. In fact, the antimicrobial activities of several nanoparticles are characterized and considered as safer and viable alternatives as antimicrobial regimens for the treatment of dysbiosis in the GI tract. The advancements in the understanding of nanotechnology and their treatment modalities lead to translational heights in drug-delivery approaches in various regions of the gastrointestinal tract.

This research is thriving on the advancements in understanding the instrumental role of nanoparticles in biological systems at each hierarchal level from cell, tissue, organ and organ system, systemic and localized effect as well as genomics, proteomics, and metabolomics processes in various pathological conditions [8]. The charisma of nanotechnological intervention is still stymied by their toxicity in a biological context and clinical evaluation. Therefore this chapter is focused on applications of nanomaterials in targeted drug delivery in the gastrointestinal tract. In addition, it also emphasizes the types of nanoparticles used for drug-delivery applications, their current bottlenecks, and future perspectives. The detailed overview of this chapter opens up the "Pandora's box" and supports successful applications of nanomediated drug delivery in various GI disorders including gastric and colon cancer, inflammatory bowel disease (IBD), and esophageal cancer [9].

2 TYPES OF NANOMEDIATED DRUG-DELIVERY SYSTEMS USED IN GASTROINTESTINAL SYSTEM

The nanomaterials presented a wealth of opportunities to explore their potential in advanced drug-delivery systems, as they have the remarkable potential to enhance the biodistribution and pharmacokinetics of the drug and enable controlled release at the target site [10]. The comprehensive library composed of nanoparticles of different shapes, sizes, and materials and possessing various surface properties are available for their widespread use. The growing continuum of nanoparticles further expanded this library for futuristic utilization. The ideal nanostructures used for drug-delivery applications should be biocompatible and nontoxic to the biological system. The literature is leveraged with various types of nanomaterials comprising different physicochemical properties that have been used to improve the efficiency of drug delivery in various regions of the

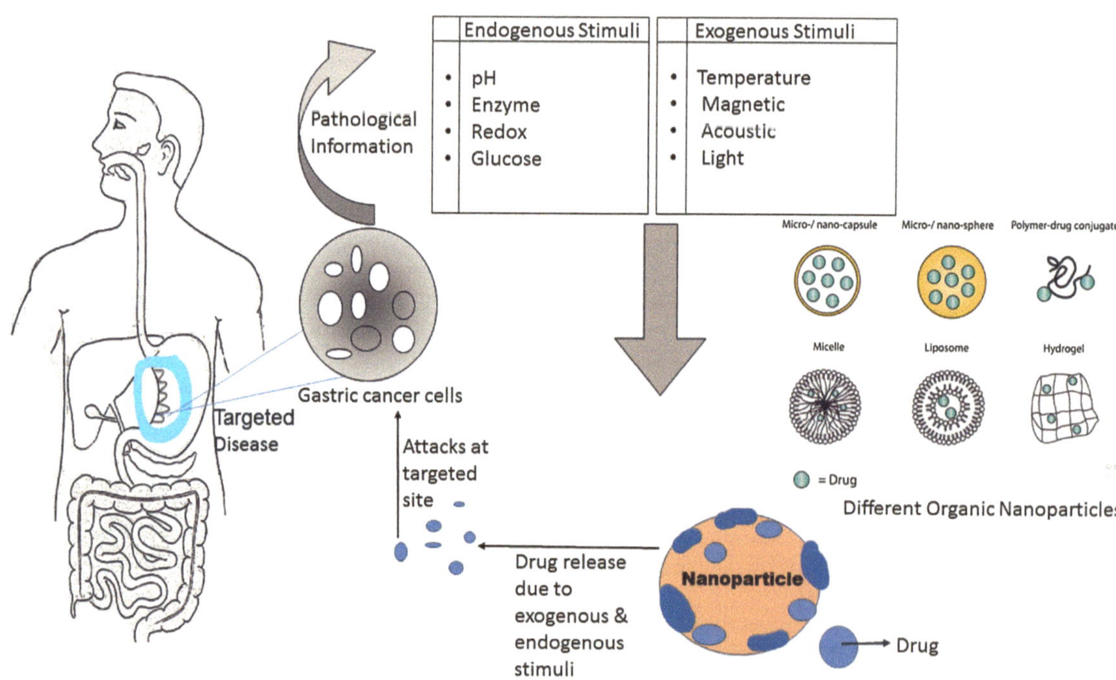

FIG. 4.1 An overview of the functionality of smart drug delivery (SDDS) in gastrointestinal tract (GIT).

gastrointestinal tract. The broad classification is depicted in Fig. 4.2 and is described as follows:

2.1 Polysaccharide-Based Drug-Delivery Systems

There has been growing interest in the development of polysaccharide-based drug-delivery systems due to their easy and ample availability, stability, accessible modifications, safety, and biodegradability. Currently, these systems are of popular choice for targeted delivery of drugs at GI. The polysaccharides differ remarkably with respect to chemical composition, positive or negative charge, source, molecular weight, and chemical structure [11]. There are numerous polymeric materials that are currently being utilized for the production of nanoparticles for drug delivery that includes poly(glycolic acid), poly(lactic acid), polysaccharides (e.g., pectin, pullulan, cellulose, hyaluronic acid (HA), dextran, alginate, and guar gum), and polycaprolactone (chitosan). Based on the location of drugs, these polymeric nanoparticles comprised both matrix systems (nanospheres) and vesicular systems (nanocapsules). Nanospheres are polymeric systems in which the dispersion of drugs occurs thoroughly in the polymer matrix, while in nanocapsules the drug is confined in the cavity that is surrounded by polymeric membranes [12]. Besides that, the diversity present in polymers leads to the development of diversified nanostructures such as polymeric nanomicelles and nanovesicles, nanogels, dendrimers, and stimulus-sensitive (single stimulus and dual-/multistimuli responsive) polymer nanostructures [13]. Currently, the immersive applications have been envisaged in the field of nanoparticles synthesized with natural polysaccharides. The presence of various hydrophilic groups including hydroxyl, carboxyl, and amino groups makes these molecules soluble in water and facilitates the noncovalent binding with mucosal membranes and biological tissues. Thus these biomaterials possess the hydrophilic properties by which the bioadhesiveness and mucoadhesive property, as well as amenability for chemical modification of the macromolecules to bind drugs or targeting agents, confer a selective advantage for the drug-delivery applications [14]. These nanoparticles are synthesized by four different methodologies such as ionic and covalent cross-linking, self-assembling of hydrophobically modified polysaccharides, and polyelectrolyte complexation, and these methods have been chosen depending on their particle size, particle size distribution, and area of application. However, comprehensive studies are still lacking about the interaction between drugs and different polysaccharides at the molecular level, their mechanism of action, and their fate in vivo that has to be resolved for their commercial applications [15].

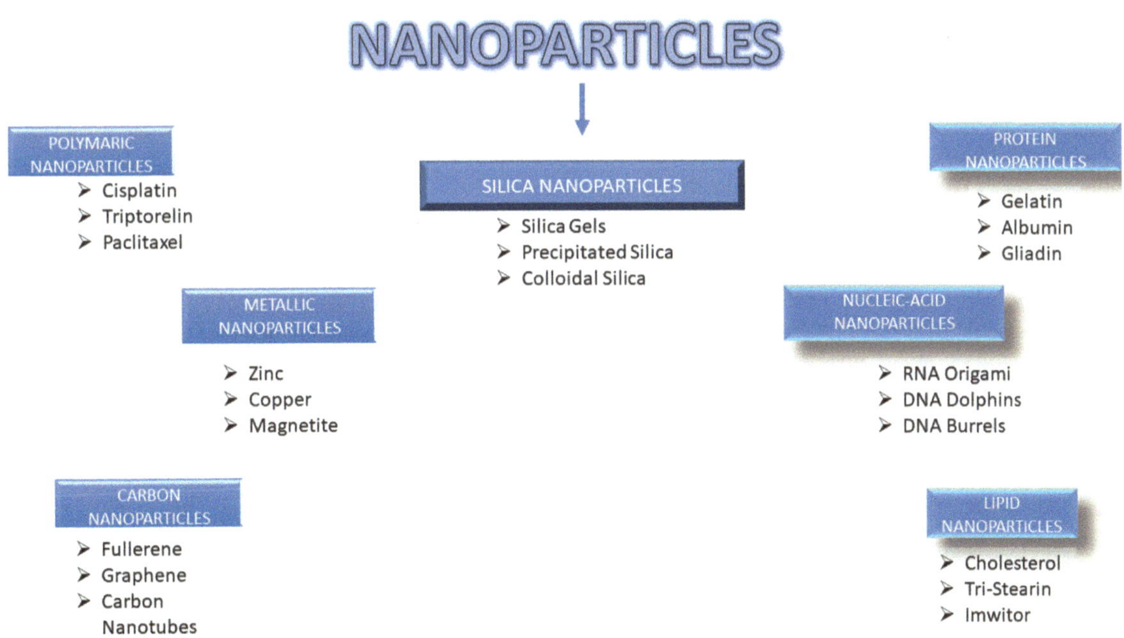

FIG. 4.2 Types of Nanomaterials used in targeted drug delivery in gastrointestinal tract (GIT).

2.2 Metallic Nanoparticle-Based Drug-Delivery Systems

Metallic nanoparticles are attracting attention in targeted drug delivery owing to their multifarious properties, and they provide the enhancement of therapeutic index and bioavailability of the drug, provide protection of drugs in biological milieu, facilitate drugs to reach their in vivo target site, prevent multidrug resistance, and ease for surface functionalization [16]. Currently, metallic nanoparticles have been served as probing molecules for imaging and tracking using magnetic resonance imaging and high-resolution quantum interference techniques. These are versatile solid colloidal nanoparticles, and their size may vary between 10 and 1000 nm in which the drug is dispersed, encapsulated, or covalently attached. The metallic nanoparticles are utilized on the basis of their magnetic and optical properties, their catalytic ability, and conductive properties [17]. Currently, the optoelectronic properties of noble metals have been explored for targeted drug delivery in the GI tract due to large surface area-to-volume ratio, manipulation of the size, and physical properties of drug particles that facilitate the targeted drug delivery. Diverse metals such as gold, silver, gadolinium, zinc, copper metal dichalcogenides, and superparamagnetic oxides such as magnetite (Fe_3O_4) and maghemite (Fe_2O_3) have been explored for efficient drug-delivery regimes in the GI tract [18].

2.3 Carbon-Based Nanostructures

Carbon-based nanomaterials are playing a pivotal role in drug-delivery applications. The remarkable electrical, mechanical, and chemical characteristics as well as possessing the ability to hybridize with a wide range of organic and inorganic materials to develop efficient and more stable composite materials make these nanoparticles suitable for targeted drug delivery [19]. The wide range of sp^2-hybridized carbon from zero- to two-dimensional nanomaterials such as fullerenes (0D), one-dimensional (1D) carbon nanotubes, and two-dimensional (2D) graphene imparts a significant role in developing smart materials for drug delivery [20]. From the past decade, there has been tremendous utilization of carbon-based nanomaterials owing to their low cytotoxicity, good biocompatibility, and high propensity to cross cellular barriers such as plasma membranes, easy fabrication, and surface modification. As these nanomaterials are water-insoluble in nature, the attachment of hydrophilic moieties on the surface of these nanostructures makes them soluble and has turned out to be able to work as excellent drug-delivery vehicles [21].

2.4 Protein-Based Nanostructures

Currently, the highly sophisticated drug-delivery system has been envisaged based on the proteins. The nanostructures derived from proteins provide remarkable credentials such as nonantigenicity, biodegradability, physiologically metabolizable, easily tunable for self-assembly process and surface functionalization, and covalent attachment of drugs and various ligands [22]. The proteins can form spatially well-defined nanostructures of differential size and shape such as spherical vesicles, nanotubes, nanofibrils, ordered molecular chains, nanowires and hybrid nanoparticles, nanoribbons, and nanotapes. The various proteins such as gelatin, albumin, gliadin, and legumin as well as self-assembled peptides have been explored for drug delivery, such as various peptide amphiphiles, diphenylalanine, and collagen mimetic self-assembled peptides [23]. These nanoparticles are preferentially used as parenteral carriers for different drugs as these are able to cross the blood–brain barriers. Among all these proteins and peptides described earlier, the mucoadhesive properties of gliadin have been exploited for the delivery of drugs in the GI tract. The solubility and hydrophobicity of gliadin lead to the strong tropism for upper regions of the gastrointestinal tract through binding with the mucus layer and are capable of protection of drugs as well as ensure its controlled release. Therefore comprehensive efforts have been made for exploring the wide world of proteins by rational design for targeted drug delivery as well as improvement in their drug loading capacities (DLC) [24]. The constant emergence of novel protein-based nanostructures with stimuli-responsive function, various peptide motifs, and triggered drug delivery at targeted sites paved the way for a promising avenue in biomedical research [25].

2.5 Nucleic Acid-Based Nanostructures

The advancements in nucleic acid nanotechnology have witnessed significant progress in nucleic acid-based nanostructures of various sizes, dimensions, geometries, and shapes and have been well studied for targeted drug delivery. They impart significant attributes such as facile fabrication through self-assembling process, less immunogenicity, high flexibility for loading of drugs, tunable for heterologous decoration of other biological molecules, and excellent controlled release of drugs to the target site. To date, there are two approaches that are being utilized for the manufacturing of DNA-based nanoparticles: (a) DNA origami and (b) rolling circle amplification [26]. In the DNA origami approach, the long-stranded DNA of bacteriophage has been folded through the use of more than 200 complementary staple strands in which the positively charged drugs/proteins, peptides are intercalated and allows to reach the target site.

By this method, a variety of two-dimensional as well as three-dimensional nanostructures such as tetrahedrons, smiley faces, DNA nanotubes, DNA "dolphins," and DNA barrels have been developed. In rolling circle amplification, a circular DNA template with repeating DNA sequences has been used for the creation of long-stranded structures through DNA polymerase. This approach also adds diversity in the formation of nanostructures such as nanoballs, nanotubes, nanobelts, and nanoclews and exploits the precise delivery of genes and drugs [27]. Apart from that, oligonucleotides have been also explored for the development of polyhedral nanostructures and DNA dendrimers that have been used for efficient drug delivery. More recently, DNA-based hybrid nanostructures of about 10 nm in the form of either hydrogels or nanofilms have been developed for drug delivery of hydrophobic drugs. Simultaneously, recently RNA-based nanostructures that rely on functionalized phi29 pRNA have been utilized in drug delivery. The methodology based on RNA origami has been utilized for the formation of nanostructures of excellent plasma stability and thermodynamic stability. Research studies supported the fact that RNA origami is more favorable as drug-delivery vehicles for controlled release of drugs as compared with its counterpart DNA origami. Still owing to great potential in drug delivery, more focused studies are still needed for in vivo pharmacokinetic bioavailability of nucleic acid-based nanostructures [28].

2.6 Silica-Based Nanostructures

The prolific growth has been observed for the development of porous nanostructures based on biocompatible inorganic compounds, such as alumina, silica, calcium phosphate, and titania. The significant advantage in the perspectives of drug delivery has been realized nowadays due to possessing nanopores size (diameter) and density of these nanostructures that can be precisely controlled to achieve a constant drug-delivery rate in the GI tract [29]. The most extensively investigated mesoporous silica nanoparticles (MSNs) possess outstanding properties such as high specific surface area, monodispersity, dynamic functionalization, and tunable pore size and diameter. The galactose or mannose, as well as folate functionalized MSNs, has been reported to depict more cytotoxicity of cancer cells as compared with unfunctionalized MSNs. A variety of drugs have been successfully delivered to their respective target site including cysteine, doxorubicin, paclitaxel, telmisartan, camptothecin, and chlorambucil through MSNs or functionalized (covalently grafted) MSNs [30]. Apart from that, recently targeted drug-delivery potential of calcium phosphate (depict similarity to bone chemical composition) systems, such as hydroxyapatite and tricalcium phosphates, has been explored. The differential stoichiometry, biocompatibility, dissolution properties, varied surface charge density, and functionality render them suitable carriers for drug delivery. In addition to that, growing developments have been made for exploring alumina and titania ceramic nanoparticles for drug-delivery applications. These particles depicted diverse spherical nanostructures like spherical flaky assemblies, mesoporous spheres as well as they have a property for binding with natural proteins. However, more in-depth knowledge is needed concerning their nonbiodegradability and bioaccumulation in the body cells followed by their toxic effects [31].

2.7 Lipid-Based Structures

These are classified as broad-spectrum nanoparticles such as colloids, liposomes, emulsions, nanostructured lipid carriers (NLCs), and solid lipid NPs (SLNs). Recently, the lipid nanoparticles have found significant applications in targeted drug delivery in the GI tract. As these are composed of lipids, they are less susceptible to degradation of gastrointestinal enzymes and depicted improved stability and specificity [32]. Among these, liposome-based nanostructures have been recognized as first drug-delivery vehicles at nanoscale dimensions and set the benchmark from laboratory to clinical applications. These are synthetic vesicles composed of phospholipids such as phosphatidylcholine, phosphatidylserine, and phosphatidylglycerol. These are classified on the basis of lipid bilayers from unilamellar vesicles (UVs) (small UVs and large UVs) and multilamellar vesicles (MLVs) [33]. Unilamellar vesicles consist of a single lipid bilayer and have a size range of less or more than 100 nm while multilamellar vesicles consist of several lipid bilayers and vary in size from few hundreds to thousands of nanometers. The inherent capability of liposomes to solubilize and compartmentalize both hydrophobic and hydrophilic drugs as well as biodegradability and biocompatibility adds dynamism in their properties and makes these entities attractive as drug-delivery vehicles [34]. Apart from that, liposomes also prevent the degradation of drug molecules, reduce side effects, and facilitate targeted delivery of drugs. Furthermore, colloids and emulsions refer to simple lipids suspended with drugs and have fewer side effects, and they become a good choice for targeted delivery however suffering from the limitations of drug degradation, low stability, and no intermembrane interactions. These nanoparticles are composed of surfactants and solid fat that can deliver both hydrophilic as well as hydrophobic drugs [35]. The structural diversity has been seen in their designing either drug-enriched shells or drug-enriched cores but both of them depicted increased stability and decreased toxicity. The designing of NLCs utilizes both solid/liquid combinations in which the drug molecules are embedded within the fatty acid chains that increase their drug retention capacity and stability. Therefore diverse composition, structure, size, and surface modification with diversified ligands lead to varied functionality that makes these vehicles as the most intelligent system for the delivery of therapeutic molecules in the GI tract [36].

3 NANOMEDIATED DRUG-DELIVERY APPROACHES IN GASTROINTESTINAL SYSTEM

The targeted delivery of nanoparticles can be mediated broadly in two ways on the basis of either physical encapsulation or chemical conjugation. These methods are known as passive and active targeting. In passive targeting, drugs are physically encapsulated into the internal cavity of the nanocarriers and stabilized by noncovalent interactions followed by disassembly at a specific site of action and release of drugs [37]. However, this method involves multiple obstacles like mucosal barriers for the drugs to reach their target site as well as accompanied by very low drug loading capacities (DLC) and nonspecific delivery. In active targeting, the surface functionalization of nanocarriers has been performed with those ligands that are specifically recognized by receptors on the surface of the target cells. The controlled release of drugs has been occurring by the cleavage of the conjugation between nanocarriers and drugs at the target site [38]. Currently, the concept of self-delivery has been come up with great potential as in this method drug molecules themselves made well-defined nanostructures by possessing the ability of self-assembly. By this approach, the content and distribution of drugs can be accurately controlled in drug-based nanostructures. There is a growing trend for the development of various self-delivering nanostructures through innovations in rational analysis, design, and fabrication methodologies in nanotechnology. The targeted drug delivery in the gastrointestinal region is quite challenging as the gastrointestinal tract (GIT) is represented as the most variable region of the body. Therefore the drug has to accustom to complex physiology, varied fluid volumes, and wide pH fluctuation and transit times. Based on this consideration, four approaches for drug-delivery systems have been designed on the basis of the enzyme environment, pH, ambient pressure, and relative constant transit time in the GI tract [39].

3.1 Enzyme-Based Approach

This strategy is based on the specific enzymes produced by colonic bacteria that degrade the nanomaterials for their growth and facilitate the release of drugs at the specific region of the colon [40]. Several polymeric (chitosan, alginate, amylopectin, poly(lactic acid)) and other biocompatible nanomaterials have been synthesized and loaded with different drugs such as anticancerous, antiinflammatory, and others for the successful release of drugs in targeted site. This strategy rules out the possibility of nonspecific delivery of drugs targeted toward the colonic area of the GI tract [41].

3.2 pH-Based Approach

This drug-delivery strategy is based on the release of drugs in pH-dependent fashion as the GI tract presents a broad range of pH variations like 2–3 in the stomach, 5–6 in the small intestine, and neutral pH 7 in the colon [42]. The strategy strongly eliminates the possibility of drug reduction during GI tract transit. A variety of nanostructures based on biodegradability and biocompatibility have been designed for drug delivery based on their pH-dependent stability and slight fluctuation in pH leads to the dramatic effects on drug-release kinetics [43]. More recently, this feature can be successfully applied for the encapsulation of prodrugs in nanocarriers that can be converted into active drug in the targeted area by pH-dependent manner [44].

3.3 Time-Based Approach

This strategy has been developed to facilitate the drug release at the site of action that is based on the transit time in the GI tract. In this approach, the drug is released after the degradation of the outer layer of nanocarriers in a time-dependent manner [45]. Various parameters such as pH, degree of swelling of the polymeric matrix, the thickness of polymeric layers, charge of polymeric molecules, as well as osmotic activity of the salt have been taken into account for designing the nanovehicles. In this way, a greater amount of drug release can be done by nanostructures in the targeted area [46].

3.4 Pressure-Based Approach

This drug-delivery strategy is based on the ability to utilize relatively strong muscular contractions present in the colon that serves the driving force to move its contents [47]. The intracolonic mechanical pressures generated during this peristaltic movement facilitate the targeted release of drugs. Therefore drug nanocarriers for example intestinal pressure-controlled colon delivery capsules (PCDCs) are designed in such a way that their outer coating is made up of water-insoluble polymeric compounds like ethyl cellulose. By acquiring optimal pressure through colonic peristalsis, these molecules get ruptured and release the required drug at the specific site [48].

3.5 Redox-Based Approach

The enormous potential has been visualized for targeting the gastrointestinal tumors through the redox-based approach for drug delivery. This strategy utilizes the reducing environment of tumors as an internal signal for releasing the drugs from nanocarriers [49]. Research studies strongly emphasized that the formation and stability of tumorigenic cells are highly dependent on the reduction and oxidation states of glutathione (GSH, GSH/GSSG) and $NADPH/NADP^+$ [50]. The major reason for reducing the environment in tumors cells due to the presence of a higher concentration of GSH than NADPH. The GSH maintains this environment due to the formation and breaking of disulfide bonds [51]. The approximate intracellular concentration of GSH has been found to be 10 mM in normal cells as compared with cancerous cells that possess a fourfold higher concentration of GSH [52]. Apart from disulfide bonds, the importance of the redox-responsive ability of diselenide bonds has been increasingly realized in targeted drug delivery through nanosystems [53]. The detailed diagrammatic representation has been depicted in Fig. 4.3. Research studies speculated that a diselenide bond offers a more efficient strategy for redox stimuli due to the presence of lower bond energy (Se—Se 172 kJ/mol) as compared with disulfide bonds (C—Se 244 kJ/mol and S—S 268 kJ/mol) [54]. In addition to that, nowadays succinimide-thioether linkage that can be cleaved through glutathione has also been visualized for assessing the reducing environment in tumor cells. Therefore the differences in the redox mechanisms between normal cells and cancer cells provide a suitable platform for targeted release of drugs through redox responsive nanovehicles [55].

4 TYPES OF INTERACTIONS OF NANOPARTICLES IN GASTROINTESTINAL SYSTEM

The diverse morphology and physiology of gut pose significant variations in drug absorption properties. The various factors are responsible for effective drug delivery of nanoparticles in the GI tract that depends upon the

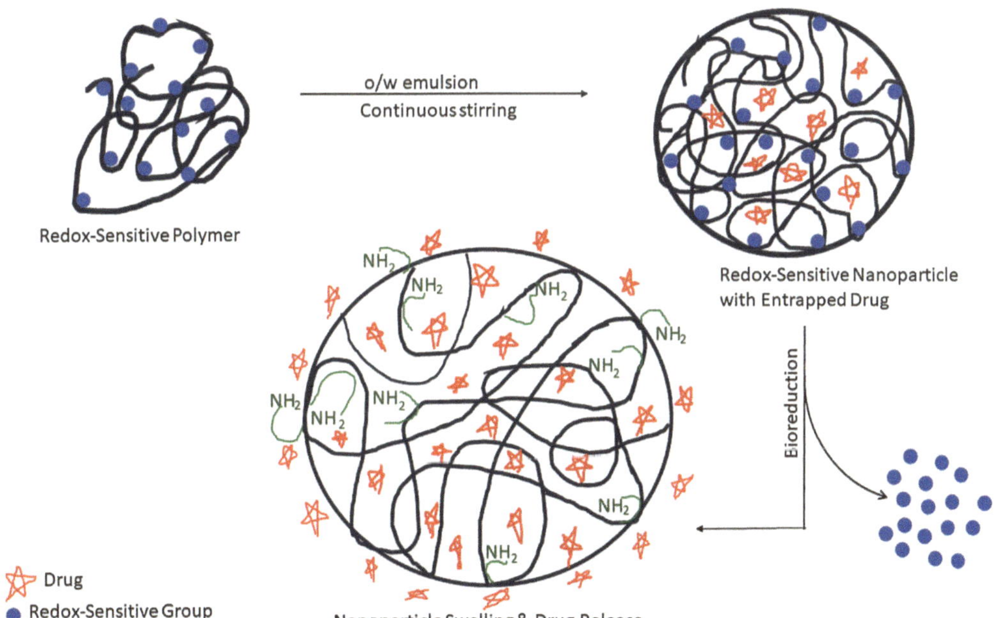

FIG. 4.3 Mechanism of redox based targeted drug delivery approach in gastrointestinal tract (GIT).

size as well as the involvement of various host pathways for their internalization [56]. More recently, the current research is aiming to comprehend the interaction of nanoparticles with our resident microbiota [57]. The different interactive avenues of nanoparticles with host and their resident microbes are described below.

4.1 Interaction With Different Host Pathways

The effective drug-delivery strategies are based on the timely release of drugs at a targeted site with the degradation of nanocarriers [58]. The variety of host pathways has been involved in the internalization of NPs into the gut. Literature studies gave speculating evidence that nanoparticles have ben interacted by clathrin-mediated endocytosis, chemically enhanced adhesion, caveolae-mediated endocytosis, and specific receptors. The internalization of these nanocarriers is highly dependent on their size that has been experimentally proven in rat colitis model [59]. The studies revealed that those NPs with diameters ≤200 nm are recognized by clathrin-mediated endocytosis, whereas NPs with diameters ≤500 nm are internalized by caveolae-mediated endocytosis. The development of mucoadhesive drug-delivery approaches had worked well in the GI tract. The overview of mucoadhesive drug delivery has been depicted in Fig. 4.4. The ionic interaction in the form of hydrogen bonds has formed between mucin chains present on the mucus layer of the GI region with polymeric chains present on drug-delivery nanosystems. The drug delivery has been facilitated by diffusion into gastrointestinal epithelial cells and also promotes enhanced accumulation and adhesion in the inflamed regions of the GI tract. It was also proven that NPs with diameters from 100 to 1000 nm displayed 5–6.5-fold enhanced accumulation in inflamed areas as compared with the healthy control group.

4.2 Interaction With Gastrointestinal Microbiota

The colonization of microbiota has been initiated from birth in the human gastrointestinal tract that imparts an outrageous impact on maintaining the homeostasis of the human body. The orchestrated molecular crosstalk between different microorganisms as well as host cells in response to different environmental signals such as genetic framework, diet, and immune signals is indeed required for modulation of host physiology and metabolism [60]. Therefore GIT is considered as one of the complex ecosystems and contains members of all three domains: *Archaea, Bacteria,* and *Eukarya*. The predominating microbial populations are as follows: *Firmicutes* (46%–60%), *Bacteroidetes* and *Actinobacteria* (8%–28%), and *Proteobacteria* (10%–30%). Therefore the

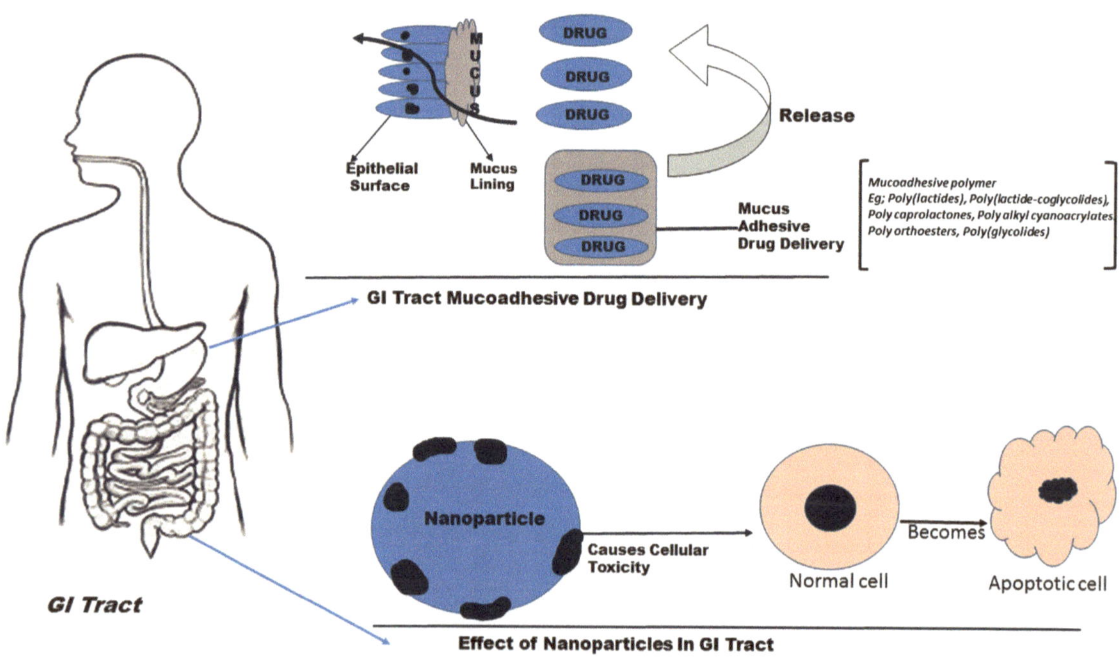

FIG. 4.4 An overview of muco-adhesive drug delivery in gastrointestinal tract (GIT).

growing continuum of nanoparticles in drug delivery has signaled the new era of understanding for the interaction of nanoparticles with GIT microbiota. The experimental studies suggested that nanoparticles confer debilitating effects against GIT microbiota through the disintegration of microbial components resulting in dysbiosis [61]. They laid significant impact on the homeostasis of the human body by altering the properties and/or composition of microbiota that ultimately perturb the finely tuned balance between health and disease. Although we are very far from understanding the exact mechanisms of action of these nanomaterials against GIT microbiota, this area of research is of prime attention due to the acquisition of antibiotic resistance in pathogenic bacteria through natural selection [62].

Therefore it is very much imperative to identify novel dysbiostatic therapeutic modalities that will prevent dysbiosis as well as minimize the incidents of natural selection that lead to the reduction of the rising global menace of antibiotic resistance. Moreover, recent research studies suggested that the composition and number of microbiota residing in GIT may also influence the effects and fate of nanoparticles within GIT, thus affecting the therapy and disease outcome. This has been observed, for example, in the case of nanostructured drug-delivery systems used to target a particular disease through the gastrointestinal route [63,64]. This bioriented interaction requires highly focused attention that clarifies the dilemma of using nanomaterials and develops a clear understanding of novel pathways and mechanisms that substantially add the growing repertoire of smart drug-delivery systems in the GI tract.

5 NANOMEDIATED TARGETED DRUG-DELIVERY IN VARIOUS GASTROINTESTINAL DISORDERS

5.1 Gastrointestinal Ulcers

The GIT represents a physical barrier and has to perform multiple functions simultaneously; therefore, maintaining its intricate functionalities requires immense efficiency. There are multiple factors that lead to the damage of gastrointestinal mucosa including various toxins, drugs, aggressive hydrochloric acid, and most chronic infection with *Helicobacter pylori* bacteria [65,66]. These factors lead to chronic gastritis accompanied by gastrointestinal ulcers thus being a major contributing factor for the pathogenesis of gastric adenocarcinoma and mucosa-associated lymphoid tissue lymphoma [67]. Besides the conventional treatment for the ulcers, the antibiotic encapsulated in nanoparticles seems to be a promising avenue for the treatment of gastrointestinal infection but major hurdles lie for the eradication of *H. pylori*. The successful eradication of this bacterium is stymied by two important factors, as the improper access of antimicrobial drugs due to the presence of *H. pylori* in the inner lining of gastric mucosa as well as effective and active concentrations of drugs is not delivered at the targeted site.

Therefore rigorous efforts have been undertaken for the development of nanomediated drug-delivery vehicles that can cross thick mucus lining of ulcerated tissues of the gastrointestinal tract as well as persist for a longer duration for delivering the drugs at proper concentration [68]. In recent years, different strategies were developed for enhancing the gastric residence time of the delivery system that includes the floating drug-delivery system, density-based systems, the swelling system, and the mucoadhesive system. Therefore first successful trails were done in 2000 by developing mucoadhesive gelatin microspheres embedded with amoxicillin for the eradication of *H. pylori*. Later on, the concept of dual therapy nanoparticles made from biodegradable and biocompatible materials have been developed for increasing the targeted drug delivery as well as ensuring the incorporation of both hydrophilic and hydrophobic substances and the feasibility of different routes of administration. In this continuation, mucoadhesive gliadin nanoparticles containing two antibiotics clarithromycin and omeprazole were prepared and evaluated against *H. pylori* [69]. These nanoparticles provide greater therapeutic efficiency and long retention in gastric mucosa as compared with conventional drug formulations. Moreover, lectin-conjugated gliadin nanoparticles showed twofold increased inhibition of the growth of *H. pylori*. Furthermore, lectin-conjugated nanoparticles containing three antibiotics, namely clarithromycin, amoxicillin, and omeprazole, have shown the higher ability for the eradication of this bacterium by delivering the drugs in a more specific manner [70]. In continuation of this, fucose-conjugated chitosan-glutamate nanoparticles loaded with triple-drug formulations were also used a standard therapy for eradication of *H. pylori*. Furthermore, for enhancing the acidic resistance, amoxicillin-loaded chitosan poly-g-glutamic acid nanoparticles were prepared and encapsulated in calcium alginate-gelatin gel that is governed by pH fluctuations in the GI tract. In addition to that, polystyrene-based fluorescent nanoparticles were evaluated in the gastric ulcer-induced mouse model [71]. The drug was delivered through nanoparticles, inflammation was monitored through myeloperoxidase activity, and particle deposition was assessed through fluorescent spectroscopy. The analysis revealed enhanced deposition of nanoparticles in ulcerated regions as compared with normal tissues.

In continuation of this, more biocompatible nanoparticles have been synthesized based on complex formed by chitosan, berberine, and heparin. These three polymers constitute both therapeutic as well as drug-delivery vehicles for the treatment of *H. pylori* as berberine is used for the treatment of gastroenteritis, heparin is known for its ability to bind to the cell receptors for drug delivery, and chitosan is utilized for its adhesive and biocompatible properties [72]. Furthermore, more nanoparticles were developed based on the heparin-chitosan complex for enhancing the gastric resistance or the heparin-genipin complex that was successfully used for drug delivery in gastric-ulcerated tissues. Later on, polymeric nanoparticles loaded with antioxidant quercetin have been used for a reduction in oxidative stress by downregulating NOS-2 and MMP-9 in an ethanol-induced gastric ulcer in mice [73]. More recently, liposome-based nanoparticles have been prepared from fucosylated glycolipids that facilitate more targeted delivery of drugs for the treatment of both coccoid and spiral forms of bacteria. In addition to that, nanocapsules were designed as a gastroretentive drug-delivery system by an alternative coating of polycation (poly(allylamine hydrochloride)) and polyanion (poly(acrylic acid)) using liposomes as the core encapsulated with amoxicillin and metronidazole [74]. This multilayered system possessed the capability of prolonged drug release that can be more specific for the eradication of *H. pylori*.

5.2 Gastric Cancer

Gastric cancer is recognized as the most complicated and multifarious process and the second leading cause of cancer-related deaths across a global scale. The various reasons attributed to the occurrence of this cancer include degradation of the extracellular matrix, production of reactive oxygen species (ROS), infection with *Helicobacter pylori*, and damage of mitochondrial components [75]. To date, the conventional treatments including chemotherapies, radiation, and surgery have been implied for eradicating early and in situ gastric cancers but not responsive to advanced metastatic stages. The growing resistance toward chemotherapy-induced apoptosis poses a major hurdle for combating this deadly disease and suffers from poor prognosis with a survival rate of less than 24% [76]. Exhaustive efforts have been undertaken for the development of therapeutic modalities for early diagnosis and treatment of gastric cancer, and among those, nano-based approaches both for diagnosis and treatment prove efficient and fast system [77].

Although our focus especially lies in targeted delivery of drugs in the GI tract, the further discussion has been mainly aligned with drug delivery only. The research studies leveraged with a variety of nanoparticles that have been used for an efficient drug-delivery system for gastric cancer cells [78]. Research studies reported that ursolic acid acts as a potential drug for the treatment of gastric cancer by inhibiting the expression of cyclooxygenase-2 (COX-2) and inducing apoptosis in cancer cell lines [79,80]. The drugs suffer from the limitation of its hydrophobicity; therefore, ursolic acid nanoparticles used as drug vehicles based on amphiphilic methoxy poly(ethylene glycol)-polycaprolactone (mPEGPCL) copolymers lead to efficient drug delivery that occurs in SGC7901 cells by membrane penetration followed by inhibition of COX-2 and caspase-3 activity and enhanced apoptosis [81]. Moreover, the side effects of paclitaxel (Ptx), a potent drug for the treatment of gastric cancer, have been reduced by the encapsulation of paclitaxel in biodegradable nanoparticles synthesized with tetrandrine. The combined therapy leads to the synergistic effect of tetrandrine-induced ROS production and paclitaxel-induced cytotoxicity by depleting the level of cellular antioxidants [82]. Therefore coadministration of two drugs through these nanostructures provides an efficient therapeutic approach for the treatment of gastric cancer. More recently, NLCs have been used for the delivery of curcumin and etoposide for the treatment of gastric cancers. As curcumin has low bioavailability and etoposide drug must be delivered in large concentrations, NLCs loaded with these two therapeutic molecules lead to enhanced selective delivery to gastric tumor sites that have been experimentally proven in mouse models [83]. Recently, the emerging trend of RNA nanoparticles confers a paradigmatic shift in the growing avenue of targeted drug delivery.

These nanoparticles are fabricated with packaging RNA (pRNA) of bacteriophage (phi29 DNA) packaging motor that forms stable dimers, trimers, and hexamers via interlocking loops through ionic interactions. The presence of thermodynamically stable three-way junction (3WJ) motif30-32 pRNA provides binding platforms for various therapeutic modules, such as miRNA, siRNA, or ribozymes and other molecules such as receptor-binding aptamers [84]. The research studies confirmed that folate receptors have been overexpressed in gastric and breast cancer cells; therefore, folic acid-conjugated silica-modified gold nanorods encapsulated with drugs have been used for imaging as well as photothermal therapy of gastric cancer (MGC803) cells [85]. In addition to that, antibody-conjugated fluorescent magnetic nanoparticles have been utilized for imaging and hyperthermal therapy of in vivo gastric cancer cells by targeting antigenic epitope SSKKQKRSHK. One of the promising molecular target *BRCAA1 (breast cancer-associated antigen1, AF208045)* exhibited overexpression in breast cancer and gastric cancer cells. Therefore FA-conjugated pRNA-3WJ nanoparticles have been used for a more targeted approach to deliver therapeutic BRCAA1 siRNA to in vivo stomach cancer cells. These nanoparticles provide a stable platform and facilitate specific binding with cell surface receptors of gastric cancer cells followed by internalization and delivery of miRNA, siRNA, and drugs for the treatment of gastric cancer [86].

5.3 Gastrointestinal Irritation and Gastric Toxicity

Currently, the nanomaterial-mediated drug-delivery approaches are not only proving their potential for targeted drug delivery but their enormous potential has been realized for reducing the burden of gastrointestinal irritation and gastric toxicity due to the consumption of various drugs orally or intravenously that is mixed with gastrointestinal milieu and imparting adverse effects to the gastrointestinal tract [87]. One of the commonly used antiproliferative drug ibuprofen in cancer therapy that is taken orally causes severe gastric toxicity.

This toxicity has been reduced by incorporating the drug into poly(DL-lactic acid) nanoparticles that minimize the dosing capacity of this drug as well as facilitate systematic circulation in the GI tract. Apart from that, dendrimers based on poly(amidoamine) polymer conjugated with SN38 have been shown to improve the oral bioavailability of persuasive anticancer drugs with reduced GI toxicity. Similarly, the GI irritation caused by commonly used immunosuppressive, antiinflammatory, and anticancerous drug triptolide has been significantly reduced by encapsulation of this drug in SLNs or liposomes [88]. The nanoencapsulation mediates the sustained release of this drug into the GI tract as well as imparts stability over gastric secretions even after 3 h of incubation. Apart from that, polymeric hyaluronan-cisplatin-conjugated nanoparticles entrapped in Eudragit S100-coated pectinate/alginate microbeads as a drug-delivery vehicle for colonic sites are helpful for the alleviation of nephrotoxicity in vivo conferred by cisplatin [89]. More recently, Ulcerative colitis overburdened by GI drugs has been circumvented by polyethyleneimine-loaded NPs with siRNA/CD98 by releasing CD98-specific siRNA to colonic cells and downregulating the CD98 in macrophage cells and the intestinal epithelial cells [90]. Thus nanomediated drug-delivery vehicles are of immense importance for minimizing gastrointestinal irritation and toxicity; therefore, more translational applications can be investigated for retrieving benefits for this purpose.

5.4 Esophageal Cancer

Esophageal cancer is recognized as one of the chronic causes of cancer mortality on a global scale. This cancer is characterized into two forms: esophageal adenocarcinoma (EAC) and squamous cell carcinoma (ESCC). Current literature studies suggested that EAC is more prevalent in western countries while higher incidences of ESCC were reported in developing countries [91]. Currently, the early prognosis of this cancer is very poor; therefore, therapeutic modalities are very limited due to diagnosis at the advanced stages. Currently, chemotherapeutic interventions such as platinum drugs, the anthracycline compounds, and epirubicin have been routinely used but suffer from significant toxicities and lacking cell specificity [92]. Therefore there is a pressing need for identifying effective treatments accompanied by targeted drug-delivery strategies to reduce the side effects of drugs. A promising possibility has been envisaged by nanomediated drug-delivery strategies that impart enhanced retention effect and improvised solubility of hydrophobic drugs and minimize drugs' side effects.

Nanoparticles have been utilized for both imaging as well as therapeutic vehicles for early diagnosis of this cancer. Initially, fluorescent nanoparticles based on either quantum dots or organic fluorophores have been utilized but suffers from the limitation of biocompatibility and photobleaching [93]. Therefore currently, peptide-based fluorescent nanoparticles that utilize the remarkable capability of self-assembly of natural amino acids and alleviate the biodegradability and biocompatibility issues have been developed. The zinc-coordinated fluorescent cyclic peptide nanoparticles have been recently explored for online monitoring of the drug release and targeted esophageal cancer cell imaging [94]. This study also suggested the possibility of tuning the fluorescence of these nanoparticles to near-infrared range that paved the way for better in vivo translational applications due to deeper tissue penetration. Apart from that, biocompatible metallic nanoparticles based on chitosan-coated gold/gold sulfide nanoparticles have also been used for photothermal ablation therapy for esophageal cancer cells without affecting the healthy cells [95]. In addition to that, cerium oxide nanoparticles also possess the radical-scavenging ability that renders them as effective drug delivery and therapeutic vehicles for the treatment of this cancer. Thus targeted drug delivery remains challenging for diagnosis as well as treatment of esophageal cancer; therefore, much more comprehensive efforts are direly needed in this direction.

5.5 Colorectal Cancer

Colorectal cancer is among the fatal cancers that are currently being impacting one millions of people per year at global scenario. The inefficiency of conventional chemotherapeutic strategies for discriminating healthy versus metastatic cells paved the way for targeted drug-delivery approaches to prevent colorectal cancer. The current scenario profusely utilizes the potential of nanoparticle-based drug-delivery systems to deliver the drugs in a specific region of the colon [96]. In the early stage of the development of nano-based drug-delivery vehicles, the passive accumulation of drugs occurs in tumors due to tumor vascularization and angiogenesis. More recently, active targeting has been preferred in which the nanocarriers have been modified with targeting ligands, such as proteins, antibodies, polysaccharides, and aptamers, that have the ability to recognize receptors that are overexpressed by colorectal cancer cells or angiogenic endothelial cells [97]. The PLGA nanosystem has been used for delivering meloxicam as an antiinflammatory agent for the effective inhibition of COX-2 enzymes in colon cancer cells (HT-29 cells) and displayed enhanced cytotoxicity at 800 μM concentrations [98]. Similarly, folate-conjugated guar gum nanoparticles encapsulated with methotrexate deliver controlled release of this drug into the colonic region while protecting the prerelease of the drug in the upper gastrointestinal tract [99]. In continuation of this, deoxycholic acid-conjugated nanoparticles (DexDA) encapsulated with retinoic acid have been explored for their anticolon cancer potential in both in vitro and in vivo conditions [100].

Apart from that, polymeric chitosan-based nanoparticles encapsulated with IL-12 (CS-TPP/IL-21) were found to significantly reduce the colorectal and subsequently liver metastasis [101]. Furthermore, more precise nano-based hydrogel systems have been used to deliver drugs in malignant and inflamed regions of the colon by utilizing the electrostatic interactions of chitosan (Ca^{2+}positive) and alginate (and SO_4^2- negative) polysaccharides [102]. More recently, a solid lipid vesicular system based on cholesteryl butyrate (SLN) has been used as a delivery system for targeting the colon cancer cells [103]. The short-chain fatty acid butyrate was used as a drug that downregulates the extracellular signal-regulated kinase (ERK) and *p*38 phosphorylation in human cancer cell lines. Apart from that, edible exosomes based on grape exosome-like nanoparticles have been recently explored for remodeling and protection of colonic epithelium cells by enhancing the Lgr5hi stem cells that also help in the alleviation of colonic malignancy [104].

In addition to that, gold nanoparticles based on the extracts of brown macroalgae *Cystoseira baccata* extracts were used for delivering these natural extracts into two colorectal cancer cell lines (Caco-2 and HT-29) and the fibroblast cell line PCS-201-010 and displayed considerable cytotoxicity in HT-29 cells [105]. In a similar fashion, the biogenic silver NPs based on fruit extracts of *Balanites aegyptiaca* were used for targeting colon and liver cancer cells. More recently, lipid core nanocapsules based on resveratrol have been developed and characterized for their ability to target the colon cancer cells and displayed enhanced therapeutic efficacy [106]. Therefore wide arena of nanostructures have been developed for targeting drug delivery for colorectal cancer, and much efforts are needed for their clinical transformations in in vivo conditions.

5.6 Inflammatory Bowel Disease

Inflammatory bowel disease (IBD) is a chronic inflammatory disease of the gastrointestinal tract that is manifested in two forms commonly known as ulcerative colitis and Crohn's disease [107]. Ulcerative colitis has restricted premises of inflammation within the colon and rectum and caused profound alteration in the structural framework of mucosal crypts. It is primarily characterized through severe neutrophil infiltration in the intestinal epithelium. Crohn's disease has a more chronic effect as it develops through the entire gastrointestinal tract and leads to the development of granulomas due to the extensive deposition of plasma cells and macrophages. The acute inflammation due to the accumulation of macrophages, neutrophils, and dendritic cells leads to alteration in the architecture of intestine and thus enhances the permeability for the migration of various cells and macromolecules. To date, the disease is considered to be idiopathic and still a matter of mystery to the researchers and clinicians. Therefore in the present scenario, no curative treatment has been available for IBD; however, a variety of antiinflammatory molecules such as corticosteroids, vascular adhesion molecules, aminosalicylates, monoclonal antibodies, and cytotoxic drugs have been given as the first line of treatment of IBD. These treatment modalities can alleviate the initial level of IBD to some extent but their prolonged use has been restricted due to the serious side effects and poor efficiency due to the nontargeted approach. Thus it is worth the alternative speculation of targeted drug-delivery vehicles that can reasonably deliver the drugs at the specific site of GI.

Recently, the rigorous application of various nanoparticles for the targeted drug delivery has been realized and currently, two administration strategies, namely intravenous injections and oral formulations, have been adopted for the treatment of moderate-to-severe episodes and mild-to-moderate symptoms of IBD, respectively. The research studies speculated that the targeted drug delivery in the inflamed colon has been explored at three levels namely organ, tissue and cell [108]. The oral formulations can only target at the organ (colon) level through pressure, pH, or bacteria-responsive mechanisms. Interestingly, the phenomenon "epithelial EPR" effect [109] that is characterized by severe inflammation in the colon due to accumulation of immune cells beneath the disrupted enterocyte layer and tissue hyperpermeability has been exploited to the preferential homing of NPs with greater efficiency and with an enhanced concentration of drugs at inflamed sites. Therefore NPs based on ROS-sensitive mechanisms can perform active targeted drug delivery at the level of colonic tissues. Apart from that, receptor-mediated targeted drug delivery has been also possible through NPs at the cellular level [110]. Although a plethora of various nanoparticles have been developed for the treatment and diagnosis of cancerous lesions, the development of nanostructures for the treatment of IBD is still in infancy. At present, two major targets for the treatment of IBD have been explored: (a) antiinflammation that includes the inhibition of synthesis and release of proinflammatory cytokines; (b) mucosal healing.

Currently, the antiinflammatory pathways such as a reduction in synthesis and production of proinflammatory cytokines have been explored for treatment avenues but fewer studies have been pursued toward mucosal healing through nanomediated interventions [111]. The inflammation has been reduced in the dextran sodium sulfate (DSS)-induced colitis mouse model by synthesizing the nanoparticles with mannosylated biocompatible cationic polymer assembled with TNF-α siRNA and tripolyphosphate [112]. As the mannose receptor is expressed on the surface of macrophages of inflamed areas of GIT, these nanoparticles can be easily internalized into macrophages by receptor-mediated endocytosis.

Therefore the developed NPs significantly reduced the expression of TNF-α, which has been confirmed through ex vivo experiments. In a similar fashion, the overexpression of galactose receptors on the surface of macrophages has also been utilized for antiinflammatory regime for the IBD [113]. More recently, nanoparticles with galactose trimethyl chitosan-cysteine have been fabricated and assembled with mitogen-activated protein kinase 4 siRNA (siMap4k4), which has been orally delivered for the treatment of DSS-induced colitis mouse models into activated macrophages. The results revealed a significant reduction in TNF-α production and improved episodes of colon length shortening, DSS-induced body weight, and myeloperoxidase activity in in vivo conditions. Similarly, the overexpression of F4/80 glycoprotein as a macrophage marker can be used for the target site for functionalized NPs. The NPs were synthesized with the Fab fragment of the F4/80 antibody and loaded with TNF-α siRNA that has been administered in vivo in the colitis regions [114] and revealed considerable improvements in various physiological symptoms and therapeutic potential as compared with nontargeted NPs.

At present, the corticosteroids especially lead the mainstay for the treatment of IBD but stymied by their long term use due to quite adverse effects such as nausea and vomiting, sleep and mood disturbance, fluid retention and weight gain, dyspepsia, and hypertension, diverse efforts have been done for the specific delivery of this drug with maximum potency associated with alleviation of side effects. Therefore dexamethasone-embedded spherical polymeric nanoconstructs were developed and tested for their efficient therapeutic potential and drug delivery in DSS-induced colitis [115] and displayed reduced inflammation at both localized and systemic regions of the GI tract. Recently, SDDS was developed through self-assembly of biocompatible polyphenols and polymers in which spherical-shaped NPs were fabricated with tannic acid and Pluronic F-68 and loaded with 10% dexamethasone as a treatment for IBD [116]. The enzyme-based release method was chosen for drug delivery, and the results showed promising improvement in inflammation and radical-scavenging capabilities.

The research efforts have been also conferred for the treatment of IBD by targeting mucosal healing. In this regard, CD98 glycoprotein was chosen as a marker that is overexpressed in macrophages and the apical plasma membrane of epithelial cells in GIT. The quantum dot-loaded NPs were fabricated with CD98 Fab′-bearing regions and administered in induced colitis mouse models. The results revealed better internalization into colonic epithelial cell lines as compared with their nontargeted counterparts (PEG-NPs) [117]. More recently, CD44 receptor that has been overexpressed on the surface of epithelial cells has been chosen as a promising target for the mucosal healing. NPs functionalized with HA that has a natural ability of binding of CD44 receptor with high affinity was targeted for the drug delivery in colonic epithelial cells for the treatment of IBD [118]. Therefore in a nutshell, great potential lies in nanomediated targeted drug delivery for the efficient treatment of IBD but still many challenges are there to overcome. Future studies should certainly seek more promising avenues for targeting antiinflammation and mucosal healing.

6 CURRENT RESEARCH PROGRESS

The rising strides have been seen in the application of nanotechnology in drug delivery that leads to the research hotspot field nowadays. In recent years, tremendous progress has been envisaged in the field of nanomediated targeted drug delivery. Recently, joint efforts of MIT researchers with Draper's bioelectronics developed ingestible biosensors based on microelectromechanical systems that have been implanted for longer duration and facilitate extended drug delivery in the GI tract [119]. The midfield transmission technology has been used to power these implantable devices in which the power has been supplied from an antenna located outside the body to an antenna placed inside the digestive tract that can sense the temperature, heart rate, and levels of gases and specific nutrients in the stomach.

These miniaturized sensors show path-breaking advancements in the field of drug delivery. Apart from that, molecular manufacturing technologies such as diamond mechanosynthesis have been utilized for the development of computerized controlled self-powered medical nanorobot systems in the form of pharmacyte that has been recently explored for precise, systematic delivery of drugs [120]. Besides that, the development of molecular clusters in the form of niosome has been under pipeline of development that is formed by the self-assembly of nonionic surfactants in an aqueous phase. This is nowadays recognized as a novel drug-delivery system that has the ability to encapsulate both hydrophobic and hydrophilic drugs [121]. They have displayed enormous potential to deliver drugs in controlled release fashion, minimize the side effects of drugs, and facilitate targeted site delivery with different routes. Furthermore, the development of a nanomedical device of the microscopic size called "microbivores" that is under process helps for the destruction of microbial pathogens found in the bloodstream of the GI tract [122]. These are recognized as artificial mechanical phagocytes comprising 610 billion finely tuned structural atoms arranged in 12.1 μm geometric volume.

The estimated power consumption of these devices is documented up to 200 pW per 2 μm of organic material per 30-s cycle. These devices have been found to be approximately 80 times more efficient and 1000 times faster than natural macrophages; therefore, they act as potential nanosystems for minimizing the sepsis as well as other GI pathogenesis. More recently, surface layer proteins from archaea and GRAS organisms have been identified and explored for the construction of biomimetic supramolecular structures [123]. These structures provide a suitable platform for the comprehensive understanding of membrane functionality as well as provide high mechanical stability, resistance toward high temperature, pH, and osmotic stress, and prospectively used as suitable drug-delivery vehicles. The self-assembling properties of three putative surface layer (S-layer) protein genes (*slp*1, *slp*2, and *slp*3) have been identified from *Viridibacillus arvi JG-B58* strain that was isolated from heavy metal-contaminated sites [124]. This self-assembly property has been used for the construction of the lipoidal vesicular system commonly known as emulsomes in which the central core is composed of lipids that are surrounded by phospholipid multilayers.

These serve as excellent drug-delivery vehicles as these nanostructures facilitate the enhancement of membrane permeability, inhibition of efflux transport, enhanced bioavailability, modulated enterocyte-based drug transport, and targeted lymphatic system transport. The inculcation of recombinant surface layer proteins (biotin-streptavidin system) in the emulsomes also includes the diverse finely tuned functional domains that may have potential applications in drug delivery [125]. Recently, the noncytotoxic effect of S-layer-coated emulsomes has been reported in in vitro cell culture studies in human liver carcinoma cells (HepG2). Furthermore, the enhancement of bioavailability of curcumin has been reported by the development of curcuemulsomes into HepG2 cells [126]. Moreover, renal toxicity has been reduced by the delivery of emulsome-coated amphotericin B that has been used for the treatment of invasive fungal infections [127]. Moreover, trilaurin and tristearin-based emulsomal formulations have been used for the targeted delivery of azidothymidine and to ensure the internalization of the encapsulated drug for the treatment of intracellular liver infections [128].

Therefore these drug-delivery vehicles are proving themselves as excellent drug-delivery vehicles for the targeted drug delivery in the GI tract. More recently, the lymphatic delivery of methotrexate in emulsome formulations has enhanced its bioavailability in the treatment of cancer [129].

More recently, the exhaustive research is going on for targeting the nanomediated drug-delivery approaches for the induction of autophagy for curing gastrointestinal malignancy by alleviating the chemoresistance of cancer cells. Recently, magnetic nanoparticles based on N-(2-hydroxy-3-trimethylammonium)propyl) chitosan chloride (HTCC)/alginate-encapsulated Fe_3O_4 has been used for targeting autophagy in a normal gastric cell line, malignant gastric cell line (SGC7901), and multidrug resistance (SGC7901/ADR) cell lines [130]. The autophagy was observed through Prussian blue iron staining in which the formation of ferric-ferrocyanide complex inside the cells enhances the uptake of HTCC-MNPs. The cell viability assay [3-(4,5-Dimethylthiazol-2-yl)-2,5-diphenyltetrazolium bromide] (MTT) was performed, and reduced viability was observed for SGC7901 and SGC7901/ADR cells with increasing the dose of HTCC-MNPs [105]. Similarly, the artocarpin NP-activated autophagic death was also observed in HepG2 cells. In continuation with this, polyvinylpyrrolidone-coated silver (Ag) NPs were found to activate autophagy in HepG2 cells [131]. Burgeoning knowledge of carbon-based nanomaterials has led to the development of nanodiamonds that include the nanosized amorphous carbon, diamondoids, fullerenes, and tubes that possess enhanced retention, selectivity, surface functionality, and stability. Therefore tremendous progress has been going on based on this vision so that the targeted nanomediated drug-delivery systems in the GI tract will revolutionize the translational application of nanotechnology in medical sciences.

7 CURRENT BOTTLENECKS

The exuberant field of targeted drug-delivery approaches based on various nanostructures is certainly on the horizon but the debate on nanotoxicology paradigms is continuously going on. This dichotomous ideology between the widespread use as well as embracing the toxicity in the biological system must be properly considered. The dynamic conditions conferred by the GI tract pose a significant challenge for the construction of robust and stable nanocarriers. The varied class and nonuniformity of various nanostructures is a prime challenge for designing the medical-oriented nanoparticles. The pharmacodynamics and pharmacokinetics of nanoparticles are quite sensitive for any alteration in their properties that also confers a significant challenge during development as drug carriers. The crucial issue of nanoparticle stability and reproducible functionality replicates with the same authenticity from in vitro laboratory-scale conditions to in vivo conditions poses dilemmatic situations and challenges among scientific and medical communities. Apart from that, the interaction of the microbial ecosystem that is thriving on the GI tract also leads to aggravate the challenge in drug-delivery regimes. Thus there is a continuing demand for comprehensive research to understand the mechanisms of nanomaterial uptake, and interaction with the biological entities in the surrounding environment and toxicity that can overcome the challenges for framing nanoentities with enhanced properties making ideal vehicles for drug delivery.

8 FUTURE RESEARCH PRIORITIES

It is conceivably envisioned that all together, the various types of nanostructures described so far are serving as good drug-delivery regimes for rectification of GI tract disorders. However, their clinical approval needs urgent attention for their commercialization. There should be well-documented guidelines on safety and risk analysis by separate regulatory Acts of International and National Agencies for the utilization of these nanoentities as drug-delivery vehicles. In addition to existing siRNA and miRNA-based nanostructures, nowadays more focus are on efficient genome-editing technologies that can overcome the off-target barriers of these technologies [132]. These editing tool comprised programmable nucleases that include the clustered regularly interspaced short palindromic repeat (CRISPR)-associated nuclease Cas9, meganucleases, transcription activation-like effector nucleases (TALENs), and zinc-finger nucleases (ZFNs) for editing genome in malignant/diseased cells to cause correction/inactivation of malfunctioning gene(s) [133]. The technological advancements in the identification and imaging of malignant and infiltrating immune-responsive cells also laid a strong foundation for the futuristic utilization of multifunctional nanoimmunocarriers for eradicating the malfunctioning of the GI tract [134]. In light of above mentioned facts huge potential of nano-mediated drug delivery regimes have been envisaged for the development of advanced drug delivery vehicles.

9 CONCLUSION

The budding age of nanotechnology has signaled a wave of entrepreneurial possibilities in drug delivery, which is serving as the most promising area in biomedical applications. This has been proved from the current studies that nanomediated drug-delivery vehicles can protect the drug components in the systemic circulation, and deliver the drugs in a sustained and controlled rate at the targeted site. The growing continuum of various SDDSs based on nanostructures for the effective drug-delivery regimes in GIT, the most complicated part of the human body, led to more fascination toward these studies. However, the development of nanomediated drug-delivery vehicles has been yanked between two important facts of their beneficial credentials as well as its toxicity. This chapter entails various aspects and embodiments related to the novel and recent development of nanostructures used in drug delivery in the GI tract. However, highly focused research must be needed for the complete elucidation of the impact and mechanisms of nanoparticles and their interaction with microbiota in the gastrointestinal route.

ABBREVIATIONS

GI	Gastrointestinal
CNTs	Carbon nanotubes
PA	Peptide amphiphiles
FF	Diphenylalanine
DLC	Drug-loading capacity
DNA	Deoxyribonucleic acid
RNA	Ribonucleic acid
MSNs	Mesoporous silica nanoparticles
NPs	Nanoparticles
NLCs	Nanostructured lipid carriers
SLNPs	Solid lipid nanoparticles
TDDS	Targeted drug-delivery system
UVs	Unilamellar vesicles
MLVs	Multilamellar vesicles
NLCs	Nanostructured lipid carriers
DLC	Drug loading capacities
GIT	Gastrointestinal tract
PCDCs	Pressure-controlled colon delivery capsules
ECM	Extracellular matrix
ROS	Reactive oxygen species
COX-2	Cyclooxygenase-2
mPEGPCL	Methoxy poly(ethylene glycol)-polycaprolactone
SDDS	Smart drug-delivery system
Ptx	Paclitaxel
H. pylori	*Helicobacter pylori*
PAH	Poly(allylamine hydrochloride)
PAA	Poly(acrylic acid)
PLA	Poly(DL-lactic acid)
PAMAM	Poly(amido amine)
SLN	Solid lipid nanoparticles
HCNPs	Hyaluronan cisplatin-conjugated nanoparticles
PAMs	Pectinate/alginate microbeads
IBU	Ibuprofen
IBD	Inflammatory bowel disease
EAC	Esophageal adenocarcinoma
EPI	Epirubicin
QDs	Quantum dots
NIR	Near infrared
CS-GGS	Chitosan-coated gold/gold sulfide
MEMS	Microelectromechanical system
NDDS	Novel drug-delivery system
GRAS	Generally regarded as safe
HA	Hyaluronic acid
CBNs	Carbon-based nanoparticles

GLOSSARY

Nanotechnology	The application of engineering and technology for the manipulation of matter at atomic, molecular, and supramolecular level.
Inflammatory bowel disease	It is a group of inflammatory conditions that occur in all parts of the GI tract or restricted to small intestine or colon.
Ulcerative colitis	It is defined as the inflammation and ulcers of colon and rectum.
Esophageal cancer	The uncontrolled growth of cells of esophagus is termed as esophageal cancer.
Nanocapsules	The manipulation of nontoxic polymeric materials to make nanoscale shell is termed as nanocapsules.
Nanomicelles	These are defined as nanosized ultramicroscopic materials comprising exterior

	hydrophilic polar heads and interior hydrophobic fatty acid chains.
Nanovesicles	These are nanosized composite vehicles used for drug delivery in targeted regions of the body.
Nanogels	These are nanoparticles comprising cross-linked framework of hydrophilic polymers.
Dendrimers	The macromolecular structures comprising highly branched polymeric molecules.
Mucoadhesive	The property of adhesion of any materials to biological mucous membranes.
Optoelectronics	The application of electronic systems and devices to control and detect the optical phenomenon in the given materials.
Therapeutics	The scientific branch that deals with the treatment of disease with the action of various therapeutic molecules or medicines.
Paramagnetic oxides	The oxides that lose their magnetic property after the removal of applied magnetic field externally is termed as paramagnetic oxides.
Maghemite	A ferrimagnetic iron oxide that is derived from the oxidation of magnetite.
Fullerenes	It is an allotropic form of carbon represented either by hollow sphere, tube, ellipsoid, and various other shapes and used for the development of nanoparticles for a variety of applications.
Carbon nanotubes	The cylindrical depiction of allotropic form of carbon is known as carbon nanotubes. These are hollow tubes composed of either single or multiple wall.
Graphene	It is an allotropic form of carbon containing hexagonally arranged single layer of carbon atoms.
Nanofibrils	Nanofibrils are nanosized filamentous structures.
Nanoribbons	Nanoribbons are ribbons with diameters less than 50 nm.
Nanotapes	The parallel arrangement of carbon nanotubes that makes them strong and sticky structures is known as nanotapes.
Immunogenicity	The ability of a molecule to evoke an immune response in the body of living being is termed as immunogenicity.
Carbon nanobelts	The belt-shaped molecules comprised fused benzene rings of carbon nanotubes.
Personalized medicine	The medicine or therapy that has been given to the individual patients based on their symptoms, case history, and predicted response.
RNA origami	The nanoscale folding of RNA for attaining different shapes for various functions is termed as RNA origami.
Mesoporous silica nanoparticles	Rod-shaped silica nanoparticles derived from the reaction of tetraethyl orthosilicates that have a regular arrangement of pores on the surface.
Liposome	A spherical vesicle composed of phospholipid bilayer and serves as a carrier for the administration of drugs or nutrients.
Emulsions	A mixture of two or more immiscible liquids is known as emulsion.
Nanostructured lipid carriers	Lipid carriers composed of both solid and liquid lipids that are used in targeted drug delivery with their size ranging from 1 to 100 nm.
Solid lipid nanoparticles	A spherical nanoparticle composed of fatty acids with an average diameter between 10 and 1000 nm.

Unilamellar vesicles	A spherical chamber vesicle composed of single amphiphilic lipid layer containing hydrophilic core inside the chamber.
Multilamellar vesicles	A spherical chamber vesicle composed of multiple amphiphilic lipid layers containing hydrophilic core inside the chamber.
Clathrin-mediated endocytosis	The process of the inward folding of plasma membrane known as clathrin that can be used for the absorption of proteins, hormones, and various metabolites is recognized as Clathrin-mediated endocytosis.
Caveolar-mediated endocytosis	The process of folding of plasma membrane in the bulb-shaped invaginations of approximately 50–60 nm known as caveolae for performing various functions of absorption.
Targeted drug delivery	The delivery of medicine or drug at the specific or required site of disease as compared with other tissues of the body followed by the enhancement of the concentration of the drug on the affected regions for the maximum effect is termed as targeted drug delivery. The targeted delivery has been mediated with the use of nanoparticles.
Extracellular matrix	The extracellular environment outside the plasma membrane of living cells that imparts the structural and biochemical support of the cell.
Reactive oxygen species	Oxygen containing chemicals that are highly reactive in nature.
Apoptosis	The programmed cell death by activation of specific genes in multicellular organisms.
Photothermal therapy	The therapy based on the utilization of electromagnetic radiation in the treatment of cancer.
SiRNA	Small sequence of RNA that prevents translation of mRNA after transcription.
miRNA	Small noncoding RNA molecule found in eukaryotes and viruses that regulates RNA silencing and gene expression.
Mucosa-associated lymphoid tissue	The small concentrations of lymphoid tissue found in various submucosal membranes in various parts of the body.
CD-98	A heterodimeric glycoprotein that is a constituent of large neutral amino acid transporter (LAT1).
Quantum dots	Nanosized semiconductors differing in optoelectronic properties from light emitting diode (LED) particles.
Photobleaching	The removal of the ability for exhibiting fluorescence from a fluorophore or dye through photochemical reactions is called photobleaching.
Endothelial barriers	The semiselective barrier that occurs between the lumen of blood vessel and its surrounding tissues. It allows the selective exchange of materials and white blood cells in the bloodstream.
Angiogenesis	The process of genesis of new blood vessels from existing ones.
Exosomes	These are defined as vesicles of cellular origin present in biological fluids.
Microbivores	Microbe-eating organisms or nanorobots is called microbivores.
Microelectromechanical system	The technology used for making microscopic electromechanical devices with moving parts.
Lymphatic system	The system is defined as an important part of

immunology consisting of lymph nodes and vessels that can deliver clean fluid lymph toward the heart.

CRISPR — It is an immune system of prokaryotes that is involved in immune response against viruses.

TALENs — Engineered restriction endonucleases named as transcription activator-like effector nucleases (TALEN) that can be modified to cut specific parts of DNA.

ZFNs: — Zinc-finger nucleases (ZFNs) are synthetic restriction enzymes that have a DNA-binding and DNA-cleaving domain.

Gastrointestinal disorders: — These are defined as the diseases that occur in the gastrointestinal tract.

Gut microbiota — The diverse population that lives in digestive tracts of organisms including humans, cattles, and insects.

Nanoparticles — The particles having the dimensions of very small size from 1 to 100 nm are termed as nanoparticles.

Gastrointestinal tract — It is an organ system connected in long twisting tube from mouth to anus in which food is digested and absorbed in the form of nutrients and energy and waste is excreted as feces.

ACKNOWLEDGMENTS

The preparation of this manuscript is entirely facilitated by Gautam Buddha University, and the authors wish to present a great sense of gratitude toward Gautam Buddha University for providing all necessary facilities for writing this chapter. All authors declare that they have no conflict of interest.

REFERENCES

[1] Bamrungsap S, Zhao Z, Chen T, Wang L, Li C, Fu T, Tan W. Nanotechnology in therapeutics: a focus on nanoparticles as a drug delivery system. Nanomedicine 2012;7(8):1253–71.

[2] Wen H, Jung H, Li X. Drug delivery approaches in addressing clinical pharmacology-related issues: opportunities and challenges. The AAPS Journal 2015;17(6): 1327–40.

[3] Singh R, Lillard Jr JW. Nanoparticle-based targeted drug delivery. Experimental and Molecular Pathology 2009; 86(3):215–23.

[4] Cong VT, Gaus K, Tilley RD, Gooding JJ. Rod-shaped mesoporous silica nanoparticles for nanomedicine: recent progress and perspectives. Expert Opinion on Drug Delivery 2018;15(9):881–92.

[5] Fell JT. Targeting of drugs and delivery systems to specific sites in the gastrointestinal tract. Journal of Anatomy 1996;189(3):517.

[6] Laroui H, Rakhya P, Xiao B, Viennois E, Merlin D. Nanotechnology in diagnostics and therapeutics for gastrointestinal disorders. Digestive and Liver Disease 2013; 45(12):995–1002.

[7] Melton SD, Genta RM, Souza RF. Biomarkers and molecular diagnosis of gastrointestinal and pancreatic neoplasms. Nature Reviews Gastroenterology & Hepatology 2010;7(11):620.

[8] Qiu K, Durham PG, Anselmo AC. Inorganic nanoparticles and the microbiome. Nano Research 2018:1–19.

[9] Patel DJ, Mistri PA, Prajapati JJ. Treatment of cancer by using nanoparticles as a drug delivery. International Journal of Drug Development & Research 2012;4(1):14–27.

[10] Bhatia S. Nanoparticles types, classification, characterization, fabrication methods and drug delivery applications. In: Natural polymer drug delivery systems. Cham: Springer; 2016. p. 33–93.

[11] Liu Z, Jiao Y, Wang Y, Zhou C, Zhang Z. Polysaccharides-based nanoparticles as drug delivery systems. Advanced Drug Delivery Reviews 2008;60(15):1650–62.

[12] Cirillo G, Iemma F, Puoci F, Parisi OI, Curcio M, Spizzirri UG, Picci N. Imprinted hydrophilic nanospheres as drug delivery systems for 5-fluorouracil sustained release. Journal of Drug Targeting 2009;17(1):72–7.

[13] Debele TA, Mekuria SL, Tsai HC. Polysaccharide based nanogels in the drug delivery system: application as the carrier of pharmaceutical agents. Materials Science and Engineering: C 2016;68:964–81.

[14] Ahmad Nor Y, Niu Y, Karmakar S, Zhou L, Xu C, Zhang J, Zhang H, Yu M, Mahony D, Mitter N, Cooper MA. Shaping nanoparticles with hydrophilic compositions and hydrophobic properties as nanocarriers for antibiotic delivery. ACS Central Science 2015; 1(6):328–34.

[15] Na K, Lee TB, Park KH, Shin EK, Lee YB, Choi HK. Self-assembled nanoparticles of hydrophobically-modified polysaccharide bearing vitamin H as a targeted anti-cancer drug delivery system. European Journal of Pharmaceutical Sciences 2003;18(2):165–73.

[16] Adeyemi OS, Sulaiman FA. Evaluation of metal nanoparticles for drug delivery systems. Journal of Biomedical Research 2015;29(2):145.

[17] Sadighi A, Faramarzi MA. Congo red decolorization by immobilized laccase through chitosan nanoparticles

on the glass beads. Journal of the Taiwan Institute of Chemical Engineers 2013;44(2):156–62.
[18] McBain SC, Yiu HH, Dobson J. Magnetic nanoparticles for gene and drug delivery. International Journal of Nanomedicine 2008;3(2):169.
[19] Cha C, Shin SR, Annabi N, Dokmeci MR, Khademhosseini A. Carbon-based nanomaterials: multifunctional materials for biomedical engineering. ACS Nano 2013;7(4):2891–7.
[20] Bianco A, Kostarelos K, Prato M. Applications of carbon nanotubes in drug delivery. Current Opinion in Chemical Biology 2005;9(6):674–9.
[21] Kong B, Seog JH, Graham LM, Lee SB. Experimental considerations on the cytotoxicity of nanoparticles. Nanomedicine 2011;6(5):929–41.
[22] Verma D, Gulati N, Kaul S, Mukherjee S, Nagaich U. Protein based nanostructures for drug delivery. Journal of Pharmaceutics 2018;2018.
[23] Fan T, Yu X, Shen B, Sun L. Peptide self-assembled nanostructures for drug delivery applications. Journal of Nanomaterials 2017;16.
[24] Peng D, Jin W, Tang C, Lu Y, Wang W, Li J, Li B. Foaming and surface properties of gliadin nanoparticles: influence of pH and heating temperature. Food Hydrocolloids 2018;77:107–16.
[25] Kobayashi N, Inano K, Sasahara K, Sato T, Miyazawa K, Fukuma T, Hecht MH, Song C, Murata K, Arai R. Self-assembling supramolecular nanostructures constructed from de Novo extender protein nanobuilding blocks. ACS Synthetic Biology 2018;7(5):1381–94.
[26] Li H, LaBean TH, Leong KW. Nucleic acid-based nanoengineering: novel structures for biomedical applications. Interface Focus 2011;1(5):702–24.
[27] Linko V, Ora A, Kostiainen MA. DNA nanostructures as smart drug-delivery vehicles and molecular devices. Trends in Biotechnology 2015;33(10):586–94.
[28] Shu Y, Cinier M, Shu D, Guo P. Assembly of multifunctional phi29 pRNA nanoparticles for specific delivery of siRNA and other therapeutics to targeted cells. Methods 2011;54(2):204–14.
[29] Bitar A, Ahmad NM, Fessi H, Elaissari A. Silica-based nanoparticles for biomedical applications. Drug Discovery Today 2012;17(19–20):1147–54.
[30] Bharti C, Nagaich U, Pal AK, Gulati N. Mesoporous silica nanoparticles in target drug delivery system: a review. International Journal of Pharmaceutical Investigation 2015;5(3):124.
[31] C Thomas S, Kumar Mishra P, Talegaonkar S. Ceramic nanoparticles: fabrication methods and applications in drug delivery. Current Pharmaceutical Design 2015; 21(42):6165–88.
[32] Puri A, Loomis K, Smith B, Lee JH, Yavlovich A, Heldman E, Blumenthal R. Lipid-based nanoparticles as pharmaceutical drug carriers: from concepts to clinic. Critical Reviews in Therapeutic Drug Carrier Systems 2009;26(6):523–80.
[33] Laurencin M, Georgelin T, Malezieux B, Siaugue JM, Ménager C. Interactions between giant unilamellar vesicles and charged core– shell magnetic nanoparticles. Langmuir 2010;26(20):16025–30.
[34] Moon JJ, Suh H, Bershteyn A, Stephan MT, Liu H, Huang B, Sohail M, Luo S, Um SH, Khant H, Goodwin JT, Ramos J, Chiu W, Irvine DJ. Interbilayer-crosslinked multilamellar vesicles as synthetic vaccines for potent humoral and cellular immune responses. Nature Materials 2011;10(3):243–51.
[35] Pfeiffer C, Rehbock C, Hühn D, Carrillo-Carrion C, de Aberasturi DJ, Merk V, Parak WJ. Interaction of colloidal nanoparticles with their local environment: the (ionic) nanoenvironment around nanoparticles is different from bulk and determines the physico-chemical properties of the nanoparticles. Journal of the Royal Society Interface 2014;11(96):20130931.
[36] Khan S, Baboota S, Ali J, Khan S, Narang RS, Narang JK. Nanostructured lipid carriers: an emerging platform for improving oral bioavailability of lipophilic drugs. International Journal of Pharmaceutical Investigation 2015; 5(4):182.
[37] Werengowska-Ciećwierz K, Wiśniewski M, Terzyk AP, Furmaniak S. The chemistry of bioconjugation in nanoparticles-based drug delivery system. Advances in Condensed Matter Physics 2015;27.
[38] Mout R, Moyano DF, Rana S, Rotello VM. Surface functionalization of nanoparticles for nanomedicine. Chemical Society Reviews 2012;41(7):2539–44.
[39] Habibi N, Kamaly N, Memic A, Shafiee H. Self-assembled peptide-based nanostructures: smart nanomaterials toward targeted drug delivery. Nano Today 2016;11(1):41–60.
[40] Vlasova II, Kapralov AA, Michael ZP, Burkert SC, Shurin MR, Star A, Kagan VE. Enzymatic oxidative biodegradation of nanoparticles: mechanisms, significance and applications. Toxicology and Applied Pharmacology 2016;299:58–69.
[41] Feng JJ, Song YY, Feng XM, Shrestha NK, Wisitruangsakul N. Biocompatible functional nanomaterials: synthesis, properties, and applications. Journal of Nanomaterials 2013;1.
[42] Gao W, Chan JM, Farokhzad OC. pH-responsive nanoparticles for drug delivery. Molecular Pharmaceutics 2010;7(6):1913–20.
[43] Thamphiwatana S, Fu V, Zhu J, Lu D, Gao W, Zhang L. Nanoparticle-stabilized liposomes for pH-responsive gastric drug delivery. Langmuir 2013;29(39):12228–33.
[44] Cheng B, Lu B, Liu X, Meng T, Tan Y, Zhu Y, Liu N, Yuan H, Huang X, Hu F. A pH-responsive glycolipid-like nanocarrier for optimising the time-dependent distribution of free chemical drugs in focal cells. International Journal of Pharmaceutics 2017;522(1–2):210–21.
[45] Zhao Y, Wang X, Wang Z, Li X, Ren Y. Nanoparticle fouling and its combination with organic fouling during forward osmosis process for silver nanoparticles removal from simulated wastewater. Scientific Reports 2016;6:25859.
[46] Quant C,A, Krishna T, Marla J. Carson Meredith, Osmotic pressure and chemical potential of silica

nanoparticles in aqueous poly(ethyleneoxid) solution. Colloids and Surfaces A: Physicochemical and Engineering Aspects 2008;317(1−3):129−35. 2008.

[47] Schroer MA, Lehmkühler F, Möller J, Lange H, Grübel G, Schulz F. Pressure-stimulated supercrystal formation in nanoparticle suspensions. The Journal of Physical Chemistry Letters 2018;9(16):4720−4.

[48] El-Habashy SE, Allam AN, El-Kamel AH. Ethyl cellulose nanoparticles as a platform to decrease ulcerogenic potential of piroxicam: formulation and in vitro/in vivo evaluation. International Journal of Nanomedicine 2016;11:2369.

[49] Chen B, Dai W, He B, Zhang H, Wang X, Wang Y, Zhang Q. Current multistage drug delivery systems based on the tumor microenvironment. Theranostics 2017;7(3):538.

[50] Wu G, Fang YZ, Yang S, Lupton JR, Turner ND. Glutathione metabolism and its implications for health. Journal of Nutrition 2004;134(3):489−92.

[51] Such GK, Yan Y, Johnston AP, Gunawan ST, Caruso F. Interfacing materials science and biology for drug carrier design. Advanced Materials 2015;27(14):2278−97.

[52] Meng F, Hennink WE, Zhong Z. Reduction-sensitive polymers and bioconjugates for biomedical applications. Biomaterials 2009;30(12):2180−98.

[53] Zhou Y, Jie K, Huang F. A redox-responsive selenium-containing pillar [5] arene-based macrocyclic amphiphile: synthesis, controllable self-assembly in water, and application in controlled release. Chemical Communications 2017;53(59):8364−7.

[54] Ji S, Cao W, Yu Y, Xu H. Dynamic diselenide bonds: exchange reaction induced by visible light without catalysis. Angewandte Chemie International Edition 2014;53(26):6781−5.

[55] Cho H, Bae J, Garripelli VK, Anderson JM, Jun HW, Jo S. Redox-sensitive polymeric nanoparticles for drug delivery. Chemical Communications 2012;48(48):6043−5.

[56] Rivas-Montoya E, Ochando-Pulido JM, López-Romero JM, Martinez-Ferez A. Application of a novel gastrointestinal tract simulator system based on a membrane bioreactor (SimuGIT) to study the stomach tolerance and effective delivery enhancement of nanoencapsulated macelignan. Chemical Engineering Science 2016;140:104−13.

[57] Fröhlich EE, Fröhlich E. Cytotoxicity of nanoparticles contained in food on intestinal cells and the gut microbiota. International Journal of Molecular Sciences 2016;17(4):509.

[58] García-Alonso J, Khan FR, Misra SK, Turmaine M, Smith BD, Rainbow PS, Valsami-Jones E. Cellular internalization of silver nanoparticles in gut epithelia of the estuarine polychaete Nereis diversicolor. Environmental Science & Technology 2011;45(10):4630−6.

[59] Kou L, Sun J, Zhai Y, He Z. The endocytosis and intracellular fate of nanomedicines: implication for rational design. Asian Journal of Pharmaceutical Sciences 2013; 8(1):1−10.

[60] Zhang S, Li J, Lykotrafitis G, Bao G, Suresh S. Size-dependent endocytosis of nanoparticles. Advanced Materials 2009;21(4):419−24.

[61] Vamanu E, Ene M, Biță B, Ionescu C, Crăciun L, Sârbu I. In vitro human microbiota response to exposure to silver nanoparticles biosynthesized with mushroom extract. Nutrients 2018;10(5):607.

[62] Jiang Z, Jacob JA, Li J, Wu X, Wei G, ArunPrasanna V, Mani R, Nainangu P, Rajadurai UM, Chen B. Influence of diet and dietary nanoparticles on gut dysbiosis. Microbial Pathogenesis 2018;118:61−5.

[63] Van Den Brûle S, Ambroise J, Lecloux H, Levard C, Soulas R, De Temmerman PJ, Palmai-Pallag M, Marbaix E, Lison D. Dietary silver nanoparticles can disturb the gut microbiota in mice. Particle and Fibre Toxicology 2015;13(1):38.

[64] Tian X, Jiang X, Welch C, Croley TR, Wong TY, Chen C, Fan S, Chong Y, Li R, Ge C, Chen C. Bactericidal effects of silver nanoparticles on lactobacilli and the underlying mechanism. ACS Applied Materials & Interfaces 2018; 10(10):8443−50.

[65] Hemeg HA. Nanomaterials for alternative antibacterial therapy. International Journal of Nanomedicine 2017; 12:8211.

[66] Kate V, Maroju NK, Ananthakrishnan N. *Helicobacter pylori* Infection and upper gastrointestinal disorders. Gastroenterology Research and Practice 2013;2013.

[67] Park JB, Koo JS. *Helicobacter pylori* infection in gastric mucosa-associated lymphoid tissue lymphoma. World Journal of Gastroenterology: WJG 2014;20(11):2751.

[68] Savkare AD, Pagar PS. Comparison of various approaches to increase the gastric residense time. International Journal of Pharmaceutical Sciences and Research 2017;8(6):2388−95.

[69] Ramteke S, Jain NK. Clarithromycin-and omeprazole-containing gliadin nanoparticles for the treatment of *Helicobacter pylori*. Journal of Drug Targeting 2008; 16(1):65−72.

[70] Ramteke S, Ganesh N, Bhattacharya S, Jain NK. Triple therapy-based targeted nanoparticles for the treatment of *Helicobacter pylori*. Journal of Drug Targeting 2008; 16(9):694−705.

[71] Hasani S, Pellequer Y, Lamprecht A. Selective adhesion of nanoparticles to inflamed tissue in gastric ulcers. Pharmaceutical Research 2009;26(5):1149−54.

[72] Chang CH, Huang WY, Lai CH, Hsu YM, Yao YH, Chen TY, Wu JY, Peng SF, Lin YH. Development of novel nanoparticles shelled with heparin for berberine delivery to treat *Helicobacter pylori*. Acta Biomaterialia 2011; 7(2):593−603.

[73] Chakraborty S, Stalin S, Das N, Choudhury ST, Ghosh S, Swarnakar S. The use of nano-quercetin to arrest mitochondrial damage and MMP-9 upregulation during prevention of gastric inflammation induced by ethanol in rat. Biomaterials 2012;33(10):2991−3001.

[74] Jain P, Jain S, Prasad KN, Jain SK, Vyas SP. Polyelectrolyte coated multilayered liposomes (nanocapsules) for

the treatment of *Helicobacter pylori* infection. Molecular Pharmaceutics 2009;6(2):593–603.

[75] Gu H, Huang T, Shen Y, Liu Y, Zhou F, Jin Y, Sattar H, Wei Y. Reactive oxygen species-mediated tumor microenvironment transformation: the mechanism of radioresistant gastric cancer. Oxidative Medicine and Cellular Longevity 2018;2018:8.

[76] Shi WJ, Gao JB. Molecular mechanisms of chemoresistance in gastric cancer. World Journal of Gastrointestinal Oncology 2016;8(9):673–81.

[77] Rugge M, Fassan M, Graham DY. Epidemiology of gastric cancer. In: Gastric cancer. Cham: Springer; 2015. p. 23–34.

[78] Lavelle EC. Targeted delivery of drugs to the gastrointestinal tract. Critical Reviews in Therapeutic Drug Carrier Systems 2001;18(4):341–86.

[79] Pang RP, Zhou JG, Zeng ZR, Li XY, Chen W, Chen MH, Hu PJ. Celecoxib induces apoptosis in COX-2 deficient human gastric cancer cells through Akt/GSK3β/NAG-1 pathway. Cancer Letters 2007;251(2):268–77.

[80] Cheng J, Fan XM. Role of cyclooxygenase-2 in gastric cancer development and progression. World Journal of Gastroenterology 2013;19(42):7361–8

[81] Danafar H. MPEG-PCL copolymeric nanoparticles in drug delivery systems. Cogent Medicine 2016;3(1):1142411.

[82] Wang SQ, Wang C, Chang LM, Zhou KR, Wang JW, Ke Y, Yang DX, Shi HG, Wang R, Shi XL, Ma LY. Geridonin and paclitaxel act synergistically to inhibit the proliferation of gastric cancer cells through ROS-mediated regulation of the PTEN/PI3K/Akt pathway. Oncotarget 2016;7(45):72990.

[83] Jiang H, Geng D, Liu H, Li Z, Cao J. Co-delivery of etoposide and curcumin by lipid nanoparticulate drug delivery system for the treatment of gastric tumors. Drug Delivery 2016;23(9):3665–73.

[84] Hill AC, Schroeder SJ. Thermodynamic stabilities of three-way junction nanomotifs in prohead RNA. RNA 2017;23(4):521–9.

[85] Zhou Z, Zhang C, Qian Q, Ma J, Huang P, Pan L, Zhi X. Folic acid-conjugated silica capped gold nanoclusters for targeted fluorescence/X-ray computed tomography imaging. Journal of Nanobiotechnology 2013;11(1):17.

[86] Cui D, Zhang C, Liu B, Shu Y, Du T, Shu D, Wang K, Dai F, Liu Y, Li C, Pan F. Regression of gastric cancer by systemic injection of RNA nanoparticles carrying both ligand and siRNA. Scientific Reports 2015;5:10726.

[87] Bergin IL, Witzmann FA. Nanoparticle toxicity by the gastrointestinal route: evidence and knowledge gaps. International Journal of Biomedical Nanoscience and Nanotechnology 2013;3(1–2).

[88] Zhang C, Gu C, Peng F, Liu W, Wan J, Xu H, Lam CW, Yang X. Preparation and optimization of triptolide-loaded solid lipid nanoparticles for oral delivery with reduced gastric irritation. Molecules 2013;18(11): 13340–56.

[89] Tsai SW, Yu DS, Tsao SW, Hsu FY. Hyaluronan–cisplatin conjugate nanoparticles embedded in Eudragit S100-coated pectin/alginate microbeads for colon drug delivery. International Journal of Nanomedicine 2013; 8:2399.

[90] Laroui H, Geem D, Xiao B, Viennois E, Rakhya P, Denning T, Merlin D. Targeting intestinal inflammation with CD98 siRNA/PEI–loaded nanoparticles. Molecular Therapy 2014;22(1):69–80.

[91] Abbas G, Krasna M. Overview of esophageal cancer. Annals of Cardiothoracic Surgery 2017;6(2):131.

[92] Wang X, Low XC, Hou W, Abdullah LN, Toh TB, Mohd Abdul Rashid M, Ho D, Chow EKH. Epirubicin-adsorbed nanodiamonds kill chemoresistant hepatic cancer stem cells. ACS Nano 2014;8(12):12151–66.

[93] Zhao MX, Zeng EZ. Application of functional quantum dot nanoparticles as fluorescence probes in cell labeling and tumor diagnostic imaging. Nanoscale Research Letters 2015;10(1):171–8.

[94] Fan Z, Chang Y, Cui C, Sun L, Wang DH, Pan Z, Zhang M. Near infrared fluorescent peptide nanoparticles for enhancing esophageal cancer therapeutic efficacy. Nature Communications 2018;9(1):2605.

[95] Li Y, Gobin AM, Dryden GW. Infrared light absorbing gold/gold sulphide nanoparticles induce cell death in oesophageal adenocarcinoma. International Journal of Nanomeidicne 2013;8:2153–61.

[96] Cisterna BA, Kamaly N, Choi WI, Tavakkoli A, Farokhzad OC, Vilos C. Targeted nanoparticles for colorectal cancer. Nanomedicine 2016;11(18):2443–56.

[97] Gulbake A, Jain A, Jain A, Jain A, Jain SK. Insight to drug delivery aspects for colorectal cancer. World Journal of Gastroenterology 2016;22(2):582.

[98] Şengel-Türk CT, Hasçiçek C, Dogan AL, Esendagli G, Guc D, Gönül N. Preparation and in vitro evaluation of meloxicam-loaded PLGA nanoparticles on HT-29 human colon adenocarcinoma cells. Drug Development and Industrial Pharmacy 2012;38(9):1107–16.

[99] Sharma M, Malik R, Verma A, Dwivedi P, Banoth GS, Pandey N, Sarkar J, Mishra PR, Dwivedi AK. Folic acid conjugated guar gum nanoparticles for targeting methotrexate to colon cancer. Journal of Biomedical Nanotechnology 2013;9(1):96–106.

[100] Jeong YI, Chung KD, Da Hye Kim YHK, Lee YS, Choi KC. All-trans retinoic acid-incorporated nanoparticles of deoxycholic acid-conjugated dextran for treatment of CT26 colorectal carcinoma cells. International Journal of Nanomedicine 2013;8:485.

[101] Soofiyani SR, Hallaj-Nezhadi S, Lotfipour F, Hosseini AM, Baradaran B. Gene therapy based on interleukin-12 loaded chitosan nanoparticles in a mouse model of fibrosarcoma. Iranian Journal of Basic Medical Sciences 2016;19(11):1238.

[102] Li P, Dai YN, Zhang JP, Wang AQ, Wei Q. Chitosan-alginate nanoparticles as a novel drug delivery system for nifedipine. International Journal of Biomedical Science 2008;4(3):221.

[103] Mukherjee S, Ray S, Thakur RS. Solid lipid nanoparticles: a modern formulation approach in drug delivery system. Indian Journal of Pharmaceutical Sciences 2009;71(4):349–58.

[104] Ju S, Mu J, Dokland T, Zhuang X, Wang Q, Jiang H, Xiang X, Deng ZB, Wang B, Zhang L, Roth M. Grape exosome-like nanoparticles induce intestinal stem cells and protect mice from DSS-induced colitis. Molecular Therapy 2013;21(7):1345–57.
[105] Ballesteros NG, Lopez SP, Rodriguez-Gonzalez B, Rodriguez-Arguelles MC. Green synthesis of gold nanoparticles using brown algae Cystoseira baccata: its activity in colon cancer cells. Colloids and Surfaces B: Biointerfaces 2017;153:190–8.
[106] Yassin AM, El-Deeb NM, Metwaly AM, El Fawal GF, Radwan MM, Elsayed E, Hafez EE. Induction of apoptosis in human cancer cells through extrinsic and intrinsic pathways by *Balanites aegyptiaca Furostanol Saponins and saponin*-coated silver nanoparticles. Applied Biochemistry and Biotechnology 2017. https://doi.org/10.1007/s12010-017-2426-3.
[107] Podolsky DK. Inflammatory bowel disease. New England Journal of Medicine 2002;347:417–29.
[108] Xiao B, Merlin D. Oral colon-specific therapeutic approaches toward treatment of inflammatory bowel disease. Expert Opinion on Drug Delivery 2012;9(11):1393–407.
[109] Collnot EM, Ali H, Lehr CM. Nano-and microparticulate drug carriers for targeting of the inflamed intestinal mucosa. Journal of Controlled Release 2012;161(2):235–46.
[110] Laroui H, Theiss AL, Yan Y, Dalmasso G, Nguyen HT, Sitaraman SV, Merlin D. Functional TNFα gene silencing mediated by polyethyleneimine/TNFα siRNA nanocomplexes in inflamed colon. Biomaterials 2011;32(4):1218–28.
[111] De Chambrun GP, Peyrin-Biroulet L, Lémann M, Colombel JF. Clinical implications of mucosal healing for the management of IBD. Nature Reviews Gastroenterology & Hepatology 2010;7(1):15–29.
[112] Xiao B, Laroui H, Ayyadurai S, Viennois E, Charania MA, Zhang Y, Merlin D. Mannosylated bioreducible nanoparticle-mediated macrophage-specific TNF-α RNA interference for IBD therapy. Biomaterials 2013;34(30):7471–82.
[113] Zhang J, Tang C, Yin C. Galactosylated trimethyl chitosan-cysteine nanoparticles loaded with Map4k4 siRNA for targeting activated macrophages. Biomaterials 2013;34(14):3667–77.
[114] Laroui H, Viennois E, Xiao B, Canup BS, Geem D, Denning TL, Merlin D. Fab'-bearing siRNA TNFα-loaded nanoparticles targeted to colonic macrophages offer an effective therapy for experimental colitis. Journal of Controlled Release 2014;186:41–53.
[115] Lee A, De Mei C, Fereira M, Marotta R, Yoon HY, Kim K, Kwon IC, Decuzzi P. Dexamethasone-loaded polymeric nanoconstructs for monitoring and treating inflammatory bowel disease. Theranostics 2017;7(15):3653.
[116] Wang X, Yan JJ, Wang L, Pan D, Yang R, Xu Y, Sheng J, Huang Q, Zhao H, Yang M. Rational design of polyphenol-poloxamer nanovesicles for targeting inflammatory bowel disease therapy. Chemistry of Materials 2018;30(12):4073–80.
[117] Xiao B, Yang Y, Viennois E, Zhang Y, Ayyadurai S, Baker MT, Laroui H, Merlin D. Glycoprotein CD98 as a receptor for colitis-targeted delivery of nanoparticles. Journal of Materials Chemistry B 2014;2(11):1499–508.
[118] Farkas S, Hornung M, Sattler C, Anthuber M, Gunthert U, Herfarth H, Schlitt HJ, Geissler EK, Wittig BM. Short-term treatment with anti-CD44v7 antibody, but not CD44v4, restores the gut mucosa in established chronic dextran sulphate sodium (DSS)-induced colitis in mice. Clinical and Experimental Immunology 2005;142(2):260–7.
[119] Abid A, O'Brien JM, Bensel T, Cleveland C, Booth L, Smith BR, Langer R, Traverso G. Wireless power transfer to millimeter-sized gastrointestinal electronics validated in a swine model. Scientific Reports 2017;7:46745.
[120] Saadeh Y, Vyas D. Nanorobotic applications in medicine: current proposals and designs. American Journal of Robotic Surgery 2014;1(1):4–11.
[121] Ag Seleci D, Seleci M, Walter JG, Stahl F, Scheper T. Niosomes as nanoparticular drug carriers: fundamentals and recent applications. Journal of Nanomaterials 2016;13.
[122] Col RDL, Brig SJ. Nanotechnology—the new Frontier of medicine. Medical Journal Armed Forces India 2006;62(3):263–8.
[123] Gaur N, Sharma A, Singhal B. Bacterial surface layer proteins: from moonlighting to biomimetics: a new horizonto lead. Advances in Bioscience and Biotechnology 2018;9(08):352–72.
[124] Suhr M, Lederer FL, Günther TJ, Raff J, Pollmann K. Characterization of three different unusual S-layer proteins from viridibacillus arvi JG-B58 that exhibits two super-imposed S-layer proteins. PLoS One 2016;11(6):e0156785.
[125] Moll D, Huber C, Schlegel B, Pum D, Sleytr UB, Sára M. S-layer-streptavidin fusion proteins as template for nanopatterned molecular arrays. Proceedings of the National Academy of Sciences 2002;99(23):14646–51.
[126] Ucisik MH, Küpcü S, Schuster B, Sleytr UB. Characterization of curcuemulsomes: nanoformulation for enhanced solubility and delivery of curcumin. Journal of Nanobiotechnology 2013;11(1):37.
[127] Gupta S, Vyas SP. Development and characterization of amphotericin B bearing emulsomes for passive and active macrophage targeting. Journal of Drug Targeting 2007;15(3):206–17.
[128] Ucisik Mehmet H, Sleytr uwe B, Schuster Bernhard. Emulsomes meet S-layer proteins: an emerging targeted drug delivery system. Current Pharmaceutical Biotechnology 2015;16(4):392–405.
[129] Paliwal R, Paliwal SR, Mishra N, Mehta A, Vyas SP. Engineered chylomicron mimicking carrier emulsome for lymph targeted oral delivery of methotrexate. International Journal of Pharmaceutics 2009;380(1–2):181–8.
[130] Li X, Feng J, Zhang R, Wang J, Su T, Tian Z, Han D, Zhao C, Fan M, Li C, Liu B. Quaternized chitosan/

alginate-Fe3O4 magnetic nanoparticles enhance the chemosensitization of multidrug-resistant gastric carcinoma by regulating cell autophagy activity in mice. Journal of Biomedical Nanotechnology 2016;12(5):948–61.

[131] Awad MA, El Dib RA, Almusayeib N, Al-Massarani S, Ortashi KM, Hendi AA. Novel Balanites aegyptiaca mesocarp synthesized silver nanoparticles: formation, characterization, antimicrobial, cytotoxicity and antiviral effects. Digest Journal of Nanomaterials and Biostructures 2013;8(4):1665–77.

[132] Ahmadzada T, Reid G, McKenzie DR. Fundamentals of siRNA and miRNA therapeutics and a review of targeted nanoparticle delivery systems in breast cancer. Biophysical reviews 2018;10(1):69–86.

[133] Baliou S, Adamaki M, Kyriakopoulos AM, Spandidos DA, Panayiotidis M, Christodoulou I, Zoumpourlis V. CRISPR therapeutic tools for complex genetic disorders and cancer. International Journal of Oncology 2018;53(2):443–68.

[134] Laroui H, Wilson DS, Dalmasso G, Salaita K, Murthy N, Sitaraman SV, Merlin D. Nanomedicine in GI. American Journal of Physiology – Gastrointestinal and Liver Physiology 2010;300(3):G371–83.

CHAPTER 5

Biocompatibility and Functionalization of Nanomaterials

FERNANDA MARIA POLICARPO TONELLI, PHD • FLÁVIA CRISTINA POLICARPO TONELLI, MSC • DANILO ROBERTO CARVALHO FERREIRA, BSC • KARINE EMANUELLE DA SILVA, BSC • HELON GUIMARÃES CORDEIRO, MSC • AMANDA TOMIE OUCHIDA, PHD • NÚBIA ALEXANDRE DE MELO NUNES, MSC

1 NANOTECHNOLOGY FIELD AND THE NANOMATERIALS

The nanotechnology field has been providing the world with a large amount of different types of nanomaterials with interesting and useful properties since the late 1950s [1]. The widespread definition for these materials states that they possess the size ranging from 1 to 100 nm at least in one dimension [2], and can be classified, regarding their chemical constitution, into two main groups: the carbon-based nanomaterials and the non-carbon-based nanomaterials (Fig. 5.1). From the first group, it is possible to highlight the carbon nanotubes (multi- or single-walled), graphene and its oxide, fullerene, and some quantum dots (which possess strong fluorescent emission being on the spotlight for diagnosis through imaging). From the second group, gold nanorods and nanoparticles as the ones composed by metallic groups (magnetic or not), polymers (like poly(lactic-co-glycolic acid) - PLGA), and others [3].

Sometimes, the combination of nanomaterials from the same group or from both groups is used to make possible desirable events, especially when it comes to the development of sensitive and precise diagnosis systems. Ultrasensitive detection of carcinoembryonic antigen (CEA) can be performed by combining graphene (acting as a nanocarrier) and carbon quantum dots (acting as luminophore) with perylenetetracarboxylic acid to generate a high sensitivity CEA immunosensor [4]. Vascular endothelial growth factor and prostate-specific antigen (PSA) can be detected with high sensitivity and selectivity using a gadget containing a gold electrode coated with graphene oxide/ssDNA and incorporated with poly-L-lactide nanoparticles, which allows early diagnosis of prostate cancer [5]. Immunosensors

FIG. 5.1 Examples of carbon-based nanomaterials (graphene, single-walled carbon nanotube, and fullerene) and non-carbon-based nanomaterials (gold nanorod and nanoparticle).

Intelligent Nanomaterials for Drug Delivery Applications. https://doi.org/10.1016/B978-0-12-817830-0.00005-9

to detect PSA can also be constructed from reduced graphene oxide with silver nanoparticles composites [6] or Fe_3O_4 nanoparticles loaded onto graphene oxide nanosheets; using the appropriate antibody, the latter was also capable to detect prostate-specific membrane antigen [7].

Some nanomaterials' characteristics made them already suitable for different interesting uses. As an example of it, we can mention that nanoparticles present the capacity of crossing an intact blood–brain barrier being useful to deliver drugs to the brain or to access the brain offering contrast for some diagnosis strategies [8]. It is also possible for some nanomaterials to be used for several different applications: these are multifunctional nanomaterials. As an example, the immunotargeted nanoshells (containing a silica core and a gold shell) can be used in bioimaging applications as well as to treat cancer patients through photothermal therapy [9].

One important characteristic is commonly found in nanomaterials: the great surface area/volume ratio. This characteristic makes them suitable for chemistry, physical, and biological functionalization improving their ability to interact with target cells, for example [10]. The possibility of modifying these materials' surface enables researchers to design them to present a set of desirable properties, avoiding undesirable off-target interactions [11].

Therefore improvements in fields such as vaccine development [12–15], gene, siRNA, and drug delivery [16–22], imaging and/or diagnosis [19,23–25], tissue engineering [26–29], energy storage [30,31], and development of some new research methodologies such as the construction of three-dimensional culture through magnetic levitation (using nanoshuttles) [32] have been achieved by using those nanomaterials (Table 5.1). However, the concerns regarding their utilization in live systems that are related to biocompatibility, toxicity, and biosafety arise [33].

For nanomaterials usage, it is also necessary to consider the information regarding their concentration, absorption of the material, distribution and capacity to enter not only cells and tissues that they are target to, but also how they would be metabolized, the time of body elimination and its route, and their toxicological profile.

2 BIOCOMPATIBILITY OF NANOMATERIALS

In 1986, the conference of the European Society of Biomaterials defined biocompatibility as the ability of a substance to present an appropriate response to the host in an application [46,47]. Therefore for nanomaterials, it would be their capacity to induce, in a specific situation, an appropriate response by the host (or target). If the cells or the body have an undesirable answer to the nanomaterial used (for example in a toxic and/or immunogenic way), this is considered bioincompatible and it should be modified before being added to any biological system.

The response elaborated by the living system to the presence of nanomaterials depends upon how they interact with some components that constitute cells (such as proteins, lipids, and carbohydrates) and, consequently, with the cellular normal functions (especially target cells and immune cells if it is an *in vivo* exposure—nanomaterials can naturally enter a living system through inhalation, ingestion, or passing through the skin).

Some nanomaterials' characteristics are known to interfere in these interactions and responses, being important to observe them before the medical use of nanomaterials, for example. Characteristics such as size, surface area, shape, chemical constitution (and consequently hydrophobicity), surface charge, surface topography, purity, the presence of functional groups, and tension [48] need to be considered before nanomaterials' use.

The nanomaterial uptake can be performed through active or passive transportation (Fig. 5.2) [49]; the preference by one of these transport types will depend upon the previously mentioned nanoparticle's characteristics.

Spherical gold nanoparticles in human dermal fibroblasts induce different pathways of internalization depending on particle's diameter; 45 nm particles entered cells via clathrin-mediated endocytosis, while 13 nm were phagocyted (Fig. 5.2) [50]. Gold nanoparticles conjugated to siRNA (targeting the expression of isocitrate dehydrogenase 1) were also internalized by U87 glioblastoma cells in a size-dependent manner; spheres containing a diameter of 50 and 40 nm were better internalized than the ones of 13 nm in diameter and could also abandon endocytic vesicles to deliver the siRNA more efficiently [51].

In the PC3 human prostate cancer cell line, the gold nanoparticles with 50 nm in diameter, without polyethylene glycol (PEG), attached to the cells and in the absence of serum proteins were better internalized [52]. In HeLa cells, colloidal gold nanoparticles with 50 nm in diameter also presented lower uptake half-life than the ones with 14 and 74 nm [53]. Gold nanoparticles coated with aptamers for surface mucin glycoprotein having 45 nm in diameter entered different cancer cells through endocytosis and presented

TABLE 5.1
Examples of Nanomaterials and Their Different Applications.

	Nanomaterial	Applications	References
Carbon-based nanomaterials	Single-walled carbon nanotubes wrapped with PEI-cholesterol derivatives	Gene delivery induced by laser	[21]
	Estradiol functionalized multi-walled carbon nanotubes	Efficient gene delivery	[22]
	Graphene	Induction of dental pulp stem cells' osteogenic differentiation in the abscence of chemical inducers for osteogenesis	[29]
	Graphene	Substrate for craniofacial bone tissue engineering research, being able to induce osteogenic differentiation in dental pulp stem cells	[29]
	Carbon nanotubes with copper	Electronic packaging: the material presents high thermal and electrical conductivity and low coefficience of thermal expansion	[34]
	Carbon nanotubes	Carbon nanotube derived from screen-printed electrochemical sensors for detection of oxidase enzymes' substrate	[35]
	Carboxylated multiwalled carbon nanotubes	Gene delivery to hard-to-transfect cells with less citotoxicity and higher transfection efficiency than cationic lipids	[36]
	Graphene oxide conjugated to aptamer (short single-stranded oligonucleotide)	Destroys bacteria's biofilm, for example *Salmonella typhimurium* 's biofilm	[37]
Non-carbon-based nanomaterials	Chitosan-coated PLGA nanoparticles containing the eukaryotic expression plasmid DNA of NDV F gene.	Controlled plasmid DNA delivery and vaccine adjuvant (inducing stronger immune response)	[12]
	N-2-hydroxypropyl trimethyl ammonium chloride chitosan - *N,O*-carboxymethyl chitosan nanoparticles loaded with chickens Newcastle disease and infectious bronchitis vaccines	Vaccine adjuvant and delivery carrier	[13]
	Chitosan/tripolyphosphate nanoparticles loaded with melatonin	Delivery of melatonin to U87MG glioblastoma cells increasing the drug's cellular uptake and anticancer efficacy	[18]
	Pluronic-stabilized magnetic nanoparticles incorporated with doxorubicin and paclitaxel	Efficient delivery of drugs with antiproliferative effect in breast cancer cells and potential to be analyzed by magnetic resonance imaging	[19]
	Gold nanoparticles with antimicrobial peptides	Gene delivery vehicle to mesenchymal stem cells with antibacterial activity	[20]
	Metal-enhanced fluorescent dye-doped silica nanoparticles (Ag-SiO_2-SiO_2-Tris(2,2-bipyridyl)dichlororuthenium(II) hexahydrate nanoparticle)	Detection of the prostate-specific antigen in a very sensitive, specific and reliable way	[24]
	Spiky gold nanoparticles that went through silver enhancement	Offer improvement in electrochemical detection	[25]
	TiO_2 nanoparticles chitosan scaffold	Biocompatible scaffold for bone tissue engineering	[27]

Continued

TABLE 5.1
Examples of Nanomaterials and Their Different Applications.—cont'd

Nanomaterial	Applications	References
PLGA nanoparticles carrying ciclosporin and added to polycaprolactone-based scaffolds	Assist in the innervation of regenerated teeth allowing local sustained action of the immunosuppressive drug	[28]
Silica nanoparticles	When placed between two polymer separators layers, the nanoparticles slow the growth of lithium dendrites extending a battery life	[31]
Gold nanorods with phosphatidylcholine	Gene delivery to hard-to-transfect cells with less cell death rate and higher transfection efficiency than cationic lipids	[36]
Silica nanoparticles functionalized with succinic anhydride to deliver Rivastigmine hydrogen tartrate	Rivastigmine hydrogen tartrate can be used to treat Alzheimer's disease	[38]
Mesoporous silica nanoparticles loaded with levofloxacin and externally functionalized with N-(2-aminoethyl)-3- aminopropyltrimethoxysilane	Nanocarrier to deliver the antibiotic efficiently penetrating bacterial biofilm	[39]
Nickel nanoparticles in microemulsions of sodium dodecyl sulphate (SDS)/ isopentanol/cyclohexane/water	Improve the performance of lubricating oils	[40]
Fatty acid and quaternary ammonium modified chitosan nanoparticles	Insulin delivery	[41]
Alginate-based nanoparticles	Insulin oral administration	[42]
Chitosan nanoparticles loaded with tea polyphenols And Elsholtzia splendens	Delivery of chemopreventive agent and inhibition of tumorigenesis in a variety of animal models of carcinogenesis	[43]
Chlorogenic acid loaded chitosan nanoparticles	Release of encapsulated chlorogenic acid without affecting the antioxidant properties	[44]
Chitosan-TPP nanoparticles to feluric acid (FA) encapsulated	Stimulation of insulin secretion by pancreatic while inhibiting lipid peroxidation	[45]

a higher uptake rate than largest ones [54]. Gold nanoparticles with 40 and 50 nm in diameter also presented an optimized effect in SK-BR-3 cells when it comes to the regulation of the desired membrane receptor internalization process [55] (Fig. 5.2). However, when ultrasmall gold nanoparticles were coated with Tiopronin, a drug used to treat cystinuria and rheumatoid arthritis, and injected intravenously in mice, the smaller ones (diameter of 2 and 6 nm) were able to penetrate deeply into tumor and presented higher accumulation levels than the ones with 15 nm in diameter [56]. Gold nanoparticles conjugated to polyethylene glycol-functionalized dithiolane ligands covalently attached to cell-penetrating peptides could penetrate COS-1 cell line better when the peptides were more displayed on nanomaterials' surface, and smaller particles (~2.4 nm in diameter) were more prone than larger ones to migrate to cells' nuclei [57]. However, in an interesting study evaluating HeLa cell behavior in the presence of single ultrasmall gold nanoparticles of different sizes, it was reported that from 4 to 17 nm in diameter, cell uptake and the unbinding force both increased with the sizes of the nanomaterial [58].

Thus nanomaterials' size can interfere not only on cellular uptake but also in the distribution through the intercellular environment. Results can be similar

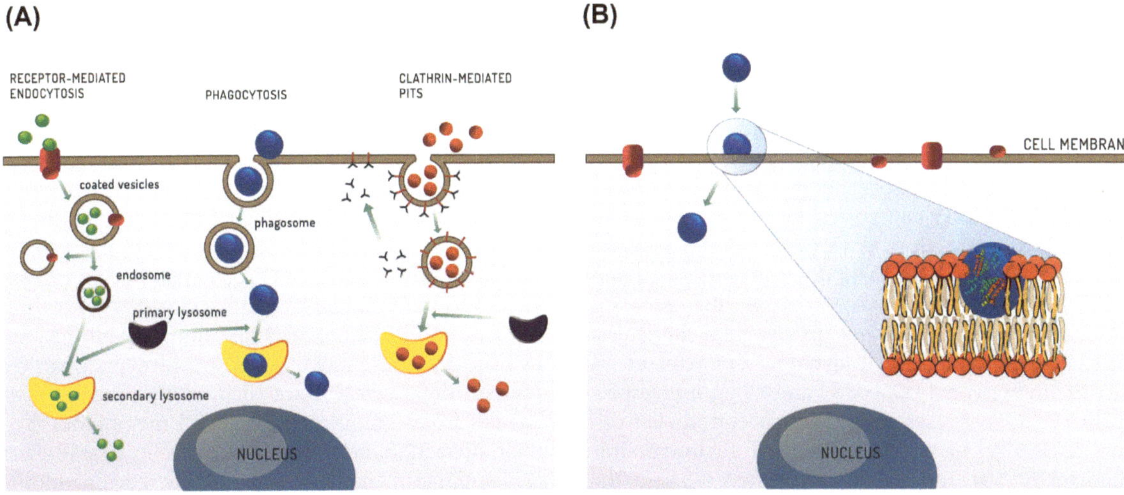

FIG. 5.2 Nanomaterial's cellular uptake can be performed through an active or passive process. In **(A)**, the most common active internalization types are presented and in **(B)**, the passive internalization.

regarding the size even when different shapes of nanomaterials are considered in the same study. Spherical gold nanoparticles of 15 and 45 nm diameter and 30 and 50 nm gold nanocages were assayed in SK-BR-3 breast cancer cells, but their surfaces were modified with polyethylene glycol, antibody anti-HER2, or polyallylamine hydrochloride. Independently of the functional group on the nanostructure surface, smaller structures presented higher uptake rates than largest ones. However, it is not true for all characteristics once the surface functional group interferes in cellular shape internalization preferences. SK-BR-3 cells internalized spherical particles more easily when modified with PEG or anti-HER2, and considering each shape separately, the internalization of polyallylamine hydrochloride-modified nanoparticles is higher than the other ones [59]. Gold nanoparticles coated with the protein transferrin in spherical and rod shape could be internalized by HeLa, SNB19, and STO cells in clathrin-dependent endocytosis, but at a lower level than the uncoated versions. Rod-shaped particles were also uptaken in a smaller amount, probably because a longer time was required by the cell to wrap a larger surface area. For particles with dimensions next to 50 nm, it was possible to enter the cell as a single structure and in a faster way; smaller particles had to cluster before uptake and larger particles also took more time due to the wrapping process [60].

Nevertheless, it is also important to highlight that the results for each nanomaterial depend on the cell types used in the experiments [36]. In macrophages, the size of the nanoparticle was not the prominent characteristic to influence uptake. The density of PEG coating the nanoparticle surface was the most important characteristic influencing cellular internalization. An increase in PEG density decreased the range of serum protein adsorption to the nanoparticle, and the cellular uptake (above ~ 0.64 PEG/nm^2) became not dependent on the presence of adsorbed proteins from serum on nanoparticle's surface. The size of the nanoparticle itself presented no influence on uptake efficiency [61]. However, when gold nanoparticles were functionalized with DNA and assayed in CaSki cells, they were internalized in a way highly dependent on the hydrodynamic diameter (smaller particles internalized much faster than larger ones), being irrelevant the effect of the DNA conformation (single- or double-stranded) and surface charge, and the DNA density played a minor role on influencing the uptake process [62].

Although the nanoparticle's size has a crucial role in biodistribution and biocompatibility, when the nanoparticles are in a physiological environment, the hydrodynamic diameter is a better parameter when the aim is to analyze the uptake and toxicity of nanoparticles [63]. Hydrodynamic diameter consists of the extent of the bilayer solvation core formed by ions and polar molecules around the nanoparticle, and that can move with it. Therefore even with a smaller size, a nanoparticle can carry a large solvation core able to influence the nanoparticle's uptake, biodistribution, and interaction with cell membranes and proteins [64]. Hydrodynamic diameter can also influence the biocompatibility of nanoparticles when it comes to their clearance pathway [65,66]. For example, in *in vivo* experiments, nanoparticles smaller than 20 nm are usually eliminated by the kidney [67]; nanoparticles between 30 and 75 nm are commonly accumulated in heart, liver, stomach, kidney [68], and bone marrow [69]; the biggest ones, above 150 nm, are usually found accumulated in liver and spleen [70]. It is possible to notice a close connection between hydrodynamic diameter and cytotoxicity when working with silica nanoparticles. In the same concentration, the smallest nanoparticles can cause more toxicity than the largest ones. Particles with a hydrodynamic diameter of 14 and 19 nm presented LC_{50} of 33 and 89 μg/mL, respectively, on *in vitro* assay using human endothelial cell line EAHY926; nanoparticles with a hydrodynamic diameter of 104 and 335 nm presented LC_{50} in the range of 1000 μg/mL. Moreover, particles with medium size, 60 nm, exhibited medium toxicity compared to the others, LC_{50} of 254 μg/mL [71].

It is also interesting to mention that most studies involving nanomaterials' surface modification determinate their dimensions before the coating process. Nonetheless, all the attached materials can also interfere with their uptake. For example, a gold nanoparticle of 20 nm in diameter after being coated with polyethyleneimine can become a 32 nm structure and after incubation in cell culture media end up being associated with protein corona and having 88 nm in diameter. Comparing its internalization rate in cells as CHO−K1, HeLa, MCF-7 with the internalization rate of 30, 50, and 80 nm in diameter (before the coating process), these 20-nm nanoparticles offered the better results. However, they were noted by the cells as 88 nm structures and not only 20 nm [72].

Besides nanomaterials' size, the surface charge can also interfere in their uptake. To characterize the surface charge of nanoparticles, a parameter called zeta potential should be used [48,73]. It indicates the electrical potential of the nanoparticle's surface and its solvation core. Therefore, this parameter is strongly influenced by the composition of the dispersing medium and the chemical composition of the nanoparticle itself. The zeta potential allows the analysis of the stability of nanoparticles' suspension in a liquid medium, indicating their tendency to agglomerate and decant [73]. The higher the superficial electrical charge of a nanoparticle (either negative or positive), the lower will be the agglomeration (once the repulsion between them will be as higher as the superficial electrical charge gets). For this reason, it is common to set a positive charge above +30 mV or a negative charge below −30 mV, avoiding the neutral charges (around 0 mV) [3,74].

The surface charge of nanoparticles has a relevant role in biological performance. The uncharged (neutral) particles have a higher opsonization rate than the charged particles [75]; in addition, the positively charged particles tend to be more internalized due to the strong interaction of these nanoparticles with the negative charges of the plasma membrane [76,77]. However, this positive charge can also impact cell survival. The nanomaterial's surface charge has also a direct influence on its cytotoxicity and biocompatibility, and usually, for mammalian cells, positive nanoparticles present the highest cytotoxicity [78]. The opposite charges from nanomaterial and cell membranes favor their interaction, increasing the possibility of side effects [79] and nanoparticle's capture by the reticuloendothelial system [80]. Positive nanoparticles can interact with negatively charged proteins in the blood creating a corona core, which increases the size of nanoparticles and stimulates the immune system [80]. However, this can be a desirable feature for cancer treatment; the tumoral microenvironment in patients has an acid pH due to the local hypoxia [81] that can make the surface charge of nanoparticles more positive, stimulating their interaction with tumor cells causing high uptake level and cytotoxicity [82]. In studies using silver nanoparticles against gram-positive bacteria (with a negative surface charge) was reported a direct influence of the nanoparticle charge over the death rate of bacteria [83]: higher death rate can be achieved with nanoparticles with a small negative surface charge or high positive surface charge. In human keratinocytes, gold nanoparticles with negative surface charge exhibited higher toxicity (inducing apoptosis at a higher level) than the positive ones [84]. In this way, the environment and the surface charge of the nanoparticle must be evaluated for each specific target cell to obtain the desired effect in each study or therapeutic strategy.

In nonphagocytic HepG2 cells, gold nanoparticles with positive surface charges were better internalized than the ones with a negative surface charge. However, the results depend upon cell type; in phagocytic RAW 264.7 cells, no difference in uptake rate was observed related to nanoparticles' surface charge [85].

Aqueous quantum dots (QDs) containing atoms of tellurium and cadmium synthetized with different thiol ligands proved to be internalized in a manner that depends most on their hydrodynamic diameter. QDs with a very similar hardcore diameter were differently internalized due to their final hydrodynamic diameter and in a way not directly related to their charge on the surface. The structure with greater hydrodynamic diameter could be accumulated in higher concentrations inside the A-427 target cells [86].

QDs made of carbon presented minimal toxicity in J774.1 murine cells in concentrations smaller than 0.5 mg/mL and the ones prepared from citric acid were easier internalized than the ones prepared from bamboo leaves. However, surface modification with phenylboronic acid could increase the uptake rate. In this particular situation, the phenylboronic acid made it easier for the nanostructure to concentrate closer to the cell surface once the acid could bind to sialic acid present on the target cell's surface [87].

When it comes to these particles, it is possible to notice that cell type also interferes in uptake capacity. Cadmium/selenide (Cd/Se) quantum dots with zinc sulfide shell and overcoated with silica were internalized in a fast way by HepG2 cells and presented no cytotoxicity at low concentrations [88]. Nonetheless, the negatively charged quantum dots (due to carboxylic acid coating) containing Cd/Se core and a zinc sulfide (ZnS) shell were successfully internalized by dendritic cells, in a sixfold smaller rate by monocytes and were not internalize by lymphocytes. Cellular uptake pathways vary with cell type and cell differentiation state; differently from HEK [89], dendritic cells used F-actin, clathrin and other substances to uptake these negatively charged quantum dots [90]. The modifications on the surface can also alter the rate of internalization of QDs. CdSe quantum dots with ZnS shell and different coatings (carboxylic acid, PEG, or amine-derivatized PEG) were presented to human MCF-10A and MCF-7 cells. After 12 h of exposure only the carboxylated version was uptaken [88]. Graphene quantum dots, in turn, were better internalized by human monocytes and granulocytes than by human lymphocytes; possessing low cytotoxicity (offering rates of ~90% cell viability) this nanostructure was well assimilated by these cells with phagocytic properties [91].

TiO_2 nanoparticles were better internalized in A549 cells when they possessed a small dimension (smaller than 80 nm) and were well dispersed in buffer/culturing media (exhibiting soft agglomerates). However, the particles with a larger diameter can end up causing pro-inflammatory responses in a size-dependent manner [92]. However, when it comes to TiO_2 nanoparticles, it is important to pay attention in their form because TiO_2 may exist in different polymorphs, such as anatase and rutile. In the anatase phase, TiO_2 nanoparticles of approximately 30 nm in diameter were assayed in the nematode *Caenorhabditis elegans* and in concentration smaller than 500 μg/mL were uptaken and caused no harm to worm's body length despite affecting its locomotion by inducing the growth of shorter axons. On the other hand, rutile TiO_2 nanoparticles induced a severe reduction in the worm population. It is due to different patterns of influence on gene expression caused by nanoparticles in genes related to metal binding or metal detoxification, fertility and reproduction, worm growth and body morphogenesis, and neuronal function [93].

Iron oxide magnetic nanoparticles coated with polyvinylpyrrolidone were better uptaken by mouse macrophage cells as diameter increased from approximately a core diameter of 7.6 nm and hydrodynamic diameter of 32.2 to a 36.8 nm core diameter and hydrodynamic diameter of 102.4 nm, which also presented a contrast enhancement in magnetic resonance imaging of hepatic lesions *in vivo*. Particles presenting larger diameter were less internalized and presented less contrast enhancement. However, all sizes showed little cytotoxicity even in high concentrations as 250 μg Fe/mL [94], but with different coatings, the results observed regarding iron nanoparticles uptake, and toxicity may vary. Polyethylenimine-coated nanoparticles are better internalized by murine macrophages and human cancer cells than PEGylated ones. For PEGylated versions, however, the internalization of smaller nanoparticles (10 nm in diameter) by these cells was higher than the internalization of larger particles (30 nm in diameter), and the cytotoxicity was only slightly observed at high concentrations while PEI-attached nanoparticles exhibit severe cytotoxicity [95].

Different cells also react differently to the exposure to polystyrene nanoparticles. Carboxylated polystyrene nanoparticles presenting a diameter of 40 nm were internalized in a faster way by 1321N1 human astrocytoma and A549 human lung carcinoma cells when compared to the internalization of nanoparticles possessing a diameter of 20 nm or 100 nm [96]. In RAW 264.7 cells, the particle of 40 nm in diameter presented uptake kinetics, which was 8, 18, 20, and 56 times higher than that observed in HCMEC D3, 1321N1, A549, and HeLa cells, respectively [97]. Carboxylated functionalized polystyrene nanoparticles of approximately 100 nm in diameter were internalized by human macrophages up to four times faster than for THP-1 cells; although when it comes to aminofunctionalized versions, they were internalized by THP-1 cells faster than by human macrophages. In the presence of serum, the uptake may occur, due to different patterns of interaction on cells surface, via phagocytosis in macrophage or via dynamin-II-dependent endocytosis in THP-1 cells [98]. A commercial version of fluorescent polystyrene nanoparticle possessing a diameter of approximately 50 nm could penetrate the chorion of developing zebrafish and decrease the heart rate, but it was not able to induce significant mortality or cause deformities [99]. Mouse neurons did not internalize polystyrene nanoparticles carboxylated or PEGylated with a diameter from 45 to 70 nm. Meanwhile, a different cell type, microglial cells, internalized large amounts of these carboxylated nanoparticles. Age-depended modifications in these nanoparticles cause different patterns of aggregation and can be responsible for an increase in toxicity and cellular uptake after storage for more than 6 months. Moreover, a particular emphasis should be placed on the medium to be used in storage, once PEGylated and carboxylated polystyrene nanoparticles proved to be able to adsorb significant amounts of bacterial lipopolysaccharide if exposed for a short period to this type of endotoxin [8]. Red blood cells can efficiently carry some polystyrene particles, as the carboxylated ones, without suffering from mechanical, osmotic or oxidative stress until high rates such as 1:50. Just rates of nanoparticles load 10-fold higher, or more, caused concerning damage [100].

Other polymeric nanoparticles, as the ones produced from poly(lactide-co-glycolide) (PLGA), can also be used to deliver substances to target cells, especially for drug oral administration. When modified with quaternary ammonium salt didodecyl dimethylammonium bromide, these charged particles were better uptaken by Caco-2 and HT-29 human cells in a size-dependent way with an optimal particle size of 100 nm [101]. Carboxymethyl chitosan-grafted nanoparticles (negatively charged) and chitosan hydrochloride-grafted nanoparticles (positively charged) from 150 to 500 nm were better phagocytized by murine macrophage when possessing larger surface area and diameter. In experiments with animals, the negatively charged particles of 150 nm in diameter were better accumulated in tumors, representing an interesting option to drug delivery [102]. Biocompatible polymeric nanoparticle synthesized to contain a liquid cored, formed by docusate sodium salt dissolved in chloroform, involved by a polyelectrolyte shell of poly-L-glutamic acid sodium salt (negatively charged) or poly-L-lysine (positively charged) could be well internalized by both murine RAW 264.7 cells and human THP-1 cell, without any significant change in the uptake level [103]. Biobased poly (thioether-ester) nanoparticles could also be an interesting alternative for drug delivery, as they presented no cytotoxicity on murine L929 cells and HeLa cells, and no hemolytic activity [104].

Silica nanoparticles from 32 to 83 nm in diameter presented no cytotoxicity or genotoxicity when assayed in Caco-2 cells. Despite the fact that both could be internalized, only the 32 nm version could reach the nucleus inner part [105]. Nanoparticles made of organically modified silica with approximately 20 nm in diameter could get inside living brains, neuronal cell bodies, and axonal projections of *Drosophila* causing no harm to its survival or to neuronal cells and processes [106].

Another interesting type of nanomaterial is the mesoporous silica nanoparticles. They possess different shapes (such as spheres, rods, and cylindrical mesopores) and can be easily chemically modified to perform a specific function. Human dendritic cells (DCs) could be penetrated by anionic surfactant-templated mesoporous silica nanoparticles and cellular functions are less affected by than as smaller nanoparticles are, and as lower dosages the cells are exposed to. Larger particles escaped from endolysosome more easily than the smaller ones [107]. The size is one of the most important characteristics to determine if mesoporous silica nanoparticles (MSNs) will be biocompatible. Adjusting the pH, it is possible to obtain MSNs with different sizes (diameters from 30 to 280 nm, for example), and in HeLa cells, the uptake is also size-dependent; the nanoparticles of 50 nm in diameter were better internalized by this cell type [108]. When it comes to red blood cells, MSNs with similar surface area and structure, but with diameters varying from 25 to 225 nm, were more prone to cause hemolysis when they were smaller in diameter and/or are present in higher dosage. Nonetheless, PEG functionalization masked the surface silanol groups attenuating the undesirable effects [109]. In this way, it is possible to conclude that not only size but also surface chemistry and charge of MSNs, as same as shape, influence on these nanoparticles' biocompatibility [110]. Spherical-shaped and tube-shaped MSNs were more efficiently internalized by CHO cells than by fibroblast cells. In CHO cells, the endocytosis of tube-shaped MSNs reached the highest rate not only in the maximum concentration; in fibroblast, the rate of endocytosis for spherical-shaped nanoparticles was faster than the one observed for tube-shaped MSNs [111]. The concentration of nanoparticle is also an important influence over biocompatibility. MSN of 80 nm in diameter could be well internalized by human dermal fibroblast cells, but an increase in concentration from 5 to 200 μg/mL led to inhibition of cell growth and proliferation in a more severe way than the effect induced by particles of 500 nm in diameter [112]. On the human vascular endothelium and platelets, the results were similar: dense and amorphous silica nanospheres with smaller diameter induced more oxidative stress, activating NF-κB and causing cell death, in a dose-dependent manner [113,114]. Moreover, larger particles were easily captured by the reticuloendothelial system and in *in vivo* assays were commonly quickly found in mice urine. Smaller particles generally accumulate in organs such as the liver, spleen, lung, kidney, and heart [115].

Approaches known as cross-linkers are widely used for the nanoparticles' synthesis, as they use the electrostatic interactions that occur between ionizable groups charged with opposite charges. Although it is a basic form, there are only small methodological variations among the different types of existing nanoparticles. In protocols using chitosan nanocarriers for insulin delivery, particles initially prepared with 100–200 nm showed the efficiency of approximately 15.7% [42]. The addition of water-soluble chitosan derivatives, such as N-trimethylated chitosan [116] and carboxymethylated chitosan [117], led to the formation of nanocomplexes with approximately 226.1 nm and provided efficiency in insulin loading of about 33%. Protocols that used cross-linkers favored the interaction between insulin, in the negatively charged form, and positively charged chitosan, which led to the formation of auto-mountable nanoparticles through electrostatic interactions. In studies that obtained particles measuring 210 nm by this method provided 39.2% efficiency in insulin loading [118]. For carrying compounds such as polyphenols, for example, chitosan nanoparticles were obtained by cross-linkers between the positively charged amine of chitosan hydrochloride and among its negatively charged carboxymethyl groups, which culminated in obtaining particles of 407 nm that were stable and showed a rate of about 70% in the sustained release of polyphenols [43].

The use of alginate nanocarriers has strongly emerged. Its use is justified once it is nontoxic, biodegradable, and low-cost, besides being mucoadhesive, biocompatible, and nonimmunogenic [42]. Alginate nanoparticles are more often obtained by complexing with calcium chloride or by mixing with an oppositely charged polyelectrolyte such as chitosan [119]. For insulin loading, nanoparticles prepared in this manner and measuring 40 nm showed a decrease in the constant release of insulin as the calcium concentration was increased (the calcium concentration evaluated was between 0.5 and 1.1 mM) [120]. In tests conducted on Swiss albino mice, there was no toxicity at the nephron or liver level, after oral administration of the nanoparticles carrying insulin, the diabetic mice presented approximately 9% reduction in glycemia [121].

The evolution in the nanotechnology field is continuous and happens in the sense of optimizing the desirable effect and decrease the undesirable ones. When it comes to suppressing the toxicity, there are also solutions in nanoscale. CuO nanoparticle cytotoxicity can be suppressed with efficiency by glutaraldehyde-cross-linked chitosan nanoparticles modified by lysine or glutamic acid, or reduced by sodium borohydride. These nanoparticles perform the chelation of Cu atoms, diminishing the intracellular level of reactive oxygen species generated by CuO nanoparticle internalization [122].

To evaluate the biocompatibility of nanomaterials in biological systems, the use of appropriate cytotoxicity experiments is necessary. However, there is no standard test to evaluate these materials' biocompatibility and toxicity [123]. It is also important to mention that cytotoxicity assays are commonly performed using cell culture because protocols tend to require the use of dye molecules, being easy to perform and possessing low cost. Nonetheless, there are different studies mentioning the interference of nanomaterials in the test results interacting with some dye molecules or influencing the values with their own absorbance/fluorescence, providing unreliable results [124,125]. In this way, it is prudent to perform not only conventional cytotoxicity assays using nanomaterials but also some other techniques such as the measurement of electrical impedance of attached cells [126].

3 NANOMATERIALS' FUNCTIONALIZATION

As it could be noticed in this chapter, the biocompatibility of nanomaterials may depend upon their surface characteristics that can allow site-directed delivery, attenuate cytotoxicity, confer solubility in some specific medium, facilitate the cellular uptake process, and avoid undesirable agglomeration [22,98,127–132], for example. Nanomaterials surfaces can be modified and through covalent bonds they can be functionalized with chemical and biomolecules to achieve a specific goal [11,133]. For example, carbon nanotubes after purification and selective oxidation can present inflammatory activity. However, by performing a benzoic acid functionalization on them, it was possible to decrease the extension of this undesirable effect, reducing the IL-1 β production and the recruitment of neutrophils and monocytes in mice [134]. Quantum dot, when PEGlyated, allowed the attachment of siRNA and tumor-homing peptides to target deliver the siRNA to

HeLa cells enabling the future development of systems to treat even metastatic cancer [135].

When the nanomaterials are submitted to direct functionalization, a single conjugation step is performed to directly attach the ligand. It acts as the modifying group itself (exhibiting the desired functionality) to the nanomaterial in the presence of a conjugating agent. When the postfunctionalization is performed, more than one compound or a bifunctional compound is required: the binding chelating group of the bifunctional compound or the first compound is reacted first being attached to the nanomaterial, and in a second step another part of the bifunctional compound is converted to the final functional group or the first compound is attached to the desired biomolecule [11] (Fig. 5.3).

There are a huge number of different protocols to functionalize different nanomaterials, each one aiming a specific function and applying a specific method. The postsynthetic grafting method or stepwise synthesis generally comprises two steps: the first one, the synthesis of the nanomaterial, and the second one, the surface modification. The cocondensation synthetic method or One-pot synthesis comprises one step with the nanomaterial synthesis and the functionalization happening at the same time [129]. The postsynthesis grafting method could be applied to functionalize silica nanoparticles using succinic anhydride, to make them able to deliver the drug Rivastigmine hydrogen tartrate [38] (a drug used in the treatment of the Alzheimer's disease). The cocondensation method allowed mesoporous silica nanoparticles to load and release aspirin attached to the nanomaterial [136].

Nanomaterials containing metals such as Ag, Cu, Pt, Hg, Fe, Si, and Au on their surface are commonly functionalized using organosulfur reagents, once sulfur compounds can easily adsorb on these metal surfaces forming strong coordination bonds [137,138]. When it comes to drug or gene delivery thiol groups can assist metal nanomaterials' conjugation to DNA, proteins/peptides, antibodies, etc. To improve tumor cell uptake of Paclitaxel in liver cancer therapy, this drug was complexed to thiol-functionalized silica nanoparticles and presented to HepG2 cells. To optimize the delivery and the time-dependent cell uptake, this complex was attached to PEG, increasing the internalization [139]. Once metal nanomaterials possess useful photo/electronic properties, sensors may be developed by functionalizing them with thiol groups present in compounds able to perform specific molecular recognition. Gold nanoparticles functionalized with thiol-modified glucose could specifically interact with a periplasmic binding protein known as maltose-binding protein allowing the development of a biodetection system [140].

Functionalization using amines is commonly performed to stabilize especially metal nanoparticles [141]. This kind of functionalization is also important to attach enzymes to the nanomaterial, for example. Amine-functionalized (using PEI, ethanolamine, and 2,2′-(ethylenedioxy)bis(ethylamine)) superparamagnetic Fe_3O_4 nanoparticles could be used as support to lipase immobilization for industrial applications. The immobilized enzyme was easy to reuse and proved to be thermally stable as well as resistant to storage [142]. Structures containing amine groups can also act as targeting agents, as it happens to N-(2-aminoethyl)-3-

(A)

(B)

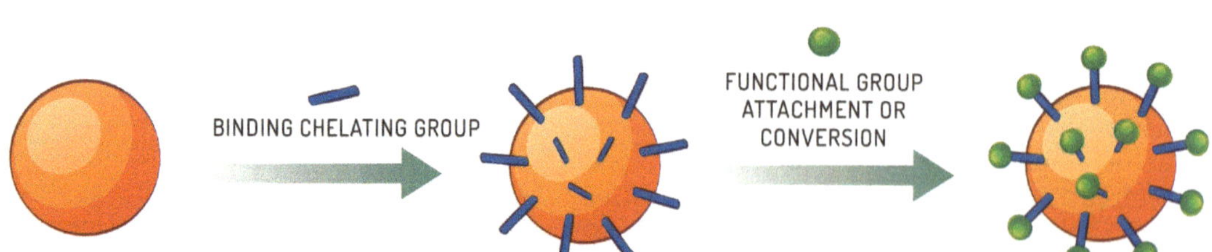

FIG. 5.3 Nanomaterial's functionalization methods. **(A)** Direct functionalization. **(B)** Postfunctionalization.

aminopropyltrimethoxysilane externally involving mesoporous silica nanoparticles loaded with levofloxacin (an antimicrobial agent); thanks to that, this delivery system could efficiently penetrate bacterial biofilm being a promising nanocarrier in the management of bone infection [39]. The N-(2-aminoethyl)-3- aminopropyltrimethoxysilane can be also used to functionalize silica-coated superparamagnetic nanoparticles allowing the attachment of antiCD34^{+} monoclonal antibodies. It made possible to separate these cells with high efficiency [143]. By applying the cocondensation method, silica nanoparticles could be functionalized using 3-aminopropyltriethoxysilane to deliver Rivastigmine hydrogen tartrate to SY5Y cells [38].

Carboxylic acids may also be employed to modify the surface of nanomaterials. The performance of graphene oxide as an additive for water-based lubricants can be dramatically improved by adding carboxyl-functionalized multiwalled carbon nanotubes. The presence of the latter helped to reduce friction and wear [144]. Carboxyl-functionalized multiwalled carbon nanotubes could also overcome polyethyleneimine-modified multiwalled carbon nanotubes capacity to act as a scavenger for toxic lead ions. In the presence of these nanotubes, freshwater zooplankton, *Daphnia magna*, could tolerate in a significantly better way lead presence, as the heavy metal adsorbed onto the nanomaterial [145].

It is also prudent to mention that the vast majority of nanoparticles need to be coated before being functionalized, which contributes to avoiding agglomeration and loss of function, and can also facilitate the functionalization process [146,147]. Nanoparticles with no coat are usually more cytotoxic than coated ones [146], thus coat biocompatibility is also an important characteristic once it is commonly in contact with the surrounding environment and different degrees of toxicity are observed for different substances that constitute it [74,148,149] (such as PLGA [150], PEG [151], poloxamer, polyoxamines, polietilenoxido [152], dimercaptosuccinic acid [153], citrate [154], dextran [155], chitosan [156]). Iron oxide nanoparticles coated with dextran, for example, induced death in immortalized primary human fibroblasts hTERT-BJ1. However, when coated with albumin, cells were induced to proliferate [148].

When it comes to biological uses of functionalized nanomaterials, it is important to highlight biofunctionalization. As previously discussed, the great relation superficial area/volume makes nanomaterials suitable to biofunctionalization and favors, after the chemical surface modification, recognition of specific cells and biomolecules and the incorporation of contrast agents to imaging diagnoses [157]. It also promotes nanomaterials' capacity of differentiation between normal and tumoral cells (increasing their uptake by these cells) [158]. As same as aptamers [159–161], antibodies [162–164], and peptides [82,146,165], amino acids can be used to chemically modify these materials' surface, and after the biofunctionalization is also possible to perform the attachment to biomolecules, for example, to be delivered. Amino acids possess at least an amino and a carboxyl group and depending on the amino acid the sulfur atom is present. Lysine could be used to functionalize Fe_3O_4 nanoparticles than bind specific antibodies and detect cancer cells [166]. Arginine could be used to modify superparamagnetic iron oxide nanoparticles to efficiently deliver the drug Paclitaxel to A549 cancer cells to induce cell death [167].

The advance in the nanotechnology field has also offered intelligent solutions even in areas that we would not immediately imagine they could offer. Smart and functional textiles' development was benefited: multiwalled carbon nanotubes could be functionalized to present oxygen-based surface groups and be incorporated in the polyester-generating fabric that presented water resistance and flame retardancy [168].

Nanotechnology/nanobiotechnology still possesses many challenges to overcome in the future, but the advances that they have been providing in diverse-related research areas are indeed a revolution. We already have examples of FDA-approved nanomedicines available to be used in the treatment of several diseases. Are relevant examples: Ferumoxytol superparamagnetic iron oxide nanoparticles (SPIONs) with polyglucose sorbitol carboxymethylether to treat iron deficiency anemia in chronic kidney disease (Feraheme/ferumoxytol (AMAG pharmaceuticals))—it decreases the number of doses necessary once it provides prolonged steady release; and albumin-bound paclitaxel nanoparticles to treat breast cancer, nonsmall cell lung cancer, and pancreatic cancer (Abraxane/ABI-007 (Celgene))—it offers a better delivery to tumor cells and an improved solubility. There are also commercial imaging agents such as SPION coated with dextran (Feridex/Endorem (AMAG pharmaceuticals)) and SPION coated with silicone (GastroMARK; umirem (AMAG pharmaceuticals)) [169].

4 FUTURE PERSPECTIVES

As mentioned in this chapter, the use of nanomaterials, especially for drug delivery, is a promising alternative for improving health and life quality. However, it is necessary to emphasize the importance of researching the safety of these materials before using them in the treatment of diseases. In this aspect, an important point to be observed is their form of entry into the human body, whether in an intentional and controlled, or accidental one. *In vitro* experiments should continue to be focused on how different cells react differently to nanomaterials, with distinct biocompatibility and toxicity rate. There are characteristics intrinsic to these materials that influence their action on living systems, such as the charge of their surface, shape, functionalization agent, etc., that should also receive attention. Due to the lack of a harmonized standard methodology to verify the toxicity of these materials in biological systems, this difficulty becomes even greater.

Therefore it is still a challenge to determine the indicated quantity of each nanomaterial to be used in each type of treatment or other biologic application of nanomaterials. It is expected continuous progress in this field to offer more reliability to the data collected through experiments. It is also expected that new nanomaterials continue to be developed as same as new methods of functionalization, aiming decrease of undesirable effects and possible damages and optimization of desirable properties that will favor their variety of applications. It is natural to expect, as nanobiotechnology advances, an increase in the number of FDA-approved nanomaterials to a variety of uses such as therapeutic and imaging agents. Thus nanotechnology is a promising field in supporting significant improvements in human life, but there is still much to be researched to apply its products safely without major concerns.

ACKNOWLEDGMENTS

Fernanda Tonelli would like to thank L'ORÉAL, UNESCO, and Academia Brasileira de Ciências for funding her research through the "For Women in Science Brazil 2017" award.

REFERENCES

[1] Drexler KE. Nanotechnology: from Feynman to funding. Bulletin of Science, Technology & Society 2004;24:21–7. https://doi.org/10.1177/0270467604263113.

[2] Weiss J, Takhistov P, McClements DJ. Functional materials in food nanotechnology. Journal of Food Science 2006;71:R107–16. https://doi.org/10.1111/j.1750-3841.2006.00195.x.

[3] Zhang Y, Yang M, Portney NG, Cui D, Budak G, Ozbay E, Ozkan M, Ozkan CS. Zeta potential: a surface electrical characteristic to probe the interaction of nanoparticles with normal and cancer human breast epithelial cells. Biomedical Microdevices 2008;10:321–8. https://doi.org/10.1007/s10544-007-9139-2.

[4] Xu L, Zhang W, Shang L, Ma R, Jia L, Jia W, Wang H, Niu L. Perylenetetracarboxylic acid and carbon quantum dots assembled synergistic electrochemiluminescence nanomaterial for ultra-sensitive carcinoembryonic antigen detection. Biosensors and Bioelectronics 2018;103:6–11. https://doi.org/10.1016/j.bios.2017.10.035.

[5] Pan LH, Kuo SH, Lin TY, Lin CW, Fang PY, Yang HW. An electrochemical biosensor to simultaneously detect VEGF and PSA for early prostate cancer diagnosis based on graphene oxide/ssDNA/PLLA nanoparticles. Biosensors & Bioelectronics 2017;89:598–605. https://doi.org/10.1016/j.bios.2016.01.077.

[6] Han L, Liu CM, Dong SL, Du CX, Zhang XY, Li LH, Wei Y. Enhanced conductivity of rGO/Ag NPs composites for electrochemical immunoassay of prostate-specific antigen. Biosensors & Bioelectronics 2017;87:466–72. https://doi.org/10.1016/j.bios.2016.08.004.

[7] Sharafeldin M, Bishop GW, Bhakta S, El-Sawy A, Suib SL, Rusling JF. Fe_3O_4 nanoparticles on graphene oxide sheets for isolation and ultrasensitive amperometric detection of cancer biomarker proteins. Biosensors & Bioelectronics 2017;91:359–66. https://doi.org/10.1016/j.bios.2016.12.052.

[8] Murali K, Kenesei K, Li Y, Demeter K, Környeia Z, Madarásza E. Uptake and bio-reactivity of polystyrene nanoparticles is affected by surface modifications, ageing and LPS adsorption: in vitro studies on neural tissue cells. Nanoscale 2015;9:4199–210. https://doi.org/10.1039/C4NR06849A.

[9] Loo C, Lowery A, Halas N, West J, Drezek R. Immunotargeted nanoshells for integrated cancer imaging and therapy. Nano Letters 2005;5:709–11. https://doi.org/10.1021/nl050127s.

[10] Albanese A, Tang PS, Chan WCW. The effect of nanoparticle size, shape, and surface chemistry on biological systems. Annual Review of Biomedical Engineering 2012;14:1–16. https://doi.org/10.1146/annurev-bioeng-071811-150124.

[11] Subbiah R, Veerapandian M, Yun KS. Nanoparticles: functionalization and multifunctional applications in biomedical sciences. Current Medicinal Chemistry 2010;17:4559–77. https://doi.org/10.2174/0929867 10794183024.

[12] Zhao K, Zhang Y, Zhang X, Shi C, Wang X, Wang X, Jin Z, Cui S. Chitosan-coated poly(lactic-coglycolic) acid nanoparticles as an efficient delivery system for Newcastle disease virus DNA vaccine. International Journal of Nanomedicine 2014;9:4609–19. https://doi.org/10.2147/IJN.S70633.

[13] Zhao K, Shanshan L, Wei L, Lu Y, Duan X, Han J, Wang X, Jin Z. Quaternized chitosan nanoparticles loaded with the combined attenuated live vaccine against Newcastle disease and infectious bronchitis elicit immune response in chicken after intranasal administration. Drug Delivery 2017;24:1574–86. https://doi.org/10.1080/10717544.2017.1388450.

[14] Saroja CH, Lakshmi P, Bhaskaran S. Recent trends in vaccine delivery systems: a review. International Journal of Pharmaceutical Investigation 2011;1:64–74. https://doi.org/10.4103/2230-973X.82384.

[15] Han J, Zhao D, Li D, Wang X, Jin Z, Zhao K. Polymer-based nanomaterials and applications for vaccines and drugs. Polymers 2018;10:1–14. https://doi.org/10.3390/polym10010031.

[16] Rizvi SAA, Saleh AM. Applications of nanoparticle systems in drug delivery technology. Saudi Pharmaceutical Journal 2018;26:64–70. https://doi.org/10.1016/j.jsps.2017.10.012.

[17] Riley MK, Vermerris W. Recent advances in nanomaterials for gene delivery—a review. Nanomaterials 2017;7:94–134. https://doi.org/10.3390/nano7050094.

[18] Kumar SY, Kumar SA, Dev A, Kaundal B, Choudhury SR, Karmakar S. Nanomelatonin triggers superior anticancer functionality in a human malignant glioblastoma cell line. Nanotechnology 2017;28:365102–13. https://doi.org/10.1088/1361-6528/aa7c76.

[19] Jain TK, Richey J, Strand M, Leslie-Pelecky DL, Flask CA, Labhasetwar V. Magnetic nanoparticles with dual functional properties: drug delivery and magnetic resonance imaging. Biomaterials 2008;29:4012–21. https://doi.org/10.1016/j.biomaterials.2008.07.004.

[20] Peng L, Huang Y, Zhang C, Niu J, Chen Y, Chu Y, Jiang Z, Gao J, Mao Z. Integration of antimicrobial peptides with gold nanoparticles as unique non-viral vectors for gene delivery to mesenchymal stem cells with antibacterial activity. Biomaterials 2016;103:137–49. https://doi.org/10.1016/j.biomaterials.2016.06.057.

[21] Kong F, Liu F, Li W, Guo X, Wang Z, Zhang H, Li Q, Luo L, Du Y, Jin Y. Smart carbon nanotubes with laser-Controlled behavior in gene delivery and therapy through a non-Digestive trafficking pathway. Small 2016;12:6753–66. https://doi.org/10.1002/smll.201601092.

[22] Jain S, Thanki K, Pandi NK, Kushwah V. Estradiol functionalized multi-walled carbon nanotubes as renovated strategy for efficient gene delivery. RSC Advances 2016;6:10792–801. https://doi.org/10.1039/C5RA26115E.

[23] Chen F, Gao W, Qiu X, Zhang H, Liu L, Liao P, Fu W, Luo Y. Graphene quantum dots in biomedical applications: recent advances and future challenges. Frontiers in Laboratory Medicine 2017;1:192–9. https://doi.org/10.1016/j.flm.2017.12.006.

[24] Xu DD, Deng YL, Li CY, Lin Y, Tang HW. Metal-enhanced fluorescent dye-doped silica nanoparticles and magnetic separation: a sensitive platform for one-step fluorescence detection of prostate specific antigen. Biosensors and Bioelectronics 2017;87:881–7. https://doi.org/10.1016/j.bios.2016.09.034.

[25] Duangkaew P, Wutikhun T, Laocharoensuk R. Triple signal amplification strategy based on size and shape transformation of ultrasmall sub-10 nm gold nanoparticles tag towards sensitivity improvement of electrochemical immunosensors. Sensors and Actuators B: Chemical 2017;239:430–7. https://doi.org/10.1016/j.snb.2016.08.037.

[26] Nobile S, Nobile L. Nanotechnology for biomedical applications: recent advances in neurosciences and bone tissue engineering. Polymer Engineering and Science 2016;2:644–50. https://doi.org/10.1002/pen.24595.

[27] Kumar P. Nano-TiO_2 doped chitosan scaffold for the bone tissue engineering applications. International Journal of Biomaterials 2018;2018:1–7. https://doi.org/10.1155/2018/6576157.

[28] Kuchler-Bopp S, Larrea A, Petry L, Idoux-Gillet Y, Sebastian V, Ferrandon A, Schwinté P, Arruebo M, Benkirane-Jessel N. Promoting bioengineered tooth innervation using nanostructured and hybrid scaffolds. Acta Biomaterialia 2017;50:493–501. https://doi.org/10.1016/j.actbio.2017.01.001.

[29] Xie H, Chua M, Islam I, Bentini R, Cao T, Viana-Gomes JC, Castro Neto AH, Rosa V. CVD-grown monolayer graphene induces osteogenic but not odontoblastic differentiation of dental pulp stem cells. Dental Materials 2017;33:e13–21. https://doi.org/10.1016/j.dental.2016.09.030.

[30] Liu Y, Zhou G, Liu K, Cui Y. Design of complex nanomaterials for energy storage: past success and future opportunity. Accounts of Chemical Research 2017;50:2895–905. https://doi.org/10.1021/acs.accounts.7b00450.

[31] Liu K, Zhuo D, Lee HW, Liu W, Lin D, Lu Y, Cui Y. Extending the life of lithium-based rechargeable batteries by reaction of lithium dendrites with a novel silica nanoparticle sandwiched separator. Advanced Materials 2017;29:1603987–92. https://doi.org/10.1002/adma.201603987.

[32] Jaganathan H, Gage J, Leonard F, Srinivasan S, Souza GR, Dave B, Godin B. Three-dimensional in vitro co-culture model of breast tumor using magnetic levitation. Scientific Reports 2014;4:6468–75. https://doi.org/10.1038/srep06468.

[33] Hu X, Sun A, Kang W, Zhou Q. Strategies and knowledge gaps for improving nanomaterial biocompatibility. Environment International 2017;102:177–89. https://doi.org/10.1016/j.envint.2017.03.001.

[34] Ferrer-Anglada N, Gomis V, El-Hachemi Z, Weglikovska UD, Kaempgen M, Roth S. Carbon nanotube based composites for electronic applications: CNT–conducting polymers, CNT–Cu. Physica Status Solidi A 2006;203:1082–7. https://doi.org/10.1002/pssa.200566188.

[35] Wang J, Musameh M. Carbon nanotube screen-printed electrochemical sensors. The Analyst 2004;129:1. https://doi.org/10.1039/b313431h.

[36] Tonelli FMP, Lacerda SMSN, Paiva NCO, Pacheco FG, Scalzo Junior SRA, de Macedo FHP, Cruz JS, Pinto MCX, Corrêa Junior JD, Ladeira LO, França LR, Guatimosim S, Resende RR. Functionalized nanomaterials: are they effective to perform gene delivery to difficult-to-transfect cells with no cytotoxicity?

Nanoscale 2015;7:18036–43. https://doi.org/10.1039/C5NR04173B.
[37] Mao B, Cheng L, Wang S, Zhou J, Deng L. Combat biofilm by bacteriostatic aptamer-functionalized graphene oxide. Applied Biochemistry and Biotechnology 2018; 65:355–61. https://doi.org/10.1002/bab.1631.
[38] Karimzadeh M, Rashidi L, Ganji F. Mesoporous silica nanoparticles for efficient rivastigmine hydrogen tartrate delivery into SY5Y cells. Drug Development and Industrial Pharmacy 2017;43:628–36. https://doi.org/10.1080/03639045.2016.1275668.
[39] Pedraza D, Díez J, Barba II, Colilla M, Vallet-Regí M. Amine-functionalized mesoporous silica nanoparticles: a new nanoantibiotic for bone infection treatment. Biomedical Glasses 2018;4:25–31. https://doi.org/10.1515/bglass-2018-0001.
[40] Qiu S, Zhou Z, Dong J, Chen G. Preparation of Ni nanoparticles and evaluation of their tribological performance as potential additives in oils. Journal of Tribology 1999;123:441–3. https://doi.org/10.1115/1.1286152.
[41] Li H, Zhang Z, Xu G, Yao P. Fatty acid and quaternary ammonium modified chitosan nanoparticles for insulin delivery. Colloids and Surfaces B: Biointerfaces 2018; 170:136–43. https://doi.org/10.1016/j.colsurfb.2018.05.063.
[42] Hu Q, Luo Y. Recents advances of polysaccharide-based nanoparticles for oral insulin delivery. International Journal of Biological Macromolecules 2018;120:775–82. https://doi.org/10.1016/j.ijbiomac.2018.08.152.
[43] Liang J, Li F, Fang Y, Yang W, Zhao L, Xin Z, Cao L, Hu Q. Synthesis, characterization and cytotoxicity studies of chitosan-coated tea polyphenols nanoparticles. Colloids and Surfaces B: Biointerfaces 2011;82:297–301. https://doi.org/10.1016/j.colsurfb.2010.08.045.
[44] Nallamuthu I, Devi A, Khanum F. Chlorogenic acid loaded chitosan nanoparticles with sustained release property, retained antioxidant activity and enhanced bioavailability. Asian Journal of Pharmaceutical Sciences 2015;10:203–11. https://doi.org/10.1016/j.ajps.2014.09.005.
[45] Panwar R, Raghuwanshi N, Srivastana A, Asvene S, Pruthi V. In-vivo sustained release of nanoencapsulated ferulic acid and its impact in induced diabetes. Materials Science and Engineering: C 2018;92:381–92. https://doi.org/10.1016/j.msec.2018.06.055.
[46] Duncan R, Izzo L. Dendrimer biocompatibility and toxicity. Advanced Drug Delivery Reviews 2005;57:2215–37. https://doi.org/10.1016/j.addr.2005.09.019.
[47] Williams D. A model for biocompatibility and its evaluation. Journal of Biomedical Engineering 1989; 11:185–91. https://doi.org/10.1016/0141-5425(89)90138-6.
[48] Adabi M, Naghibzadeh M, Adabi M, Zarrinfard MA, Esnaashari SS, Seifalian AM, Faridi-Majidi R, Aiyelabegan HT, Ghanbari H. Biocompatibility and nanostructured materials: applications in nanomedicine. Artif. Cells Nanomed. Biotechnol. 2016;45:833–42. https://doi.org/10.1080/21691401.2016.1178134.
[49] Shang L, Nienhaus K, Nienhaus GU. Engineered nanoparticles interacting with cells: size matters. Journal of Nanobiotechnology 2014;12:5–9. https://doi.org/10.1186/1477-3155-12-5.
[50] Mironava T, Hadjiargyrou M, Simon M, Jurukovski V, Rafailovich MH. Gold nanoparticles cellular toxicity and recovery: effect of size, concentration and exposure time. Nanotoxicology 2010;4:120–37. https://doi.org/10.3109/17435390903471463.
[51] Yue J, Feliciano TJ, Li W, Lee A, Odom TW. Gold nanoparticle size and shape effects on cellular uptake and intracellular distribution of siRNA nanoconstructs. Bioconjugate Chemistry 2017;28:1791–800. https://doi.org/10.1021/acs.bioconjchem.7b00252.
[52] Arnida MA, Ghandehari H. Cellular uptake and toxicity of gold nanoparticles in prostate cancer cells: a comparative study of rods and spheres. Journal of Applied Toxicology 2010;30:212–7. https://doi.org/10.1002/jat.1486.
[53] Chithrani BD, Ghazani AA, Chan WC. Determining the size and shape dependence of gold nanoparticle uptake into mammalian cells. Nano Letters 2006;6:662–8. https://doi.org/10.1021/nl052396o.
[54] Wang SH, Lee CW, Chiou A, Wei PK. Size-dependent endocytosis of gold nanoparticles studied by three-dimensional mapping of plasmonic scattering images. Journal of Nanobiotechnology 2010;8:33–9. https://doi.org/10.1186/1477-3155-8-33.
[55] Jiang W, Kim BYS, Rutka JT, Chan WCW. Nanoparticle-mediated cellular response is size-dependent. Nature Nanotechnology 2008;3:145–50. https://doi.org/10.1038/nnano.2008.30.
[56] Huang K, Ma H, Liu J, Huo S, Kumar A, Wei T, Zhang X, Jin S, Gan Y, Wang PC, He S, Zhang X, Liang XJ. Size-dependent localization and penetration of ultrasmall gold nanoparticles in cancer cells, multicellular spheroids, and tumors in vivo. ACS Nano 2012;6:4483–93. https://doi.org/10.1021/nn301282m.
[57] Oh E, Delehanty JB, Sapsford KE, Susumu K, Goswami R, Blanco-Canosa JB, Dawson PE, Granek J, Shoff M, Zhang Q, Goering PL, Huston A, Medintz IL. Cellular uptake and fate of PEGylated gold nanoparticles is dependent on both cell-penetration peptides and particle size. ACS Nano 2011;5:6434–48. https://doi.org/10.1021/nn201624c.
[58] Shan Y, Ma S, Nie L, Shang X, Hao X, Tang Z, Wang H. Size-dependent endocytosis of single gold nanoparticles. Chemical Communications 2011;47:8091–3. https://doi.org/10.1039/c1cc11453k.
[59] Cho EC, Au L, Zhang Q, Xia Y. The effects of size, shape, and surface functional group of gold nanostructures on their adsorption and internalization by cells. Small 2010;6:517–22. https://doi.org/10.1002/smll.200901622.
[60] Chithrani BD, Chan WC. Elucidating the mechanism of cellular uptake and removal of protein-coated gold

nanoparticles of different sizes and shapes. Nano Letters 2007;7:1542–50. https://doi.org/10.1021/nl070363y.

[61] Walkey CD, Olsen JB, Guo H, Emili A, Chan WC. Nanoparticle size and surface chemistry determine serum protein adsorption and macrophage uptake. Journal of the American Chemical Society 2012;134:2139–47. https://doi.org/10.1021/ja2084338.

[62] Wong AC, Wright DW. Size-dependent cellular uptake of DNA functionalized gold nanoparticles. Small 2016;12: 5592–600. https://doi.org/10.1002/smll.201601697.

[63] Veiseh O, Gunn JW, Zhang M. Design and fabrication of magnetic nanoparticles for targeted drug delivery and imaging. Advanced Drug Delivery Reviews 2010;62: 284–304. https://doi.org/10.1016/j.addr.2009.11.002.

[64] Laurent S, Boutry S, Muller RN. Metal oxide particles and their prospects for applications. In: Mahmoudi M, Laurent S, editors. Iron oxide nanoparticles for biomedical applications. Elsevier; 2018. p. 3–42. https://doi.org/10.1016/B978-0-08-101925-2.09986-1. ISBN 9780081019252.

[65] Longmire M, Choyke PL, Kobayashi H. Clearance properties of nano-sized particles and molecules as imaging agents: considerations and caveats. Nanomedicine 2008;3:703–17. https://doi.org/10.2217/17435889.3.5.703.

[66] Bareford LM, Swaan PW. Endocytic mechanisms for targeted drug delivery. Advanced Drug Delivery Reviews 2007;59:748–58. https://doi.org/10.1016/j.addr.2007.06.008.

[67] Choi HS, Liu W, Misra P, Tanaka E, Zimmer JP, Ipe BI, Bawendi MG, Frangioni JV. Renal clearance of quantum dots. Nature Biotechnology 2007;25:1165–70. https://doi.org/10.1038/nbt1340.

[68] Banerjee T, Mitra S, Singh AK, Sharma RK, Maitra A. Preparation, characterization and biodistribution of ultrafine chitosan nanoparticles. International Journal of Pharmaceutics 2002;243:93–105. https://doi.org/10.1016/S0378-5173(02)00267-3.

[69] Moghimi SM. Exploiting bone marrow microvascular structure for drug delivery and future therapies. Advanced Drug Delivery Reviews 1995;17:61–73. https://doi.org/10.1016/0169-409X(95)00041-5.

[70] Moghimi SM. Mechanisms of splenic clearance of blood cells and particles: towards development of new splenotropic agents. Advanced Drug Delivery Reviews 1995;17: 103–15. https://doi.org/10.1016/0169-409X(95)00043-7.

[71] Napierska D, Thomassen LC, Rabolli V, Lison D, Gonzalez L, Kirsch-Volders M, Martens JA, Hoet PH. Size-dependent cytotoxicity of monodisperse silica nanoparticles in human endothelial cells. Small 2009; 5:846–53. https://doi.org/10.1002/smll.200800461.

[72] Elbakry A, Wurster EC, Zaky A, Liebl R, Schindler E, Bauer-Kreisel P, Blunk T, Rachel R, Goepferich A, Breunig M. Layer-by-layer coated gold nanoparticles: size-dependent delivery of DNA into cells. Small 2012; 8:3847–56. https://doi.org/10.1002/smll.201201112.

[73] Couvreur P, Barratt G, Fattal E, Legrand P, Vauthier C. Nanocapsule technology: a review. Critical Reviews In Therapeutic Drug Carrier Systems 2002;19:99–134. https://doi.org/10.1615/CritRevTherDrugCarrierSyst.v19.i2.10.

[74] Singh N, Jenkins GJS, Asadi R, Doak SH. Potential toxicity of superparamagnetic iron oxide nanoparticles (SPION). Nano Reviews 2010;1:5358–96. https://doi.org/10.3402/nano.v1i0.5358.

[75] Owens III DE, Peppas NA. Opsonization, biodistribution, and pharmacokinetics of polymeric nanoparticles. International Journal of Pharmaceutics 2006;307:93–102. https://doi.org/10.1016/j.ijpharm.2005.10.010.

[76] Chen L, McCrate JM, Lee JC, Li H. The role of surface charge on the uptake and biocompatibility of hydroxyapatite nanoparticles with osteoblast cells. Nanotechnology 2011;22:105708–19. https://doi.org/10.1088/09574484/22/10/105708.

[77] Zhu ZJ, Carboni R, Quercio MJ, Yan B, Miranda OR, Anderton DL, Arcaro KF, Rotello VM, Vachet RW. Surface properties dictate uptake, distribution, excretion, and toxicity of nanoparticles in fish. Small 2010;6:2261–5. https://doi.org/10.1002/smll.201000989.

[78] Ran S, Downes A, Thorpe PE. Increased exposure of anionic phospholipids on the surface of tumor blood vessels. Cancer Research 2002;62:6132–40.

[79] Kreuter J. Nanoparticulate systems for brain delivery of drugs. Advanced Drug Delivery Reviews 2001;47: 65–81. https://doi.org/10.1016/S0169-409X(00)00122-8.

[80] Yang R, Han X, Shi K, Cheng G, Shim CK, Cui F. Cationic formulation of paclitaxel-loaded poly D, L-lactic-co-glycolic acid (PLGA) nanoparticles using an emulsion-solvent diffusion method. Asian Journal of Pharmaceutics 2009;4:89–95.

[81] Mahoney BP, Raghunand N, Baggett B, Gillies RJ. Tumor acidity, ion trapping and chemotherapeutics: I. Acid pH affects the distribution of chemotherapeutic agents in vitro. Biochemical Pharmacology 2003;66:1207–18. https://doi.org/10.1016/S0006-2952(03)00467-2.

[82] Cotin G, Piant S, Mertz D, Felder-Flesch D, Begin-Colin S. Iron oxide nanoparticles for biomedical applications: synthesis, functionalization, and application. In: Mahmoudi M, Laurent S, editors. Iron oxide nanoparticles for biomedical applications. Elsevier; 2018. p. 43–88. https://doi.org/10.1016/B978-0-08-101925-2.00002-4. ISBN 9780081019252.

[83] El Badawy AM, Silva RG, Morris B, Scheckel KG, Suidan MT, Tolaymat TM. Surface charge-dependent toxicity of silver nanoparticles. Environmental Science & Technology 2010;45:283–7. https://doi.org/10.1021/es1034188.

[84] Schaeublin NM, Braydich-Stolle LK, Schrand AM, Miller JM, Hutchison J, Schlager JJ, Hussain SM. Surface charge of gold nanoparticles mediates mechanism of toxicity. Nanoscale 2011;3:410–20. https://doi.org/10.1039/c0nr00478b.

[85] Liu X, Huang N, Li H, Jin Q, Ji J. Surface and size effects on cell interaction of gold nanoparticles with both

phagocytic and nonphagocytic cells. Langmuir 2013;29: 9138–48. https://doi.org/10.1021/la401556k.

[86] Dong C, Irudayaraj J. Hydrodynamic size-dependent cellular uptake of aqueous QDs probed by fluorescence correlation spectroscopy. The Journal of Physical Chemistry B 2012;116:12125–32. https://doi.org/10.1021/jp305563p.

[87] Thoo L, Fahmi MZ, Zulkipli IN, Keasberry N, Idris A. Interaction and cellular uptake of surface-modified carbon dot nanoparticles by J774.1 macrophages. Central-European Journal of Immunology 2017;42: 324–30. https://doi.org/10.5114/ceji.2017.70978.

[88] ElZorkany HE, El-Shorbagy HM, Farroh KY, Youssef T, Sabet S, Salaheldin TA. Monitoring the cellular uptake of silica-coated CdSe/ZnS quantum dots by time lapse confocal laser scanning microscopy. Journal of Applied Pharmaceutical Science 2018;8:1–8. https://doi.org/10.7324/JAPS.2018.8301.

[89] Zhang LW, Monteiro-Riviere NA. Mechanisms of quantum dot nanoparticle cellular uptake. Toxicology Sciences 2009;110:138–55. https://doi.org/10.1093/toxsci/kfp087.

[90] Zhang LW, Bäumer W, Monteiro-Riviere NA. Cellular uptake mechanisms and toxicity of quantum dots in dendritic cells. Nanomedicine 2011;6:777–91. https://doi.org/10.2217/nnm.11.73.

[91] Fasbender S, Allani S, Wimmenauer C, Cadeddu RP, Raba K, Fischer JC, Bulat B, Luysberg M, Seidel CAM, Heinzel T, Haas R. Uptake dynamics of graphene quantum dots into primary human blood cells following in vitro exposure. RSC Advances 2017;7:12208–16. https://doi.org/10.1039/c6ra27829a.

[92] Andersson PO, Lejon C, Ekstrand-Hammarstrom B, Akfur C, Ahlinder L, Bucht A, Osterlund L. Polymorph- and size-dependent uptake and toxicity of TiO_2 nanoparticles in living lung epithelial cells. Small 2011;7: 514–23. https://doi.org/10.1002/smll.201001832.

[93] Hu CC, Wu GH, Hua TE, Wagner OI, Yen TJ. Uptake of TiO_2 nanoparticles into *C. elegans* neurons negatively affects axonal growth and worm locomotion behavior. ACS Applied Materials and Interfaces 2018;10: 8485–95. https://doi.org/10.1021/acsami.7b18818.

[94] Huang J, Bu L, Xie J, Chen K, Cheng Z, Li X, Chen X. Effects of nanoparticle size on cellular uptake and liver MRI with polyvinylpyrrolidone-coated iron oxide nanoparticles. ACS Nano 2010;4:7151–60. https://doi.org/10.1021/nn101643u.

[95] Feng Q, Liu Y, Huang J, Chen K, Huang J, Xiao K. Uptake, distribution, clearance, and toxicity of iron oxide nanoparticles with different sizes and coatings. Scientific Reports 2018;8:2082–9. https://doi.org/10.1038/s41598-018-19628-z.

[96] Varela JA, Bexiga MG, Åberg C, Simpson JC, Dawson KA. Quantifying size-dependent interactions between fluorescently labeled polystyrene nanoparticles and mammalian cells. Journal of Nanobiotechnology 2012;10:39–47. https://doi.org/10.1186/1477-3155-10-39.

[97] Santos T, Varela J, Lynch I, Salvati A, Dawson KA. Quantitative assessment of the comparative nanoparticle-uptake efficiency of a range of cell lines. Small 2011;7: 3341–9. https://doi.org/10.1002/smll.201101076.

[98] Lunov O, Syrovets T, Loos C, Beil J, Delacher M, Tron K, Nienhaus GU, Musyanovych A, Mailänder V, Landfester K, Simmet T. Differential uptake of functionalized polystyrene nanoparticles by human macrophages and a monocytic cell line. ACS Nano 2011;5: 1657–69. https://doi.org/10.1021/nn2000756.

[99] Pitt JA, Kozal JS, Jayasundara N, Massarsky A, Trevisan R, Geitner N, Wiesner M, Levin ED, Di Giulio RT. Uptake, tissue distribution, and toxicity of polystyrene nanoparticles in developing zebrafish (*Danio rerio*). Aquatic Toxicology 2018;194:185–94. https://doi.org/10.1016/j.aquatox.2017.11.017.

[100] Pan D, Vargas-Morales O, Zern B, Anselmo AC, Gupta V, Zakrewsky M, Mitragotri S, Muzykantov V. The effect of polymeric nanoparticles on biocompatibility of carrier red blood cells. PLoS One 2016;11:e0152074. https://doi.org/10.1371/journal.pone.0152074.

[101] Xu A, Yao M, Xu G, Ying J, Ma W, Li B, Jin Y. A physical model for the size-dependent cellular uptake of nanoparticles modified with cationic surfactants. International Journal of Nanomedicine 2012;7:3547–54. https://doi.org/10.2147/IJN.S32188.

[102] He C, Hu Y, Yin L, Tang C, Yin C. Effects of particle size and surface charge on cellular uptake and biodistribution of polymeric nanoparticles. Biomaterials 2010;31: 3657–66. https://doi.org/10.1016/j.biomaterials.2010.01.065.

[103] Łukasiewicz S, Szczepanowicz K, Błasiak E, Dziedzicka-Wasylewska M. Biocompatible polymeric nanoparticles as promising candidates for drug delivery. Langmuir 2015;31:6415–25. https://doi.org/10.1021/acs.langmuir.5b01226.

[104] Cardoso PB, Machado TO, Feuser PE, Claudia Sayer, Meier MAR, Araújo PHH. Biocompatible polymeric nanoparticles from Castor oil derivatives via Thiol-Ene miniemulsion polymerization. European Journal of Lipid Science and Technology 2016;120. https://doi.org/10.1002/ejlt.201700212.

[105] Schübbe S, Schumann C, Cavelius C, Koch M, Müller T, Kraegeloh A. Size-dependent localization and quantitative evaluation of the intracellular migration of silica nanoparticles in Caco-2 cells. Chemistry of Materials 2012;24:914–23. https://doi.org/10.1021/cm2018532.

[106] Barandeh F, Nguyen PL, Kumar R, Iacobucci GJ, Kuznicki ML, Kosterman A, Bergey EJ, Prasad PN, Gunawardena S. Organically modified silica nanoparticles are biocompatible and can be targeted to neurons in vivo. PLoS One 2012;7:e29424. https://doi.org/10.1371/journal.pone.0029424.

[107] Vallhov H, Gabrielsson S, Strømme M, Scheynius A, Garcia-Bennett AE. Mesoporous silica particles induce size dependent effects on human dendritic cells. Nano Letters 2007;7:3576–82. https://doi.org/10.1021/nl0714785.

[108] Lu F, Wu SH, Hung Y, Mou CY. Size effect on cell uptake in well-suspended, uniform mesoporous silica nanoparticles. Small 2009;5:1408–13. https://doi.org/10.1002/smll.200900005.

[109] Lin YS, Haynes CL. Impacts of mesoporous silica nanoparticle size, pore ordering, and pore integrity on hemolytic activity. Journal of the American Chemical Society 2010;132:4834–42. https://doi.org/10.1021/ja910846q.

[110] Shi Y, Miller ML, Di Pasqua AJ. Biocompatibility of mesoporous silica nanoparticles? Comments on Inorganic Chemistry 2015;36:61–80. https://doi.org/10.1080/02603594.2015.1088439.

[111] Trewyn BG, Nieweg JA, Zhao Y, Lin VSY. Biocompatible mesoporous silica nanoparticles with different morphologies for animal cell membrane penetration. Chemical Engineering Journal 2008;137:23–9. https://doi.org/10.1016/j.cej.2007.09.045.

[112] Zhang Y, Hu L, Yu D, Gao C. Influence of silica particle internalization on adhesion and migration of human dermal fibroblasts. Biomaterials 2010;31:8465–74. https://doi.org/10.1016/j.biomaterials.2010.07.060.

[113] Corbalan JJ, Medina C, Jacoby A, Malinski T, Radomski MW. Amorphous silica nanoparticles trigger nitric oxide/peroxynitrite imbalance in human endothelial cells: inflammatory and cytotoxic effects. International Journal of Nanomedicine 2011;6:2821–35. https://doi.org/10.2147/IJN.S25071.

[114] Corbalan JJ, Medina C, Jacoby A, Malinski T, Radomski MW. Amorphous silica nanoparticles aggregate human platelets: potential implications for vascular homeostasis. International Journal of Nanomedicine 2012;7:631–9. https://doi.org/10.2147/IJN.S28293.

[115] He Q, Zhang Z, Gao F, Li Y, Shi J. In vivo biodistribution and urinary excretion of mesoporous silica nanoparticles: effects of particle size and PEGylation. Small 2011;7:271–80. https://doi.org/10.1002/smll.201001459.

[116] Song H, Ma X, Xiong F, Hong H, Li C, Li L, Wu S, Zhan X, Zhang J, Hu J. Preparation and evaluation of insulin-loaded nanoparticles based on hydroxypropyl-β-cyclodextrin modified carboxymethyl chitosan for oral delivery. Journal of Wuhan University of Technology-Mater. Sci. Ed. 2016;31:1394–400. https://doi.org/10.1007/s11595-016-1544-z.

[117] Fan W, Xia D, Zhu Q, Li X, He S, Zhu C, Guo S, Hovgaard L, Yang M, Gan Y. Functional nanoparticles exploit the bile acid pathway to overcome multiple barriers of the intestinal epithelium for oral insulin delivery. Biomaterials 2018;151:13–23. https://doi.org/10.1016/j.biomaterials.2017.10.022.

[118] Mukhopadhyay P, Sarkar K, Chakraborty M, Bhattacharya S, Mishra R, Kundu PP. Oral insulin delivery by self-assembled chitosan nanoparticles: *In vitro* and *in vivo* studies in diabetic animal model. Materials Science and Engineering: C 2013;33:376–82. https://doi.org/10.1016/j.msec.2012.09.001.

[119] Sæther HV, Holme HK, Maurstad G, Smidsrød O, Stokke BT. Polyelectrolyte complex formation using alginate and chitosan. Carbohydrate Polymers 2008;74:813–21. https://doi.org/10.1016/j.carbpol.2008.04.048.

[120] Goswami S, Bajpai J, Bajpai AK. Calcium alginate nanocarriers as possible vehicle for oral delivery of insulin. Journal of Experimental Nanoscience 2014;9:337–56. https://doi.org/10.1080/17458080.2012.661472.

[121] Bhattacharyya A, Mukherjee D, Mishra R, Kundu PP. Development of pH sensitive polyurethane–alginate nanoparticles for safe and efficient oral insulin delivery in animal models. RSC Advances 2016;48:41835–46. https://doi.org/10.1039/C6RA06749B.

[122] Zhang Y, Xu Y, Xi X, Shrestha S, Jiang P, Zhang W, Gao C. Amino acid-modified chitosan nanoparticles for Cu^{2+} chelation to suppress CuO nanoparticle cytotoxicity. Journal of Materials Chemistry B 2017;5:3521–30. https://doi.org/10.1039/C7TB00344G.

[123] Dobrovolskaia MA, McNeil SE. Immunological properties of engineered nanomaterials. Nature Nanotechnology 2007;2:469–78. https://doi.org/10.1038/nnano.2007.223.

[124] Ong KJ, MacCormack TJ, Clark RJ, Ede JD, Ortega VA, Felix LC, Dang MKM, Ma G, Fenniri H, Veinot JGC, Goss GG. Widespread nanoparticle-assay interference: implications for nanotoxicity testing. PLoS One 2014;9:e90650–61. https://doi.org/10.1371/journal.pone.0090650.

[125] Casey A, Herzog E, Davoren M, Lyng FM, Byrne HJ, Chambers G. Spectroscopic analysis confirms the interactions between single walled carbon nanotubes and various dyes commonly used to assess cytotoxicity. Carbon 2007;45:1425–32. https://doi.org/10.1016/j.carbon.2007.03.033.

[126] Lee YH, Chuang SM, Huang SC, Tan X, Liang RY, Yang GCC, Chueh PJ. Biocompatibility assessment of nanomaterials for environmental safety screening. Environmental Toxicology 2016;32:1170–82. https://doi.org/10.1002/tox.22313.

[127] Doty RC, Tshikhudo TR, Brust M, Fernig DG. Extremely stable water-soluble Ag nanoparticles. Chemistry Materials 2005;17:4630–7. https://doi.org/10.1021/cm0508017.

[128] Grancharov SG, Zeng H, Sun S, Wang SX, O'Brien S, Murray CB, Kirtley JR, Held GA. Bio-functionalization of monodisperse magnetic nanoparticles and their use as biomolecular labels in a magnetic tunnel junction based sensor. The Journal of Physical Chemistry B 2005;7:13030–5. https://doi.org/10.1021/jp051098c.

[129] Asefa T, Tao Z. Biocompatibility of mesoporous silica nanoparticles. Chemical Research in Toxicology 2012;25:2265–84. https://doi.org/10.1021/tx300166u.

[130] Dai L. Functionalization of graphene for efficient energy conversion and storage. Accounts of Chemical Research 2012;46:31–42. https://doi.org/10.1021/ar300122m.

[131] Guazzo R, Gardin C, Bellin G, Sbricoli L, Ferroni L, Ludovichetti FS, Piattelli A, Antoniac I, Bressan E, Zavan B. Graphene-based nanomaterials for tissue engineering in the dental field. Nanomaterials 2018;8:349–58. https://doi.org/10.3390/nano8050349.

[132] Slowing I, Trewyn BG, Lin VSY. Effect of surface functionalization of MCM-41-type mesoporous silica nanoparticles on the endocytosis by human cancer cells. Journal of the American Chemical Society 2006;128: 14792–3. https://doi.org/10.1021/ja0645943.

[133] Sperling RA, Parak WJ. Surface modification, functionalization and bioconjugation of colloidal inorganic nanoparticles. Philosophical Transactions of the Royal Society A 2010;368(1915):1333–83. https://doi.org/10.1098/rsta.2009.0273.

[134] Yang M, Flavin K, Kopf I, Radics G, Hearnden CHA, McManus GJ, Moran B, Villalta-Cerdas A, Echegoyen LA, Giordani S, Lavelle EC. Functionalization of carbon nanoparticles Modulates inflammatory cell recruitment and NLRP3 inflammasome activation. Small 2013;9:4194–206. https://doi.org/10.1002/smll.201300481.

[135] Derfus AM, Chen AA, Min DH, Ruoslahti E, Bhatia SN. Targeted quantum dot conjugates for siRNA delivery. Bioconjugate Chemistry 2007;18:1391–6. https://doi.org/10.1021/bc060367e.

[136] Datt A, Ndiege N, Larsen SC. Development of porous nanomaterials for applications in drug delivery and imaging. Nanomaterials for Biomedicine 2012: 239–58. https://doi.org/10.1021/bk-2012-1119.ch011.

[137] Gooding JJ, Mearns F, Yang W, Liu J. Self-assembled monolayers into the 21st century: recent advances and applications. Electroanalysis 2003;15:81–96. https://doi.org/10.1002/elan.200390017.

[138] Dong TY, Wu HH, Lin MC. Superlattice of octanethiol-protected copper nanoparticles. Langmuir 2006;22: 6754–6. https://doi.org/10.1021/la053438r.

[139] Xia QQ, Li L, Zhao L. Silica nanoparticle-based dual-responsive nanoprodrug system for liver cancer therapy. Experimental and Therapeutic Medicine 2017; 14:2071–7. https://doi.org/10.3892/etm.2017.4768.

[140] Spampinato V, Parracino MA, La Spina R, Rossi F, Ceccone G. Surface analysis of gold nanoparticles functionalized with thiol-modified glucose SAMs for biosensor applications. Frontiers in Chemistry 2016;4: 8–19. https://doi.org/10.3389/fchem.2016.00008.

[141] Neouze MA, Schubert U. Surface modification and functionalization of metal and metal oxide nanoparticles by organic ligands. Monatshefte Für Chemie - Chemical Monthly 2008;139:183–95. https://doi.org/10.1007/s00706-007-0775-2.

[142] Sahoo B, Dutta S, Dhara D. Amine-functionalized magnetic nanoparticles as robust support for immobilization of Lipase. Journal of Chemical Sciences 2016;128: 1131–40. https://doi.org/10.1007/s12039-016-1115-z.

[143] Chen W, Shen H, Li X, Jia N, Xu J. Synthesis of immunomagnetic nanoparticles and their application in the separation and purification of $CD34^+$ hematopoietic stem cells. Applied Surface Science 2006;253:1762–9. https://doi.org/10.1016/j.apsusc.2006.03.012.

[144] Min C, Zhang Q, Shen C, Liu D, Shen X, Song H, Li S, Xu D, Lin X, Zhang K. Graphene oxide/carboxyl-functionalized multiwalled carbon nanotube hybrids: powerful additives for water-based lubrication. RSC Advances 2017;7:32574–80. https://doi.org/10.1039/c7ra04730d.

[145] Jang MH, Hwang YS. Effects of functionalized multi-walled carbon nanotubes on toxicity and bio-accumulation of lead in Daphnia magna. PLoS One 2018;13:e0194935–42. https://doi.org/10.1371/journal.pone.0194935.

[146] Babič M, Horák D, Lukash LL, Ruban TA, Kolomiets YN, Shpylova SP, Grypych OA. Influence of surface-modified maghemite nanoparticles on in vitro survival of human stem cells. Beilstein Journal of Nanotechnology 2014;5: 1732–7. https://doi.org/10.3762/bjnano.5.183. eCollection 2014.

[147] Berry CC, Charles S, Wells S, Dalby MJ, Curtis AS. The influence of transferrin stabilised magnetic nanoparticles on human dermal fibroblasts in culture. International Journal of Pharmaceutics 2004;269:211–25. https://doi.org/10.1016/j.ijpharm.2003.09.042.

[148] Berry CC, Wells S, Charles S, Curtis ASG. Dextran and albumin derivatised iron oxide nanoparticles: influence on fibroblasts in vitro. Biomaterials 2003;24:4551–7. https://doi.org/10.1016/S0142-9612(03)00237-0.

[149] Auffan M, Decome L, Rose J, Orsiere T, De Meo M, Briois V, Chaneac C, Olivi L, Berge-lefranc J, Botta A, Wiesner MR, Bottero J. In vitro interactions between DMSA-coated maghemite nanoparticles and human fibroblasts: a physicochemical and cyto-genotoxical study. Environmental Science & Technology 2016;40: 4367–73. https://doi.org/10.1021/es060691k.

[150] Kellar KE, Fujii DK, Gunther WH, Briley-Saebø K, Bjørnerud A, Spiller M, Koenig SH. NC100150 injection, a preparation of optimized iron oxide nanoparticles for positive-contrast MR angiography. Journal of Magnetic Resonance Imaging 2000;11:488–94. https://doi.org/10.1002/(SICI)1522-2586(200005)11:5<488. AID-JMRI4>3.0.CO;2-V.

[151] Yue-Jian C, Juan T, Fei X, Jia-Bi Z, Ning G, Yi-Hua Z, Ye D, Liang G. Synthesis, self-assembly, and characterization of PEG-coated iron oxide nanoparticles as potential MRI contrast agent. Drug Development and Industrial Pharmacy 2010;36:1235–44. https://doi.org/10.3109/03639041003710151.

[152] Lacava LM, Lacava ZG, Da Silva MF, Silva O, Chaves SB, Azevedo RB, Pelegrini F, Gansau C, Buske N, Sabolovic D, Morais PC. Magnetic resonance of a dextran-coated magnetic fluid intravenously administered in mice. Biophysical Journal 2001;80:2483–6. https://doi.org/10.1016/S0006-3495(01)76217-0.

[153] Valois CR, Braz JM, Nunes ES, Vinolo MA, Lima EC, Curi R, Kuebler WM, Azevedo RB. The effect of DMSA-functionalized magnetic nanoparticles on transendothelial migration of monocytes in the murine lung via a β2 integrin-dependent pathway. Biomaterials 2010;31: 366–74. https://doi.org/10.1016/j.biomaterials.2009.09.053.

[154] Morais PC, Santos RL, Pimenta ACM, Azevedo RB, Lima ECD. Preparation and characterization of ultra-

stable biocompatible magnetic fluids using citrate-coated cobalt ferrite nanoparticles. Thin Solid Films 2006;515:266–70. https://doi.org/10.1016/j.tsf.2005.12.079.

[155] Ayala V, Herrera AP, Latorre-Esteves M, Torres-Lugo M, Rinaldi C. Effect of surface charge on the colloidal stability and in vitro uptake of carboxymethyl dextran-coated iron oxide nanoparticles. Journal of Nanoparticle Research 2013;15:1874–82. https://doi.org/10.1007/s11051-013-1874-0.

[156] Boca SC, Potara M, Gabudean AM, Juhem A, Baldeck PL, Astilean S. Chitosan-coated triangular silver nanoparticles as a novel class of biocompatible, highly effective photothermal transducers for in vitro cancer cell therapy. Cancer Letters 2011;311:131–40. https://doi.org/10.1016/j.canlet.2011.06.022.

[157] Cisterna BA, Kamaly N, Choi WI, Tavakkoli A, Farokhzad OC, Vilos C. Targeted nanoparticles for colorectal cancer. Nanomedicine 2016;11:2443–56. https://doi.org/10.2217/nnm-2016-0194.

[158] Villanueva A, Cañete M, Roca AG, Calero M, Veintemillas-Verdaguer S, Serna CJ, Morales Mdel P, Miranda R. The influence of surface functionalization on the enhanced internalization of magnetic nanoparticles in cancer cells. Nanotechnology 2009;20:115103. https://doi.org/10.1088/0957-4484/20/11/115103.

[159] Wu Y, Sefah K, Liu H, Wang R, Tan W. DNA aptamer–micelle as an efficient detection/delivery vehicle toward cancer cells. Proceedings of the National Academy of Sciences of the United States of America 2010;107:5–10. https://doi.org/10.1073/pnas.0909611107.

[160] Hwang DW, Ko HY, Lee JH, Kang H, Ryu SH, Song IC, Lee DS, Kim S. A nucleolin-targeted multimodal nanoparticle imaging probe for tracking cancer cells using an aptamer. The Journal of Nuclear Medicine 2010;51:98–105. https://doi.org/10.2967/jnumed.109.069880.

[161] Chen LQ, Xiao SJ, Peng L, Wu T, Ling J, Li YF, Huang CZ. Aptamer-based silver nanoparticles used for intracellular protein imaging and single nanoparticle spectral analysis. The Journal of Physical Chemistry B 2010;114:3655–9. https://doi.org/10.1021/jp9104618.

[162] Adams GP, Weiner LM. Monoclonal antibody therapy of cancer. Nature Biotechnology 2005;23:1147–57. https://doi.org/10.1038/nbt1137.

[163] Pan X, Wu G, Yang W, Barth RF, Tjarks W, Lee RJ. Synthesis of cetuximab-immunoliposomes via a cholesterol-based membrane anchor for targeting of EGFR. Bioconjugate Chemistry 2007;18:101–8. https://doi.org/10.1021/bc060174r.

[164] Yang J, Lee CH, Ko HJ, Suh JS, Yoon HG, Lee K, Huh YM, Haam S. Multifunctional magneto-polymeric nanohybrids for targeted detection and synergistic therapeutic effects on breast cancer. Angewandte Chemie International Edition in English 2007;46:8836–9. https://doi.org/10.1002/anie.200703554.

[165] Zhang N, Chittasupho C, Duangrat C, Siahaan TJ, Berkland C. PLGA nanoparticle– peptide conjugate effectively targets intercellular cell-adhesion molecule-1. Bioconjugate Chemistry 2008;19:145–52. https://doi.org/10.1021/bc700227z.

[166] Antal I, Koneracka M, Kubovcikova M, Zavisova V, Khmara I, Lucanska D, Jelenska L, Vidlickova I, Zatovicova M, Pastorekova S, Bugarova N, Micusik M, Omastova M, Kopcansky P. D,L-lysine functionalized Fe_3O_4 nanoparticles for detection of cancer cells. Colloids and Surfaces B: Biointerfaces 2018;163:236–45. https://doi.org/10.1016/j.colsurfb.2017.12.022.

[167] Rehana D, Haleel AK, Rahiman AK. Hydroxy, carboxylic and amino acid functionalized superparamagnetic iron oxide nanoparticles: synthesis, characterization and in vitro anti-cancer studies. Journal of Chemical Sciences 2015;127:1155–66. https://doi.org/10.1007/s12039-015-0876-0.

[168] Gonçalves AG, Jarrais B, Pereira C, Morgado J, Freire C, Pereira MFR. Functionalization of textiles with multi-walled carbon nanotubes by a novel dyeing- like process. Journal of Materials Science 2012;47:5263–75. https://doi.org/10.1007/s10853-012-6412-4.

[169] Bobo D, Robinson KJ, Islam J, Thurecht KJ, Corrie SR. Nanoparticle-based Medicines: a review of FDA-approved materials and clinical trials to date. Pharmaceutical Research 2016;33:2373–87. https://doi.org/10.1007/s11095-016-1958-5.

CHAPTER 6

Noninvasive/Minimally Invasive Nanodiagnostics

KANWAL AKHTAR, MS • YASIR JAVED, PHD • NAVEED AKHTAR SHAD, PHD

1 INTRODUCTION

In the recent past, the field of nanomedicine has seen tremendous growth in medical technologies. Fields of controlled drug release, targeted drug delivery, tissue engineering, and in vitro diagnostics have been progressed widely. At the nanoscale, several benefits have been achieved such as enhanced sensitivity, controlled distribution, and multifunctionality; development in this field totally changes the landscape of nanomedicine [1]. Nanotechnology provides immense utility in diagnostic imaging. This usually refers to a wide slew of technologies, which are used to examine the inside human body for the diagnosis of different pathologies. These techniques provide a diagnostic portfolio to physicians. Different techniques such as positron emission tomography, magnetic resonance imaging, and computed tomography are being used as routine procedures that provide unparalleled information about the body for diagnostic purposes [2,3].

To visualize the abnormalities through diagnostic imaging, different materials as contrast agents are employed, which show interactions with various incident radiations to produce clear visible changes on obtained images. Contrast agents greatly enhance the sensitivity of imaging techniques that allow to detect many undetectable pathologies [4,5]. The development of improved and novel contrast agents based on nanomaterials is providing a major impact in the diagnostic imaging field. Manipulations at the nanoscale level provide the scientists to enhance the biocompatibility, biodistribution, and sensitivity profile of many contrast materials (Fig. 6.1). Contrast agents used at the nanoscale level opened new domains in imaging technologies along with their applications in fluorescence imaging [6,7].

In this chapter, we will briefly discuss the synthesis methods, pharmacokinetics, and various diagnostic technologies using nanomaterials as contrast agents and fluorescence. We will discuss briefly all the possible challenges that need to be addressed for the development of nanomaterials for their promising use in nanomedicine with all possible biological interactions.

2 INSIGHTS OF NANODIAGNOSTICS

2.1 Biological Interactions

To address various challenges and design considerations for the preparation of nanoparticle (NP)-based contrast agents, the realization of key mechanisms responsible for the interactions of these NPs with the human body is necessary. When NP-based contrast agents are introduced into the human body, the immune system takes these injected NPs as foreign substances and activates different clearance mechanisms [9]. The body tries to remove the exogenous species from blood circulation quickly by using different clearance routes through effective uptake by different types of cells and components of the mononuclear phagocyte system (MPS) present in lymph nodes, spleen, and liver [10,11]. Systems based on modified NPs offer prolonged circulation time by retaining their technologically relevant properties and have a tendency of elimination from the body after performing their clinical goals. The primary factors involved in the interaction of NPs with the biological systems are their surface properties and size [12].

2.2 Clearance Routes

In clearance of NPs from the body, especially, from the complement system and MPS elements, the immune system plays the key defense system. In the presence of some foreign substances, the complement system, which is the family of proteins, is usually activated. They initiated a cascade that leads toward opsonization and protein adsorption at the NPs surface, and that marks for the clearance through the uptake of different

Intelligent Nanomaterials for Drug Delivery Applications. https://doi.org/10.1016/B978-0-12-817830-0.00006-0

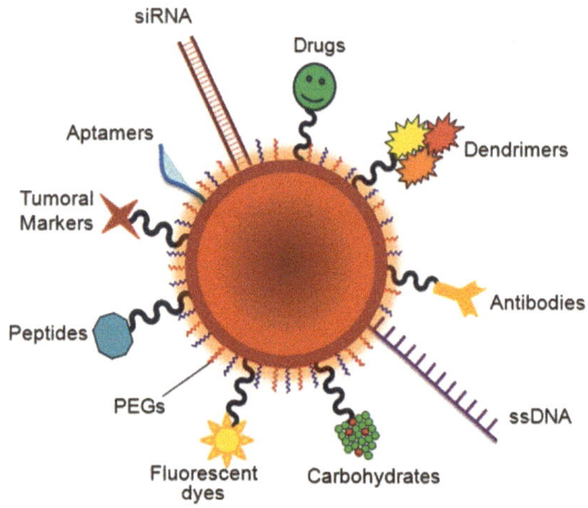

FIG. 6.1 Schematic representation of multifunctional nanodiagnostics. (Adopted from Conde J, et al. Revisiting 30 years of biofunctionalization and surface chemistry of inorganic nanoparticles for nanomedicine. Frontiers in Chemistry 2014;2(48).)

MPS cells [10,11]. To prevent opsonization of antibodies and complement proteins, NPs are desired to maintain a surface to increase the circulation time. In the recent past, it was reported that highly charged and hydrophobic surfaces possessed more opsonization tendency whereas neutral and hydrophilic surfaces were not and act stealthily. The particle size also plays a major role in the clearance process. Rapid clearance was reported for smaller particles than larger ones. Particles having a hydrodynamic size smaller than 5.5 nm possessed the rapid clearance tendency through kidneys [12,13]. To achieve long circulation time, the optimization of signal strength is usually achieved using larger particles. The particle size has a major impact on circulation and thus on tumors through different types of tissues. NPs must be removed by degradation processes from the body after their useful lifetime to prevent the body from the long-term toxicity [14,15]. Particle size usually depends on synthesis protocols used to prepare the nanomaterials, and it cannot be reduced afterward. However, for the modification of surface properties to decrease the opsonization process, different strategies can be employed [16]. Clearance from the body can depend on many parameters, and researchers are investing much time to elaborate on the clearance mechanisms. The biodistribution of carbon-based dots is shown in Fig. 6.2, where the effect of different administration routes on clearance was evaluated. The majority of the species eliminate from the mice's body in 24 h. There seemed to be no evident effect due to different injection routes [17].

2.3 Coating Materials

As discussed briefly in the previous section, the surface properties of NPs can be highly dictated by their biological interactions with the host cell. The different types of nanomaterials including quantum dots, silica, iron oxide, and some polymers possessed promising surface abilities such as hydrophobicity and high charge. The negative charge on silica NPs is because of its lower isoelectric point (pH = 2) [18,19]. In the biological environment, it experiences the rapid adsorption of proteins having positive charges that result in fast clearance. To prevent NPs from the aggregation, coating agents are required for stabilization. Under the physiological environment, iron oxide NPs are not considered colloidally stable. Hence, to increase the colloidal stability, coatings with specific stabilizing agents are required. To transmit colloidal stability, biocompatibility, and stealth properties to the NPs, different types of coating materials are used. One of the most commonly used class of coating compounds are natural and synthetic hydrophilic polymers including poly(Llysine) [20], poly(vinyl alcohol) [21–23], dextran [24–29], poly(ethylene glycol) [19,30], alginate [31–33], gum arabic [34], starch [35–37], and chitosan [34,38]. To create hydrophilic and neutral surface against protein adsorption, these materials are used for coatings. However, the major drawback is an increase in overall hydrodynamic size. For effective coating surfaces, small molecules such as zwitterionic and neutral molecules on silica and thiol consisting molecules for the quantum dots are promisingly used [18]. The major advantage of such coating molecules is that they slightly increase the hydrodynamic size of the particles; however, the disadvantage includes less-stable coatings and less-effective stabilization. One of the major selection criteria of such coating agents is to consider the functional groups present at the surface of NPs because these are important for conjugation mechanisms with other molecules including targeting ligands and external drug payloads [39].

2.4 Targeting Abilities

Identification of specific tissues and selective accumulation are the most important issues that nanotechnology is currently facing. Toxic chemotherapeutic agents at higher doses through selective targeting are delivered with minimum side effects associated with chemotherapy. Contrast agents through specific targeting

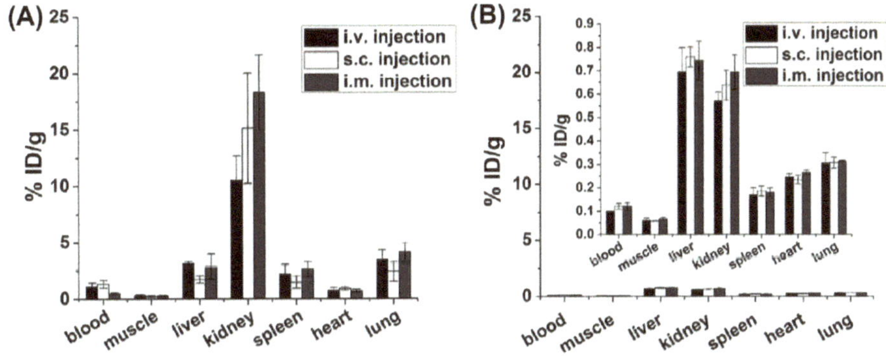

FIG. 6.2 Biodistribution studies of ^{64}Cu-Carbon dots injected through three routes (intravenous (i.v.), subcutaneous (s.c.), and intramuscular (i.m.)) after 1 h **(A)** and 24 h **(B)**. The majority of NPs accumulated in the kidney, and after 24 h, clearance showed no real dependence on the injection route. (Reproduced with permission from Huang X, et al. Effect of injection routes on the biodistribution, clearance, and tumor uptake of carbon dots. ACS Nano 2013;7(7):5684–5693, Copyright (2013) American Chemical Society.)

provide more detailed and useful information about different tissues and help to identify the abnormal and malignant tissues much earlier as compared to the systematic contrast agents [40,41]. This specific targeting makes it possible to lower the given dose, and thus reduce potential exposure of toxic materials and save expenses. Furthermore, to enhance the signal-to-noise ratio, there should be less contrast agents floating freely inside the body [42,43].

Targeting methods usually categories into two major classes: active targeting and passive targeting (Fig. 6.3). To provide selective accumulation, passive targeting depends on different types of physiological phenomena and abnormalities that evolved during diseased conditions. In solid tumors, this type of targeting is referred as enhanced retention effect and permeability [44,45]. In tumors, selective entrance and accumulation of macromolecules depend on the combination formed between poor lymphatic drainage system and permeable blood vessels. Another promising example of passive targeting involves the enhanced contrast through Feridex (commercially used liver contrast agent), which is not taken up by the cancerous tissues but by the spleen and healthy liver tissues [46–49]. In active targeting, targeting ligands recognize and show binding for specific biomarkers through conjugation to the nanomaterials surface. Biomarkers are recognized either as exclusively or in higher numbers on specified tissues with respect to normal tissues. In addition to the passive effects of native targeting, some extra layers of specificities are added to initiate the receptor-mediated endocytosis process [50,51]. At the cellular receptor surface, specific bindings of these materials on the surface of

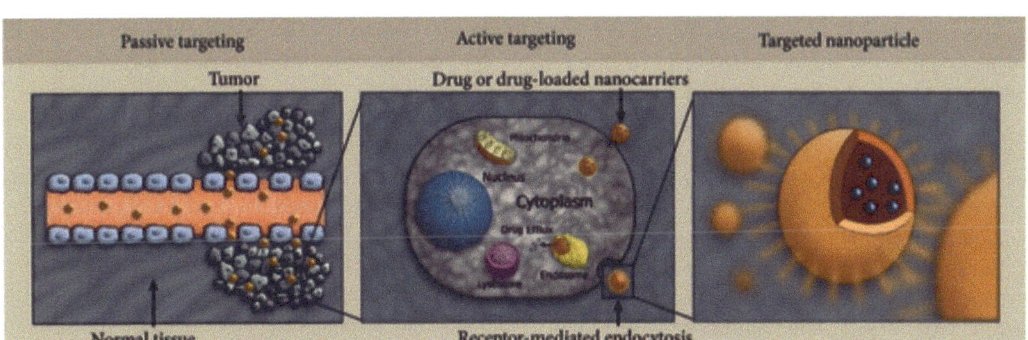

FIG. 6.3 Schematic representation of active and passive targeting. (Reproduced from Sutradhar KB, Amin ML. Nanotechnology in cancer drug delivery and selective targeting. ISRN Nanotechnology 2014;2014:12.)

cellular receptors are required. This process results in intracellular accumulation and internalization of the contrast agents in the body. Active targeting is explored for imaging of a large number of conditions such as inflamed endothelial cells, myocardial infections, and cancer [52,53].

3 SYNTHESIS PROTOCOLS

Nanomaterials can be prepared through different methods that are usually classified into a top-down approach and a bottom-up approach. The bottom-up approach is a constructive method in which NPs are made up of atoms/molecules [55]. This includes chemical vapor deposition (CVD) [56], sol–gel [57], pyrolysis [58], spinning [59], and biosynthesis [60]; whereas, in top-down method, nanometric-sized particles are obtained through reduction of bulk materials. Top-down approach includes nanolithography [61], thermal decomposition [62], laser ablation [63], mechanical milling [59], and sputtering [64] methods. In the following section, we will discuss a few synthesis protocols being frequently used for the preparation of nanodiagnostics.

3.1 Thermal Decomposition

It is an endothermic process, which utilizes heat to break the chemical bonds present in the chemical precursors at elevated temperatures. At a specific temperature, NPs are formed due to the decomposition of metal precursors that produce secondary products because of chemical reactions in the solvent [65]. The thermal decomposition process generally involves two major routes for the synthesis of NPs with great control overgrowth and nucleation mechanisms. The first route involves hot surfactant solutions in which organometallic compounds are injected directly, which results in the formation of nuclei instantaneously. The second route involves nuclei formation through the controlled heating process of injected compounds in a surfactant solution [66]. This method is very common to produce nanodiagnostics with good control over their shape and morphology. The major limitation of thermal decomposition is the use of organic solvents, and phase change is required to make water-soluble suspensions of NPs. Kim et al. synthesized the iron oxide nanocubes with the size varied from 20 to 160 nm with the use of oleic acid, benzyl ether, and iron acetylacetonate [67]. Roca et al. reported iron oleate and iron pentacarbonyl as alternative precursors [68].

3.2 Sol–Gel

Sol is referred to the colloidal solution of solid that suspended in the liquid phase. Sol, which is known as the dispersion phase, can be obtained from organic/inorganic-based precursors. Small dispersion phases of suspensions ($\sim$1–1000 nm) result in short-range and negligible gravitational forces because of dominant interactions [69]. Solid macromolecules submersed in a solvent to a form gel. Sol–gel is considered as the most cost-effective and simplified synthesis method. It is the wet chemical method in which the chemical solution acts as a precursor for the integrated system, which consists of discrete particles. In the sol–gel process, chlorides and metal oxides are commonly used as precursors [70]. The precursor is dispersed through stirring, shaking, or sonication in host liquid that forms a resultant system consisting of solid and liquid phases. To recover NPs, phase separation is carried out through different methods such as centrifugation, filtration, and sedimentation, and drying is done to remove the moisture [71,72]. Sugimoto et al. prepared TiO_2 NPs from a condensed gel of $Ti(OH)_4$ with the size ranging from 5 to 30 nm formed through hydrolysis of Ti-triethanolamine complex [73]. Rashidy et al. also prepared the silver-doped bioactive glass NPs with the conventional sol–gel process [74].

3.3 Pyrolysis

Pyrolysis is used for the large-scale production of NPs. It requires the burning of precursors through the flame. The precursor used in pyrolysis is either vapor or liquid, which is supplied through a small hole inside the furnace under high pressure where it burns. For easy evaporation, plasma is used in some furnaces at high temperature through easy evaporation [55,75]. Chen et al. reported the fluorine and nitrogen co-doped TiO_2 NPs by the pyrolysis of $(NH_4)_2TiF_6$. Decomposition of $(NH_4)_2TiF_6$ into NH_4TiOF_3 and $TiOF_2$ result in the formation of anatase-based TiO_2 with thin-shaped sheet-like morphology [76].

3.4 Laser Ablation

Laser ablation is a commonly used method for the synthesis of NPs from bulk metal sources in different solvents. A laser beam is used to irradiate the metal submersed in liquid, and a plasma plume that produces the NPs is formed. This technique provides an alternative to a conventional metal reduction method to form the metal-based NPs. NPs produced by this

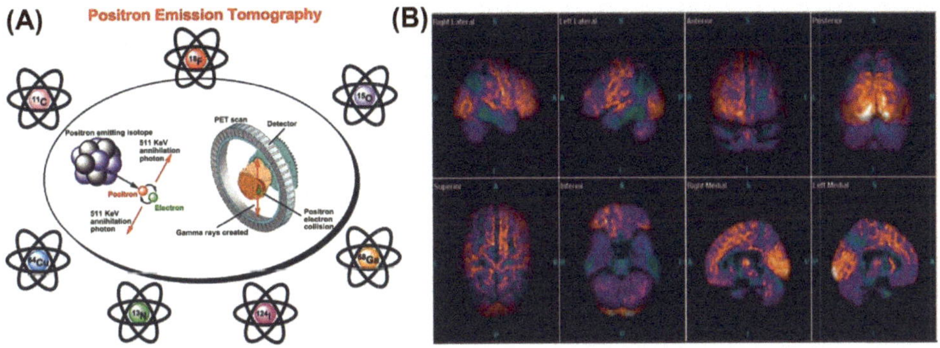

FIG. 6.4 **(A)** Schematic working of positron emission tomography. **(B)** PET images of the brain indicating hypometabolism in different hemispheres. (**(A)** Reproduced with permission from Li Z, Conti PS. Radiopharmaceutical chemistry for positron emission tomography. Advanced Drug Delivery Reviews 2010; 62(11):1031–1051, **(B)** Reproduced from Shigaeff N, et al. An interdisciplinary approach aiding the diagnosis of primary progressive aphasia: a case report. Dementia & Neuropsychologia 2017;11(1):100–102.)

method requires no chemical/stabilizing agents during the synthesis [77]. Tsuji et al. prepared silver NPs with the size varied from 12 to 29 nm by using a laser beam with different wavelengths such as 1064, 532, or 355 nm [78].

3.5 Sputtering

Sputtering is the process of NPs deposition on a suitable surface by ejecting from the target through the ion beam. Sputtering is referred to as the deposition of NPs' thin layer followed by the process of annealing. The size and shape of NPs depend on the substrate type, temperature, layer thickness, and annealing time [79]. Torimoto et al. reported the deposition of gold on ionic liquid, which results in the formation of highly uniform and dispersed gold NPs in the absence of any reducing/stabilizing agent. The reported average diameter for the gold NPs in the 1-ethyl-3-methylimidazolium tetrafluoroborate was 5.5 nm [64].

4 NONINVASIVE AND MINIMALLY INVASIVE DIAGNOSTIC TOOLS

Noninvasive diagnostic techniques usually refer to surgical techniques that are performing without any type of incision in the skin. These techniques involve magnetic resonance imaging, X-rays, and ultrasound, while the minimally invasive techniques refer to perform small practical incisions in the skins for diagnosis of certain diseases.

4.1 Positron Emission Tomography Scanning

Positron emission tomography (PET) is a frequently used technique for the diagnosis in modern medical science. In contrast to other diagnostic techniques such as X-ray machines, magnetic resonance imaging, and computed tomography scanning, which usually provide the structural-based information, functional body information can be obtained through PET (Fig. 6.4) [80]. PET is used for the quantitative analysis of the amount of contrast agents present in specified tissues or organs for a given time span. Glucose metabolism, oxygen, and blood flow can also be measured through PET [81,82]. Scans acquired through PET provide enhanced tissue penetration in comparison to other diagnostic imaging instruments such as ultrasound, fluorescence, and optical imaging. Good structural resolution cannot be obtained through PET scanning, that is why PET is used often with other imaging techniques to provide good functional and structural body information. To measure the metabolism and distribution of biomolecules through PET scans, short-life radiotracers such as $_{13}$N, $_{18}$F, $_{11}$C, $_{15}$O, and $_{64}$Cu are incorporated in biomolecules [83]. To measure glucose metabolism through PET scan, fluorodeoxyglucose as a radiotracer is used. For tracking the biodistribution of nanomaterial-based formulations, PET scanning is being used as an important tool by the researchers. For example, radiolabeled single-walled carbon nanotubes have been employed for tracking in vivo biodistribution, as reported in the literature [84]. Nanomaterials act as carriers for the radiotracer molecules to perform the task of drug delivery. Inaccurate results can be obtained through PET scanning after administration of two compounds because PET can only detect radiolabels not NPs. As mentioned earlier, PET imaging is combined with other techniques to create composite images that provide both functional and structural information. NPs due to their enhanced surface area-to-volume ratio act as a platform for multimodal contrast agents to obtain the images of conjugated therapeutic ligands. Chelated metal ($_{64}Cu$) is a radiolabeled agent used for macromolecules; however, the major drawback of these chelated metals includes their charged nature and large size that influence theoretically the pharmacokinetics and structure of macromolecules [83,85]. For the utilization of radionuclides, effective radiolabels have been developed, such as $_{18}F$, to avoid the structural influence of macromolecules. In both research and diagnosis fields, PET scanning is considered an important tool. To develop more efficient multimodal contrasts agents, new nanomaterials are needed to utilize for providing functional and structural data simultaneously through PET [86,87].

4.2 Magnetic Resonance Imaging

Magnetic resonance imaging (MRI) is one of the powerful diagnostic tools for the imaging of soft tissues. It usually possesses high temporal and spatial resolution, considered noninvasive and safe, because ionizing radiations are absent. MRI is an effective imaging technique for tumor detection and is used often in connection with the targeted cancer therapies to accomplish the real-time monitoring and accumulation of drugs in the specified infected areas. It can be adopted to footmark and image various cells in vivo [89,90]. The strong applied magnetic field interacts abundantly with water protons present in the soft tissues. Magnetic moments present in protons aligned along the applied field direction. Through the tissues, the radiofrequency pulse is applied, which makes protons to align in a specified direction at a specific frequency range. When the radiofrequency pulse is removed, protons returned to ground states known as the relaxation process [24,91]. Relaxation time consists of two major components that correspond to the decay of angular momentum vectors named longitudinal relaxation time (T_1) and transverse relaxation time (T_2). Relaxation values (T_1 and T_2) lead toward the difference in contrast which result in composite images and detected relaxation signals through computer-aided reconstruction mechanisms [92,93]. Enhanced contrast can be produced on MRI images by the difference between the relaxation time of two tissues. The use of contrast agents to enhance the imaging capabilities is a crucial factor for many applications including tracking, cellular imaging, and tumor diagnosis. With the alternation of relaxation time, surrounded water molecules can produce a pronounced image. As a result, the accumulation of contrast agents in specified tissues enhances the visibility of different lesions and structures on the MR images [94,95]. This relaxation depends on many factors in which particle size and surface functionalized organic polymers are more influential factors (Fig. 6.5) [96,97].

4.2.1 Contrast agents used in MRI

Currently, Gadolinium(III)-based chelates are used clinically as contrast agents. Gadolinium ions are paramagnetic in nature and due to the unpaired electrons, possessed large magnetic moments. On T_1-weighted images, these agents provide bright spots and positive contrast. However, these ions possessed high toxicity; therefore, proper ligand coatings are required to obtain gadolinium chelates that are highly stable and nontoxic in the body for a specific administration period [98,99]. Clinically used chelate is gadolinium-diethylenetriamine pentaacetic acid. Many types of chelates are regularly used in clinics with different biodistribution profiles for liver imaging [100,101] (Table 6.1). In targeted drug delivery, gadolinium chelates are considered as promising candidates. Part et al. conjugated gadolinium-tetraazacyclododecane tetraacetic acid to get contrast agents with enhanced tumor targeting abilities for MRI [102]. Different formulations based on gadolinium NPs have been used for creating enhanced contrast agents in MRI [103].

4.3 Computed Tomography Scanning

Computed tomography (CT) scanning is based on X-rays that can produce detailed cross-sectional micrographs of different tissues. It is accomplished by acquiring multiple images by using X-rays at distinct angles and forming CT images. It is used for taking images of the abdomen, pelvis, chest, and head. For the internal structure analysis, conventional CT scanning is used to measure the attenuation. For this purpose, the X-ray beam is projected through the body of the patient. Apparatus is rotated through the patient at the same time longitudinally. From the attenuation of the beam when rotated in each direction of the plane, data are recorded. These recorded data can be used to compute internal structure at a specific plane of the patient. Apparatus or the patient indexed along an axis and further plane is investigated. From the various slices, the three-dimensional photograph is constructed. The main problem is the specified duration of time for which the patient is not motionless to obtain the necessary slices. To avoid this problem, currently, contrast agents based on gadolinium and iodine are used for CT scans [106–108]. However, these contrast agents face many problems such as half-lives based on short circulation time (for conventional iodinated compounds, it is less than 10 min), nonspecific biodistribution, poor contrast, and slight renal toxicity

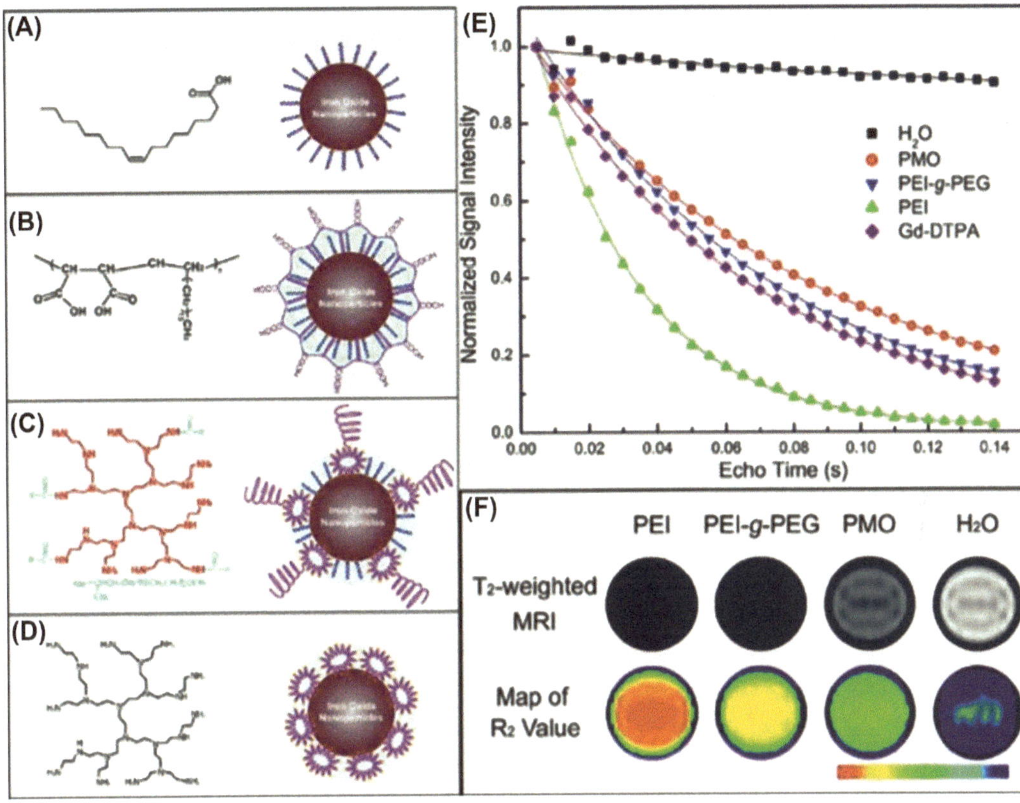

FIG. 6.5 **(A–D)** Schematic representation of different organic polymers: (**(A)** Oleic acid, **(B)** copolymers, **(C)** polyethylene glycol-polyethylenimine (PEI), **(D)** hyperbranched PEI) coatings on iron oxide NPs, **(E)** T2 relaxation comparison of pure water, gadolinium complex, and polymer-coated iron oxide NPs, **(F)** T2 contrasts and R2 color-coded maps for different sizes and coatings of iron oxide nanocrystals. (**(F)** Reproduced with permission from Duan H, et al. Reexamining the effects of particle size and surface chemistry on the magnetic properties of iron oxide nanocrystals: new insights into spin disorder and proton relaxivity. The Journal of Physical Chemistry C 2008;112(22):8127–8131.)

issues in a large number of patients [109]. To overcome such problems, contrast agents based on different nanomaterials are developed traditionally. Contrast agents based on conventional iodine including iodinated NPs, bismuth sulfide NPs, and gold NPs, as well as liposomal and polymeric carriers, are tremendously used [110–112] (Table 6.2).

4.3.1 NPs used for CT

In CT imaging, contrast agents based on polymer-coated Bi_2O_3 NPs have been developed recently. These NPs can prepare typically in a two-step synthesis mechanism; through precipitation of sodium sulfide and bismuth citrate, Bi_2S_3 cores are produced and then coating with polyvinylpyrrolidone is done to impart biocompatibility and colloidal stability [114]. These particles increase fivefold X-ray absorption as compared to contrast agents based on iodine and longer circulation half-life (more than 2 h). In the mouse model, in vivo clinical utility of these NPs enhances the imaging of vasculature, lymph nodes, and liver. The major problem that these contrast agents face includes control on the size and shape of Bi_2S_3 NPs. The second major class of contrast agents used to a greater extent is based on gold NPs [113,115]. Excellent CT contrast agents based on gold NPs greatly enhance the X-ray absorption (2.7–5.7 times) with the longer blood circulation time (in rats approximately 4 h) as compared to iodine, which is 10 min. The biocompatibility of these prolonged circulation time of gold NPs increases through polymer coatings at the surface of the NPs, which rapidly reduce the renal

TABLE 6.1
Short Summary of Different Types of Clinically Preclinically Developed Contrast Media.

Imaging Technique	Base Contrast Material	Specific Type	Trade name of Clinically Approved Agents	Current Status	References
MRI	Gadolinium nanostructures	Liposomes, polymers, inorganic NPs	–	Preclinical	[101,104,105]
MRI	Gadolinium salts and chelates		Gadofosveset trisodium (ABLAVAR); gadodiamide (OMNISCAN); gadobutrol (Gadavist); gadoxetate disodium (EOVIST); gadopentetate dimeglumine (MAGNEVIST®); gadobenate dimeglumine (MultiHance); gadoversetamide (OptiMARKTM); gadoteridol (ProHance)	Clinical use	[98–100,103]
MRI	Iron oxide NPs	Iron oxide NPs coated with polymers	Ferumoxsil (GastroMARK®, Lumirem®); ferumoxide (Feridex®, Endorem) ferrixan (Resovist®, Cliavist™);	Clinically used and under some clinical trials	[24,43,93]

clearance. Another major advantage of these contrast agents is an enhancement in contrast to micrographs with X-ray energies ranging from 80 to 100 keV [116,117]. These energy ranges reduce the absorption in bones and soft tissues when subjected at lower doses. For studying mammary carcinoma, commercially available gold NPs with size 1.6 nm have been evaluated in vivo to the group of mice [107]. These small-sized particles result in small accumulation in spleen and liver and show clear neovasculature imaging, which allowed the effective tumor diagnosis on CT scans. Their circulation and contrast properties are mainly size-dependent, and the appropriate size is required to utilize nanoparticles more efficiently (Fig. 6.6) [118]. Scanning taken through contrast agents based on iodine did not diagnose tumors clearly [119]. Kim et al. studied PEG-coated gold NPs with a size range of 30 nm on the rodent model and revealed the efficient blood pool [110,120]. Earlier studies revealed the low in vivo toxicity. Iodine is considered by far the most common contrast agent used for CT. Iodine-based contrast agents possessed attenuated X-ray capabilities; however, such contrast agents suffered from short circulation time and lack of tissue specificities. To overcome these issues, iodine is usually incorporated into different types of engineered nanostructures [81]. Different strategies have been utilized for nanoscale polymers that are able of taking large payloads and can encapsulate iodine in liposomes with the use of polymeric iodinated NPs. Therefore, longer circulation time can be achieved through these strategies as compared to CT

TABLE 6.2
Different Materials Under Consideration as Contrast Agents in CT.

Imaging Technique	Base Contrast Material	Specific Type	Trade name of Clinically Approved Agents	Current Status	References
CT	Iodinated nanostructure	Polymeric carriers, inorganic Nanostructures, liposomal Carriers,		Preclinical	[108,109,111,112]
CT	Conventional iodine	Ionic dimers	Ioxaglate (Hexabrix)	Clinical use	[113]
		Nonionic monomers	Ioversol (Optiray); iohexol (Omnipaque); iopromide (Ultravist); iopamidol (Isovue);	Clinical use	[113]
		High osmolarity	Diatrizoate anion (Renografin); iothalamate anion (Conray); diatrizoate anion (Hypaque);	Clinical use	[113]
		Nonionic dimers	Iodixanol and iotrol (Visipaque)	Clinical use	[113]
CT	Gold NPs			Clinical trials	[107,110]
CT	Bismuth sulfide (Bi2S3) NPs			Preclinical	[106]

conventional imaging. It leads toward the development of contrast agents for CT that are capable of targeting and molecular imaging. Through these strategies, an increase of iodine concentration in tissues with significantly enhanced image contrast can be achieved [121].

4.4 Fluorescence Imaging

Fluorescent proteins and organic fluorophores have been used widely in the biological research field to study extensively the different molecular and cellular processes that usually occurred both in vivo and in vitro noninvasively. Incorporation of these small entities into different types of chemical formulations is used in various techniques such as flow cytometry and confocal microscopy to perform in vitro study through the interactions present between cells and macromolecules as well as fate and in vivo biodistribution. For cell tracking and cell labeling, fluorescent molecules are tremendously used. These molecules are also employed on animal models to study the cell uptake, NPs retention time, and tracking the biodistribution [122,123]. Fluorescence imaging faces many limitations as compared to conventional fluorophores such as wide absorption and emission bands, poor tissue/skin penetration, high photobleaching susceptibility, short in vivo imaging duration, and poor signal strength. Because of these limitations, fluorescence imaging techniques have limited clinical use. New advancement has been made in florescent nanomaterials to overcome such issues that are linked with the conventional fluorophores to open the possible use of this fluorescence imaging for the diagnosis of many diseases and their progression on cellular, systemic, and molecular level [124,125]. Fluorescence tagging is one of the most common methodologies for different types of nanomaterial such as iron oxide-based MRI contrast agents and polymer vehicles for controlled drug delivery to obtain the information about tumoral accumulation, cellular uptake, and biodistribution [126,127].

4.4.1 NPs used for fluorescence imaging

Semiconductor quantum dots are generally the nanometric sized particles made up of heavy metal combinations (InP, CdSe, InAs, CdTe, etc.). Distinctive optical

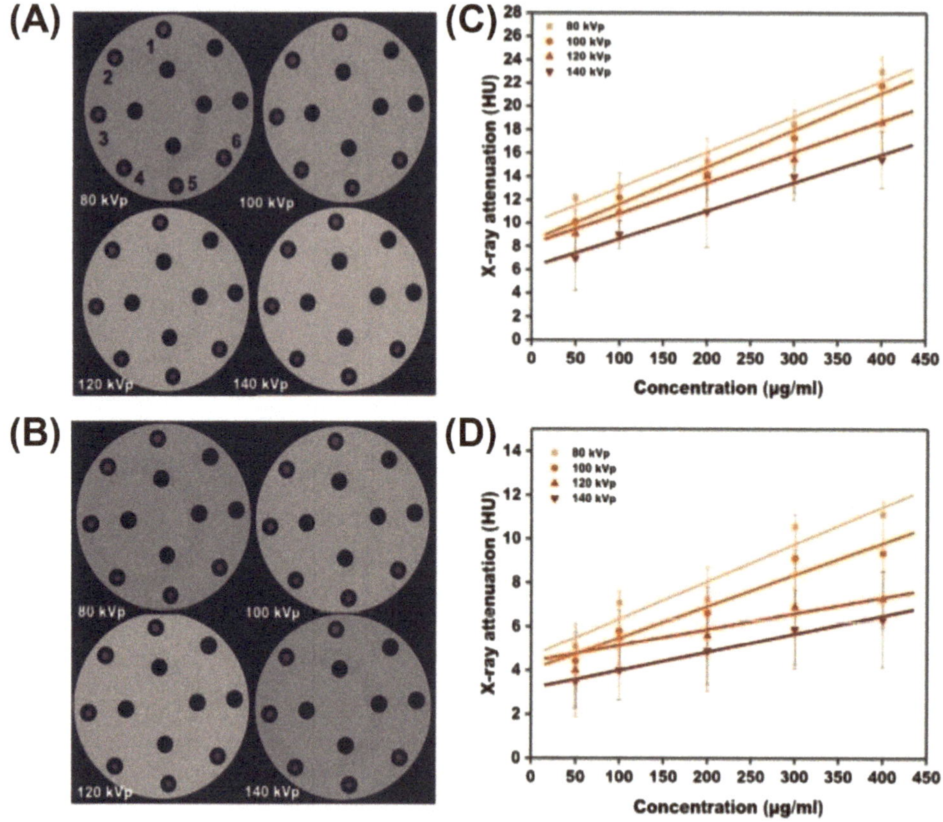

FIG. 6.6 Axial CT images of different size gold NPs **(A)** 13 nm **(B)** 60 nm. X-ray attenuation signal of **(C)** 13 nm and **(D)** 60 nm gold NPs at different concentrations and tube potentials. (Reproduced with permission from Khademi S, et al. Evaluation of size, morphology, concentration, and surface effect of gold nanoparticles on X-ray attenuation in computed tomography. Physica Medica 2018;45:127–133.)

properties of quantum dots (QDs) have emerged as promising candidates for fluorescence-based imaging that replace conventional imaging based on organic fluorophore [128,129]. QDs absorb light from the large spectral range with a narrow emission spectrum, as a result, emit a well-defined wavelength of light despite the fact that what their input wavelengths are. Great control over the emission of specified wavelengths of the particles can be achieved by tuning the composition and wavelength parameters. Due to these promising properties, QD fluorescent probes can be used as optical multiplexing where the same light source can be illuminated but with the emission of different colors [126,130]. Large extinction molar coefficients and high photostability of QDs make them an efficient candidate for the in vivo imaging because of low tissue depth sensitivity. QDs have been widely used for tracking cell movements, image vasculature, labeling of cellular components, and visualization of tissue distribution and lymph nodes of nanomaterials. The major concern associated with the QD fluorescent tags is its possible toxic effects because generally QDs are made up of heavy metals. Cadmium is a heavy metal and considered toxic in its free form (Cd^{2+}) [131,132]. QDs with the size greater than the renal excretion threshold (hydrodynamic size = 5.5 nm) show in vivo degradation mechanism. Therefore, such types of QDs raise the toxicity concerns because these are accumulated in different organs for the extended time (Fig. 6.7) [133]. Studies made in the recent past revealed that proper coatings on the QDs increase the retention time in body up to 2 years, while their fluorescent properties remain preserved [134].

4.4.2 Encapsulation of fluorophore materials

Florescent nanomaterials can be synthesized by encapsulation of conventional organic fluorophores inside the nanocarriers. It greatly enhances the fluorophore

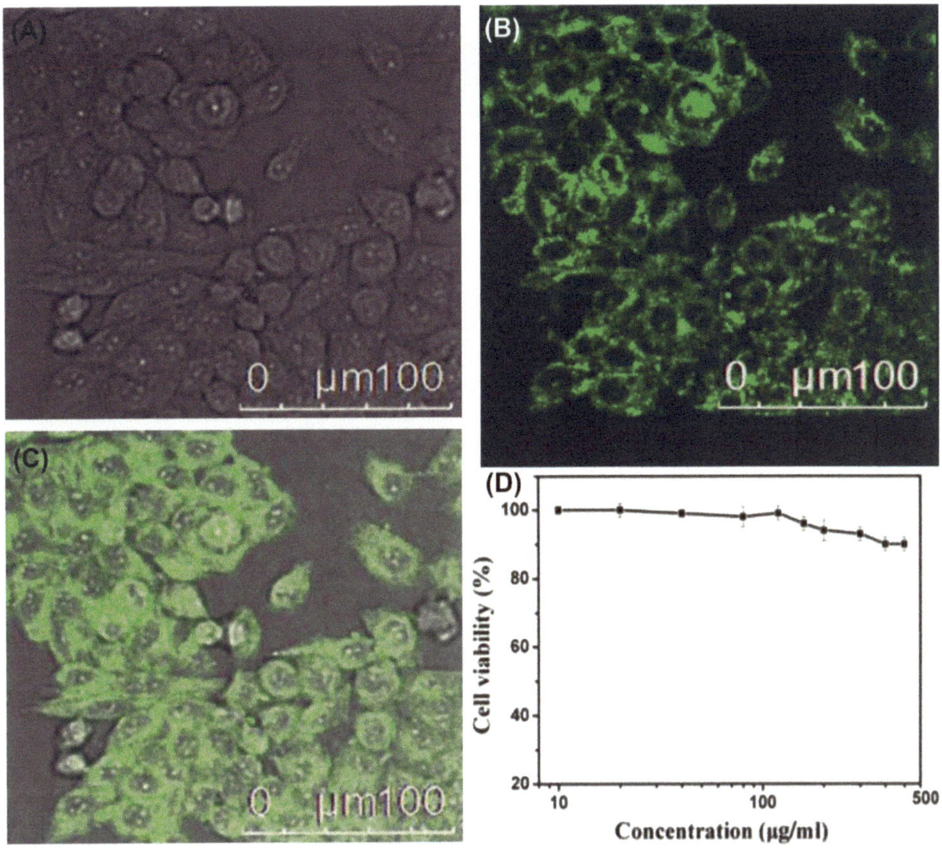

FIG. 6.7 Two photon fluorescent images of HeLa cells in **(A)** bright field **(B)** excited by 800 nm fs laser. **(C)** Combined image of **(A)** and **(B)**. **(D)** Cell viability for different concentrations of nitrogen-rich carbon quantum dots for 24 h. (Reproduced with permission from Yuan F, et al. Nitrogen-rich D-π-A structural carbon quantum dots with a bright two-photon fluorescence for deep-tissue imaging. ACS Applied Bio Materials 2018, copyright. (2018) American Chemical Society.)

properties with the increase in signal strength. Therefore, assay sensitivity enhances the use of a large number of entrapped dye molecules. Resistance to photobleaching improves as compared to the encapsulated dyes. This allows great control over the biodistribution mechanism and clearance rates as compared to the organic fluorophores based on small molecules [135] (Table 6.3). One of the commonly used encapsulation methods involves the organic fluorophores trapped inside silica NPs. These particles can be prepared through different synthesis routes. However, the entrapment of hydrophobic dye molecules in the hydrophilic environment of silica-based NPs is considered as a major challenge [136,137]. Covalent linking of the silica NPs with the dye molecules can be accomplished. It can be used during the synthesis of silica nanoparticles in the classic Stober method [138]. These synthesis protocols can create highly consistent core–shell-structured particles. The microemulsion method can be used for the synthesis through polar dye molecules. In the oil emulsion method, the particles are created inside water droplets and the dye is entrapped physically inside the NPs. The different types of methods are available to entrap the hydrophobic dyes through the microemulsion technique that can produce more control over the monodispersity of the NPs as compared to the Stober method [139]. Silica-based nanosystems are tremendously used due to high surface potential modification and lower toxicity. The encapsulation of dye molecules can be done through different methods inside polymer particles; however, there are more issues related to dye molecules, that is, dye aggregation and leakage as compared to the silica NPs [140].

TABLE 6.3
Materials Under Consideration as Contrast Agents in Fluorescence Imaging.

Imaging Technique	Base Contrast Material	Specific Type	Trade Name of Clinically Approved Agents	Current Status	References
Optical/ Fluorescence Imaging	Quantum dots	–	–	Preclinical	[123,124,128,132]
	Nanostructure-based radionuclides	–	–	Preclinical	[83,84]
	Fluorophores encapsulated with polymers	–	–	Preclinical	[135,136]
	Fluorescent silica particles	–	–	Preclinical	[135,137,139,140]

5 FUTURE PERSPECTIVES

In medical diagnosis, efficient and proper pharmaceutical delivery and artificial cell development are the major fields where nanomaterials have been found of practical implementation. Nanotechnology applications to nanomedicine involve the three major mutually progressive and overlapping molecular technologies. The fabrication of nanomaterial-based devices has shown great importance in smart drugs, targeted drug delivery, advanced diagnostics, immunoisolation therapies, and biosensors. Bionanotechnology offers the major benefits of molecular medicine through genomics, artificially engineered microbes, and proteomics. Medical nanorobots and molecular machine systems can allow the instant detection of pathogens, chromosome replacement, extermination, individual in vivo cell surgery, and improvement of the physiological functions. In the next coming decades, diagnostic devices based on nanotechnology will be available, which can make many measurements inexpensive and quick. In the future, diagnostic trends will continue in reducing the biochip-based technologies to the nanoscale range. In response to the localized infection or injury, estimation of these devices' capabilities can be used for the evaluation of their typical performance of chemical-released mechanism into the blood by tissues.

These devices can indicate single-cell differentiation through chemical sources from the in vivo chemical concentration and, as a result, can provide a high-resolution diagnosis in both space and time. Currently used techniques for blood analysis such as chemical sources when diluted in blood volume are much difficult to differentiate from the background and withdraw them as blood samples. This trend will be helpful in diagnosis devices through bottom-up approaches, starting with smaller building blocks. Another trend is the limited use of fluorescent labeling because it reduces the signal strength, but fluorescent labeling methods can be improved through viable methods with NPs.

6 CONCLUSIONS

Over the past years, nanotechnology is applied practically in creating artificial tissues, cells, and organs. Artificial cells are actively being investigated for the replacement of incorrectly or defectively functioning organs and cells, majorly related to the metabolic functions. To extend the current limits of molecular diagnosis and integration of diagnosis with therapeutics through the development of personalized medicines is the promise of nanotechnology. Important clinical applications of nanotechnology include the DNA detection assay, cancer detection, biomarker discovery, and diagnosis of infectious microorganisms.

REFERENCES

[1] Gu FX, et al. Targeted nanoparticles for cancer therapy. Nano Today 2007;2(3):14–21.
[2] Petkar KC, et al. Nanostructured materials in drug and gene delivery: a review of the state of the art. Critical Reviews in Therapeutic Drug Carrier Systems 2011;28(2).
[3] Rosen J, et al. Nanotechnology and diagnostic imaging: new advances in contrast agent technology. Journal of Nanomedicine & Nanotechnology 2011;2(5):1–12.

[4] Shi J, et al. Nanotechnology in drug delivery and tissue engineering: from discovery to applications. Nano Letters 2010;10(9):3223–30.
[5] Power S, Slattery MM, Lee MJ. Nanotechnology and its relationship to interventional radiology. Part II: drug delivery, thermotherapy, and vascular intervention. Cardiovascular and interventional radiology 2011;34(4): 676–90.
[6] Erickson D, et al. Nanobiosensors: optofluidic, electrical and mechanical approaches to biomolecular detection at the nanoscale. Microfluidics and Nanofluidics 2008; 4(1–2):33–52.
[7] Power S, Slattery MM, Lee MJ. Nanotechnology and its relationship to interventional radiology. Part I: imaging. Cardiovascular and Interventional Radiology 2011;34(2):221–6.
[8] Conde J, et al. Revisiting 30 years of biofunctionalization and surface chemistry of inorganic nanoparticles for nanomedicine. Frontiers in Chemistry 2014;2(48).
[9] Meerasa A, Huang JG, Gu FX. CH50: a revisited hemolytic complement consumption assay for evaluation of nanoparticles and blood plasma protein interaction. Current Drug Delivery 2011;8(3):290–8.
[10] Vega-Villa KR, et al. Clinical toxicities of nanocarrier systems. Advanced Drug Delivery Reviews 2008;60(8): 929–38.
[11] Zahr AS, Davis CA, Pishko MV. Macrophage uptake of core – shell nanoparticles surface modified with poly (ethylene glycol). Langmuir 2006;22(19):8178–85.
[12] Moghimi SM, Hunter AC, Murray JC. Long-circulating and target-specific nanoparticles: theory to practice. Pharmacological Reviews 2001;53(2):283–318.
[13] Alexis F, et al. Factors affecting the clearance and biodistribution of polymeric nanoparticles. Molecular Pharmaceutics 2008;5(4):505–15.
[14] Owens III DE, Peppas NA. Opsonization, biodistribution, and pharmacokinetics of polymeric nanoparticles. International Journal of Pharmaceutics 2006;307(1):93–102.
[15] Choi HS, et al. Renal clearance of quantum dots. Nature Biotechnology 2007;25(10):1165.
[16] Vonarbourg A, et al. Parameters influencing the stealthiness of colloidal drug delivery systems. Biomaterials 2006;27(24):4356–73.
[17] Huang X, et al. Effect of injection routes on the biodistribution, clearance, and tumor uptake of carbon dots. ACS Nano 2013;7(7):5684–93.
[18] Forge D, et al. Optimization of the synthesis of superparamagnetic contrast agents by the design of experiments method. The Journal of Physical Chemistry C 2008; 112(49):19178–85.
[19] Park JY, et al. Highly water-dispersible PEG surface modified ultra small superparamagnetic iron oxide nanoparticles useful for target-specific biomedical applications. Nanotechnology 2008;19(36):365603.
[20] Babic M, et al. Poly (L-lysine)-modified iron oxide nanoparticles for stem cell labeling. Bioconjugate Chemistry 2008;19(3):740–50.
[21] Bajpai A, Gupta R. Synthesis and characterization of magnetite (Fe_3O_4)—polyvinyl alcohol-based nanocomposites and study of superparamagnetism. Polymer Composites 2010;31(2):245–55.
[22] Mahmoudi M, et al. Optimal design and characterization of superparamagnetic iron oxide nanoparticles coated with polyvinyl alcohol for targeted delivery and imaging. The Journal of Physical Chemistry B 2008; 112(46):14470–81.
[23] Lin H, et al. Preparation of magnetic poly (vinyl alcohol) (PVA) materials by in situ synthesis of magnetite in a PVA matrix. Journal of Applied Polymer Science 2003; 87(8):1239–47.
[24] Qiao R, Yang C, Gao M. Superparamagnetic iron oxide nanoparticles: from preparations to in vivo MRI applications. Journal of Materials Chemistry 2009; 19(35):6274–93.
[25] Sirlin CB, et al. Gadolinium-DTPA-dextran: a macromolecular MR blood pool contrast agent1. Academic Radiology 2004;11(12):1361–9.
[26] Sirlin CB, et al. Gadolinium-DTPA-dextran: a molecular MR blood pool contrast agent. Academic Radiology 2005;12(5):S80–2.
[27] Hifumi H, et al. Dextran coated gadolinium phosphate nanoparticles for magnetic resonance tumor imaging. Journal of Materials Chemistry 2009;19(35): 6393–9.
[28] Earhart C, et al. Synthesis of carbohydrate-conjugated nanoparticles and quantum dots. Langmuir 2008; 24(12):6215–9.
[29] Hong R, et al. Preparation and characterization of magnetite/dextran nanocomposite used as a precursor of magnetic fluid. Chemical Engineering Journal 2009; 150(2–3):572–80.
[30] Yu MK, et al. Drug-loaded superparamagnetic iron oxide nanoparticles for combined cancer imaging and therapy in vivo. Angewandte Chemie 2008;120(29):5442–5.
[31] Ma HL, et al. Superparamagnetic iron oxide nanoparticles stabilized by alginate: pharmacokinetics, tissue distribution, and applications in detecting liver cancers. International Journal of Pharmaceutics 2008; 354(1–2):217–26.
[32] Grant GT, et al. Biological interactions between polysaccharides and divalent cations: the egg-box model. FEBS Letters 1973;32(1):195–8.
[33] Gombotz WR, Wee S. Protein release from alginate matrices. Advanced Drug Delivery Reviews 1998;31(3): 267–85.
[34] Tsai Z-T, et al. In situ preparation of high relaxivity iron oxide nanoparticles by coating with chitosan: a potential MRI contrast agent useful for cell tracking. Journal of Magnetism and Magnetic Materials 2010;322(2): 208–13.

[35] Kim DK, et al. Starch-coated superparamagnetic nanoparticles as MR contrast agents. Chemistry of Materials 2003;15(23):4343–51.
[36] Saboktakin MR, Maharramov A, Ramazanov MA. Synthesis and characterization of superparamagnetic nanoparticles coated with carboxymethyl starch (CMS) for magnetic resonance imaging technique. Carbohydrate Polymers 2009;78(2):292–5.
[37] Jiang J-S, et al. A novel magnetic fluid based on starch-coated magnetite nanoparticles functionalized with homing peptide. Journal of Nanoparticle Research 2009;11(6):1321–30.
[38] López-Cruz A, et al. Water dispersible iron oxide nanoparticles coated with covalently linked chitosan. Journal of Materials Chemistry 2009;19(37):6870–6.
[39] Rosen JE, Gu FX. Surface functionalization of silica nanoparticles with cysteine: a low-fouling zwitterionic surface. Langmuir 2011;27(17):10507–13.
[40] Maeda H, et al. Tumor vascular permeability and the EPR effect in macromolecular therapeutics: a review. Journal of Controlled Release 2000;65(1–2):271–84.
[41] Noguchi Y, et al. Early phase tumor accumulation of macromolecules: a great difference in clearance rate between tumor and normal tissues. Japanese Journal of Cancer Research 1998;89(3):307–14.
[42] Bannon-Peppas L, Blanchette J. Nanoparticle and targeted system for cancer therapy. Advanced Drug Delivery Reviews 2004;56:1649–59.
[43] Laurent S, et al. Magnetic iron oxide nanoparticles: synthesis, stabilization, vectorization, physicochemical characterizations, and biological applications. Chemical Reviews 2008;108(6):2064–110.
[44] Zhang C, et al. Specific targeting of tumor angiogenesis by RGD-conjugated ultrasmall superparamagnetic iron oxide particles using a clinical 1.5-T magnetic resonance scanner. Cancer Research 2007;67(4):1555–62.
[45] Sonvico F, et al. Folate-conjugated iron oxide nanoparticles for solid tumor targeting as potential specific magnetic hyperthermia mediators: synthesis, physicochemical characterization, and in vitro experiments. Bioconjugate Chemistry 2005;16(5):1181–8.
[46] Allen TM. Ligand-targeted therapeutics in anticancer therapy. Nature Reviews Cancer 2002;2(10):750.
[47] Chen TJ, et al. Targeted folic acid-PEG nanoparticles for noninvasive imaging of folate receptor by MRI. Journal of Biomedical Materials Research Part A: An Official Journal of The Society for Biomaterials, The Japanese Society for Biomaterials, and The Australian Society for Biomaterials and the Korean Society for Biomaterials 2008;87(1):165–75.
[48] Iyer AK, et al. Exploiting the enhanced permeability and retention effect for tumor targeting. Drug Discovery Today 2006;11(17–18):812–8.
[49] Rosen JE, et al. Iron oxide nanoparticles for targeted cancer imaging and diagnostics. Nanomedicine: Nanotechnology, Biology and Medicine 2012;8(3):275–90.
[50] Fan C, et al. Tumor selectivity of stealth multi-functionalized superparamagnetic iron oxide nanoparticles. International Journal of Pharmaceutics 2011;404(1–2):180–90.
[51] Weissleder R, et al. Antimyosin-labeled monocrystalline iron oxide allows detection of myocardial infarct: MR antibody imaging. Radiology 1992;182(2):381–5.
[52] Kang HW, et al. Magnetic resonance imaging of inducible E-selectin expression in human endothelial cell culture. Bioconjugate Chemistry 2002;13(1):122–7.
[53] Islam T, Josephson L. Current state and future applications of active targeting in malignancies using superparamagnetic iron oxide nanoparticles. Cancer Biomarkers 2009;5(2):99–107.
[54] Sutradhar KB, Amin ML. Nanotechnology in cancer drug delivery and selective targeting. ISRN Nanotechnology 2014;2014:12.
[55] Ealia SAM, Saravanakumar M. A review on the classification, characterisation, synthesis of nanoparticles and their application. In: IOP conference series: materials science and engineering. IOP Publishing; 2017.
[56] Malesevic A, et al. Synthesis of few-layer graphene via microwave plasma-enhanced chemical vapour deposition. Nanotechnology 2008;19(30):305604.
[57] Mackenzie JD, Bescher EP. Chemical routes in the synthesis of nanomaterials using the sol–gel process. Accounts of Chemical Research 2007;40(9):810–8.
[58] Nemade K, Waghuley S. Synthesis of MgO nanoparticles by solvent mixed spray pyrolysis technique for optical investigation. International Journal of Metals 2014;2014.
[59] Yadav TP, Yadav RM, Singh DP. Mechanical milling: a top down approach for the synthesis of nanomaterials and nanocomposites. Nanoscience and Nanotechnology 2012;2(3):22–48.
[60] Mohanpuria P, Rana NK, Yadav SK. Biosynthesis of nanoparticles: technological concepts and future applications. Journal of Nanoparticle Research 2008;10(3):507–17.
[61] Glass R, Möller M, Spatz JP. Block copolymer micelle nanolithography. Nanotechnology 2003;14(10):1153.
[62] Salavati-Niasari M, Davar F. Synthesis of copper and copper (I) oxide nanoparticles by thermal decomposition of a new precursor. Materials Letters 2009;63(3–4):441–3.
[63] Zeng H, et al. Nanomaterials via laser ablation/irradiation in liquid: a review. Advanced Functional Materials 2012;22(7):1333–53.
[64] Torimoto T, et al. Sputter deposition onto ionic liquids: simple and clean synthesis of highly dispersed ultrafine metal nanoparticles. Applied Physics Letters 2006;89(24):243117.
[65] Salavati-Niasari M, Davar F, Mir N. Synthesis and characterization of metallic copper nanoparticles via thermal decomposition. Polyhedron 2008;27(17):3514–8.
[66] Lu AH, Salabas EL, Schüth F. Magnetic nanoparticles: synthesis, protection, functionalization, and application. Angewandte Chemie International Edition 2007;46(8):1222–44.
[67] Kim D, et al. Synthesis of uniform ferrimagnetic magnetite nanocubes. Journal of the American Chemical Society 2008;131(2):454–5.

[68] Roca A, et al. Structural and magnetic properties of uniform magnetite nanoparticles prepared by high temperature decomposition of organic precursors. Nanotechnology 2006;17(11):2783.
[69] Laurent S, et al. Magnetic iron oxide nanoparticles for biomedical applications. Future Medicinal Chemistry 2010;2(3):427–49.
[70] Sivakumar P, et al. Synthesis and characterization of $NiFe_2O_4$ nanoparticles and nanorods. Journal of Alloys and Compounds 2013;563:6–11.
[71] Mann S, et al. Sol–gel synthesis of organized matter. Chemistry of Materials 1997;9(11):2300–10.
[72] Trewyn BG, et al. Synthesis and functionalization of a mesoporous silica nanoparticle based on the sol–gel process and applications in controlled release. Accounts of Chemical Research 2007;40(9):846–53.
[73] Sugimoto T, Zhou X, Muramatsu A. Synthesis of uniform anatase TiO_2 nanoparticles by gel–sol method: 3. Formation process and size control. Journal of Colloid and Interface Science 2003;259(1):43–52.
[74] El-Rashidy AA, et al. Preparation and in vitro characterization of silver-doped bioactive glass nanoparticles fabricated using a sol-gel process and modified Stöber method. Journal of Non-Crystalline Solids 2018;483:26–36.
[75] D'Amato R, et al. Synthesis of ceramic nanoparticles by laser pyrolysis: from research to applications. Journal of Analytical and Applied Pyrolysis 2013;104:461–9.
[76] Chen D, et al. A facile method to synthesize nitrogen and fluorine co-doped TiO_2 nanoparticles by pyrolysis of $(NH_4)_2TiF_6$. Journal of Nanoparticle Research 2009;11(2):303–13.
[77] Amendola V, Meneghetti M. Laser ablation synthesis in solution and size manipulation of noble metal nanoparticles. Physical Chemistry Chemical Physics 2009;11(20):3805–21.
[78] Tsuji T, et al. Preparation of silver nanoparticles by laser ablation in solution: influence of laser wavelength on particle size. Applied Surface Science 2002;202(1–2):80–5.
[79] Okazaki K-i, et al. Single-step synthesis of gold–silver alloy nanoparticles in ionic liquids by a sputter deposition technique. Chemical Communications 2008;(6):691–3.
[80] Shigaeff N, et al. An interdisciplinary approach aiding the diagnosis of primary progressive aphasia: a case report. Dementia & Neuropsychologia 2017;11(1):100–2.
[81] Cai W, Chen X. Nanoplatforms for targeted molecular imaging in living subjects. Small 2007;3(11):1840–54.
[82] Herth MM, et al. Radioactive labeling of defined HPMA-based polymeric structures using [18F] FETos for in vivo imaging by positron emission tomography. Biomacromolecules 2009;10(7):1697–703.
[83] Hong H, et al. Molecular imaging and therapy of cancer with radiolabeled nanoparticles. Nano Today 2009;4(5):399–413.
[84] Liu Z, et al. In vivo biodistribution and highly efficient tumour targeting of carbon nanotubes in mice. Nature Nanotechnology 2007;2(1):47.
[85] Alavi A, et al. Positron emission tomography imaging of regional cerebral glucose metabolism. In: Seminars in nuclear medicine. Elsevier; 1986.
[86] Markus R, et al. Topography and temporal evolution of hypoxic viable tissue identified by 18F-fluoromisonidazole positron emission tomography in humans after ischemic stroke. Stroke 2003;34(11):2646–52.
[87] Arnaud L, et al. Is 18F-fluorodeoxyglucose positron emission tomography scanning a reliable way to assess disease activity in Takayasu arteritis? Arthritis & Rheumatism: Official Journal of the American College of Rheumatology 2009;60(4):1193–200.
[88] Li Z, Conti PS. Radiopharmaceutical chemistry for positron emission tomography. Advanced Drug Delivery Reviews 2010;62(11):1031–51.
[89] Jain TK, et al. Magnetic nanoparticles with dual functional properties: drug delivery and magnetic resonance imaging. Biomaterials 2008;29(29):4012–21.
[90] Smirnov P, et al. In vivo single cell detection of tumor-infiltrating lymphocytes with a clinical 1.5 Tesla MRI system. Magnetic Resonance in Medicine: An Official Journal of the International Society for Magnetic Resonance in Medicine 2008;60(6):1292–7.
[91] Javed Y, et al. MRI based on iron oxide nanoparticles contrast agents: effect of oxidation state and architecture. Journal of Nanoparticle Research 2017;19(11):366.
[92] Shapiro EM, et al. In vivo detection of single cells by MRI. Magnetic Resonance in Medicine: An Official Journal of the International Society for Magnetic Resonance in Medicine 2006;55(2):242–9.
[93] Shapiro EM, et al. MRI detection of single particles for cellular imaging. Proceedings of the National Academy of Sciences 2004;101(30):10901–6.
[94] Banerjee SS, Chen D-H. Multifunctional pH-sensitive magnetic nanoparticles for simultaneous imaging, sensing and targeted intracellular anticancer drug delivery. Nanotechnology 2008;19(50):505104.
[95] Santra S, et al. Drug/dye-loaded, multifunctional iron oxide nanoparticles for combined targeted cancer therapy and dual optical/magnetic resonance imaging. Small 2009;5(16):1862–8.
[96] Duan H, et al. Reexamining the effects of particle size and surface chemistry on the magnetic properties of iron oxide nanocrystals: new insights into spin disorder and proton relaxivity. The Journal of Physical Chemistry C 2008;112(22):8127–31.
[97] Estelrich J, Sánchez-Martín MJ, Busquets MA. Nanoparticles in magnetic resonance imaging: from simple to dual contrast agents. International Journal of Nanomedicine 2015;10:1727–41.
[98] Zhu D, et al. Biocompatible nanotemplate-engineered nanoparticles containing gadolinium: stability and relaxivity of a potential MRI contrast agent. Journal of

Nanoscience and Nanotechnology 2006;6(4):996–1003.

[99] Helm L. Optimization of gadolinium-based MRI contrast agents for high magnetic-field applications. Future Medicinal Chemistry 2010;2(3):385–96.

[100] Khemtong C, Kessinger CW, Gao J. Polymeric nanomedicine for cancer MR imaging and drug delivery. Chemical Communications 2009;(24):3497–510.

[101] Zhang WL, et al. Fabrication of polymer-gadolinium (III) complex nanomicelle from poly (ethylene glycol)-polysuccinimide conjugate and diethylenetriaminetetraacetic acid-gadolinium as magnetic resonance imaging contrast agents. Journal of Applied Polymer Science 2011;120(5):2596–605.

[102] Park JA, et al. Gd-DOTA conjugate of RGD as a potential tumor-targeting MRI contrast agent. ChemBioChem 2008;9(17):2811–3.

[103] Caravan P, et al. Gadolinium (III) chelates as MRI contrast agents: structure, dynamics, and applications. Chemical Reviews 1999;99(9):2293–352.

[104] Karfeld-Sulzer LS, et al. Multivalent protein polymer MRI contrast agents: controlling relaxivity via modulation of amino acid sequence. Biomacromolecules 2010;11(6):1429–36.

[105] Kamaly N, Miller AD. Paramagnetic liposome nanoparticles for cellular and tumour imaging. International Journal of Molecular Sciences 2010;11(4):1759–76.

[106] Rabin O, et al. An X-ray computed tomography imaging agent based on long-circulating bismuth sulphide nanoparticles. Nature Materials 2006;5(2):118.

[107] Hainfeld J, et al. Gold nanoparticles: a new X-ray contrast agent. British Journal of Radiology 2006;79(939):248–53.

[108] Pan D, et al. Detecting vascular biosignatures with a colloidal, radio-opaque polymeric nanoparticle. Journal of the American Chemical Society 2009;131(42):15522–7.

[109] Mukundan Jr S, et al. A liposomal nanoscale contrast agent for preclinical CT in mice. American Journal of Roentgenology 2006;186(2):300–7.

[110] Kim D, et al. Antibiofouling polymer-coated gold nanoparticles as a contrast agent for in vivo X-ray computed tomography imaging. Journal of the American Chemical Society 2007;129(24):7661–5.

[111] Dekrafft KE, et al. Iodinated nanoscale coordination polymers as potential contrast agents for computed tomography. Angewandte Chemie 2009;121(52):10085–8.

[112] Hyafil F, et al. Noninvasive detection of macrophages using a nanoparticulate contrast agent for computed tomography. Nature Medicine 2007;13(5):636.

[113] Hahn MA, et al. Nanoparticles as contrast agents for in-vivo bioimaging: current status and future perspectives. Analytical and Bioanalytical Chemistry 2011;399(1):3–27.

[114] Galperin A, et al. Radiopaque iodinated polymeric nanoparticles for X-ray imaging applications. Biomaterials 2007;28(30):4461–8.

[115] Elrod DB, et al. An iodinated liposomal computed tomographic contrast agent prepared from a diiodophosphatidylcholine lipid. Nanomedicine: Nanotechnology, Biology and Medicine 2009;5(1):42–5.

[116] Kweon S, et al. Liposomes coloaded with iopamidol/lipiodol as a RES-targeted contrast agent for computed tomography imaging. Pharmaceutical Research 2010;27(7):1408–15.

[117] Zheng J, et al. Liposome contrast agent for CT-based detection and localization of neoplastic and inflammatory lesions in rabbits: validation with FDG-PET and histology. Contrast Media & Molecular Imaging 2010;5(3):147–54.

[118] Khademi S, et al. Evaluation of size, morphology, concentration, and surface effect of gold nanoparticles on X-ray attenuation in computed tomography. Physica Medica 2018;45:127–33.

[119] Chrastina A, Schnitzer JE. Iodine-125 radiolabeling of silver nanoparticles for in vivo SPECT imaging. International Journal of Nanomedicine 2010;5:653.

[120] Kim D, et al. Antibiofouling polymer-coated gold nanoparticles as a contrast agent for in vivo X-ray computed tomography imaging [J. Am. Chem. Soc. 2007, 129, 7661–7665]. Journal of the American Chemical Society 2007;129(41):12585.

[121] Aillon KL, et al. Iodinated NanoClusters as an inhaled computed tomography contrast agent for lung visualization. Molecular Pharmaceutics 2010;7(4):1274–82.

[122] Nune SK, et al. Nanoparticles for biomedical imaging. Expert Opinion on Drug Delivery 2009;6(11):1175–94.

[123] Smith AM, et al. Multicolor quantum dots for molecular diagnostics of cancer. Expert Review of Molecular Diagnostics 2006;6(2):231–44.

[124] Pan J, Wang Y, Feng SS. Formulation, characterization, and in vitro evaluation of quantum dots loaded in poly (lactide)-vitamin E TPGS nanoparticles for cellular and molecular imaging. Biotechnology and Bioengineering 2008;101(3):622–33.

[125] Frangioni JV. In vivo near-infrared fluorescence imaging. Current Opinion in Chemical Biology 2003;7(5):626–34.

[126] Larson DR, et al. Water-soluble quantum dots for multiphoton fluorescence imaging in vivo. Science 2003;300(5624):1434–6.

[127] Gustafsson MG. Nonlinear structured-illumination microscopy: wide-field fluorescence imaging with theoretically unlimited resolution. Proceedings of the National Academy of Sciences 2005;102(37):13081–6.

[128] Cai W, et al. Are quantum dots ready for in vivo imaging in human subjects? Nanoscale Research Letters 2007;2(6):265.

[129] Rao J, Dragulescu-Andrasi A, Yao H. Fluorescence imaging in vivo: recent advances. Current Opinion in Biotechnology 2007;18(1):17–25.

[130] Gao X, et al. In vivo cancer targeting and imaging with semiconductor quantum dots. Nature Biotechnology 2004;22(8):969.

[131] Yang L, et al. Single chain epidermal growth factor receptor antibody conjugated nanoparticles for in vivo tumor targeting and imaging. Small 2009;5(2):235–43.
[132] Ballou B, et al. Sentinel lymph node imaging using quantum dots in mouse tumor models. Bioconjugate Chemistry 2007;18(2):389–96.
[133] Yuan F, et al. Nitrogen-rich D-π-A structural carbon quantum dots with a bright two-photon fluorescence for deep-tissue imaging. ACS Applied Bio Materials 2018;1(3):853–8.
[134] Ballou B, Ernst LA, Waggoner AS. Fluorescence imaging of tumors in vivo. Current Medicinal Chemistry 2005; 12(7):795–805.
[135] Wang L, et al. Watching silica nanoparticles glow in the biological world. ACS Publications; 2006.
[136] Yan J, et al. Dye-doped nanoparticles for bioanalysis. Nano Today 2007;2(3):44–50.
[137] Burns AA, et al. Fluorescent silica nanoparticles with efficient urinary excretion for nanomedicine. Nano Letters 2008;9(1):442–8.
[138] Gholami T, et al. Synthesis and characterization of spherical silica nanoparticles by modified Stöber process assisted by organic ligand. Superlattices and Microstructures 2013;61:33–41.
[139] Ow H, et al. Bright and stable core–shell fluorescent silica nanoparticles. Nano Letters 2005;5(1):113–7.
[140] Choi J, et al. Core-shell silica nanoparticles as fluorescent labels for nanomedicine. Journal of Biomedical Optics 2007;12(6):064007.

CHAPTER 7

Nanopharmacology Intervention in Human Pathological Diseases

ISHITA MATAI, PHD* • DEEPA GARG, MTECH* • UPASANA GUPTA, BTECH • ABHAY SACHDEV, PHD

1 INTRODUCTION

The advent of nanoscience and bioengineering has revolutionized our understanding and capability to tailor the existing biological process and manipulate them to improve healthcare. Nanomedicine is associated with pharmaceutical formulations of nanoscale-sized particles to develop better drug molecules and technologies for early diagnosis and treatment of diseases with the objective to enhance the efficacy and specificity so as to improve the life span and quality of patient lives [1]. Basically, it is an interdisciplinary field with the interaction of nanoscience, nanotechnology, and nanoengineering with life sciences and aims to target various domains of medical sciences including vaccine and drug development, diagnostics and imaging, antibacterial, wearable devices, tissue engineering, and drug delivery [2]. Nanomaterials (NMs) used in medicine can potentially mimic the activity of biological molecules, thereby replacing or modifying them using nanotechnology to enhance their therapeutic efficiency. Such NMs include lipid micelles [3], carbon dots (CDs) [4], graphene [5], carbon nanotubes (CNTs) [6], polymeric nanostructures [7], nanodiamonds (ND), silica nanoparticles (NPs), and magnetic NPs (such as superparamagnetic iron oxide NPs (SPIONs)) [1]. Fig. 7.1 depicts diverse NMs and NPs used in nanopharmacology with their potential targets and interactions.

These NMs have unique structure, shape, and size, which further leads to improved optical, magnetic, biological, and optical properties depending upon the process used to fabricate, purify, and characterize them. Even a small shift in the shape or size of a nanomaterial can significantly affect its properties. Therefore a great precision is necessary to synthesize NPs with focused distribution to achieve specific functions [1,8]. Nanomaterials have been demonstrated to improve the diagnosis and detection of a wide variety of diseases by acting as contrast agents for anatomical and functional imaging. They have also been used as drugdelivery carriers due to their ability to move across biological barriers [9,10]. Their application and related mechanism for drug delivery will be discussed in detail in the next section of this chapter. Besides diagnostics and drugdelivery applications, nanotechnology-based innovations have been applied to fabricate scaffolds that aid in tissue regeneration. Although most of them are still in their early stages of research and development, some have been translated into clinical practice [1].

With this perspective, this chapter mainly focuses on the overview of the nanotechnology-based drug and gene delivery and their controlled release. Different types and structures of NMs developed based on their application, mechanism, and mode of interaction with the biological system will be discussed. Other areas of medicine such as *in vivo* imaging, *in vitro* diagnostics, and therapeutic techniques are also described. Further, the challenges and upcoming prospective in the field of nanomedicine have been highlighted.

2 NANOPHARMACEUTICS FOR DRUG AND GENE DELIVERY

Medicine largely depends on the use of drugs to treat disease. The effect of a drug is directly related to the process by which it is delivered or administered as these drugs are not inherently effective. The manner in which drug delivery is carried out, in turn, affects the pharmacokinetics (PKs), distribution, excretion, toxicity, and duration of therapeutic effect [11]. Therefore along with the discovery and invention of new drug entities, there is a dire need to develop modalities to effectively deliver

*Authors contributed equally to the manuscript.

Intelligent Nanomaterials for Drug Delivery Applications. https://doi.org/10.1016/B978-0-12-817830-0.00007-2

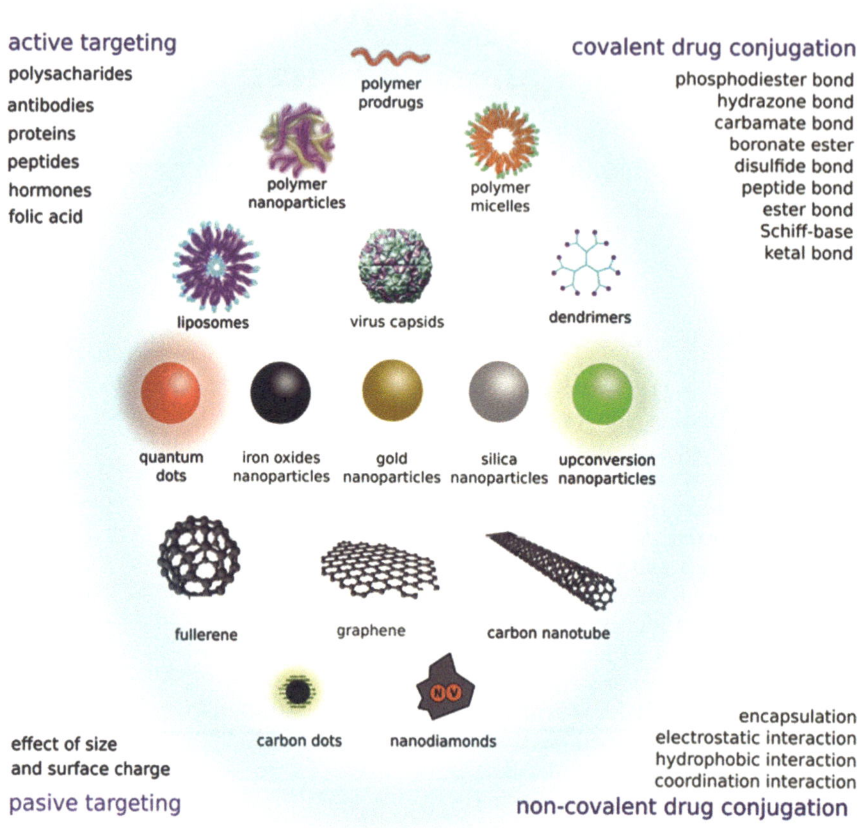

FIG. 7.1 Schematic representation of different nanocarriers as drug delivery systems along with various possibilities of drug targeting and its conjugation. (Reprinted with permission from Ulbrich K, Holá K, Šubr V, Bakandritsos A, Tuček J, Zbořil R. Targeted drug delivery with polymers and magnetic nanoparticles: covalent and noncovalent approaches, release control, and clinical studies. Chemical Reviews 2016;116(9): 5338e431. (Copyright (2016) American Chemical Society.)

them to the specific site. In this perspective, fabrication and utilization of biocompatible materials to package and deliver drugs to the target sites in the living systems are the main goals of nanomedicine [11,12].

To augment the conventional drug-delivery systems, nanocarriers need to possess certain essential features such as biocompatibility, specificity, and easy clearance. The main approach to achieve these is to modulate the PKs of nanocarriers by improving their physicochemical properties (such as shape, size, surface chemistry, and composition) and administration routes (such as intravenous intramuscular and subcutaneous). Subsequently, the improved PKs determine the pharmacodynamics (PDs) (pharmacological response of the drugs in the body) and biocompatibility (toxicity profile) of nanocarriers [13].

The first factor to be considered while designing nanocarriers is the "**size**." As NPs possess a high aspect ratio due to their small size, they are capable of penetrating the cell membrane and delivering the drugs by crossing the biological barriers that larger-sized particles cannot. In addition, NPs of dimensions 10–12 nm are preferred as they offer optimum cell permeation and nominal tissue accumulation. The size of the NPs defines the manner in which they will be eliminated. For example, small-sized NPs (hydrodynamic diameter (HD) < 5.5 nm) prefer the route of renal clearance while the larger NPs (HD > 5.5 nm) are excreted through the hepatic pathway [14]. In general, the pathway of renal clearance is preferred as it allows minimum involvement of cellular metabolism and thereby reduced body interaction with the NPs [1,15].

The "**shape**" of the NPs also influences their cell permeation and therefore affects their biodistribution and circulation time in the physiological environment. The rod-shaped NPs show better cell permeation than the spherical ones [15,16]. As elongated NPs have higher surface areas than spherical NPs, they have more number of target-specific binding sites that allow multivalent interaction with the cell surface receptors and higher cellular uptake [17]. In this context, Zhao and group demonstrated that long rod NPs possess improved ability to overcome rapid clearance by reticuloendothelial system and therefore offer better *in vivo* blood circulation time and bioavailability in comparison to other NPs such as mesoporous silica NPs (MSNPs), short rod NPs, and spherical NPs [18].

In addition, alteration of the "**surface properties**" of the NPs significantly affects their interaction with the biological system. The induction of charges on the surface of NPs influences their absorption and biodistribution. For example, NPs with the positively charged surfaces have faster and greater cellular uptake due to enhanced electrostatic interactions with the negatively charged cell membrane [15]. The coating of NPs with neutral polymers such as polyethylene glycol (PEG), poly(lactic-coglycolic acid) (PLGA), etc. also improves their stability by increasing the blood circulation time and half-life due to decreased opsonization. For instance, PEGylation (coating surface with PEG) of NPs such as proteins, quantum dots (QDs), and liposomes enhance their circulation time, solubility, and reduce toxic effects [1,19]. Moreover, charge reversible (zwitterionic) polymer-based NPs mimicking proteins (such as plasma, serum albumin) can be synthesized by exploiting their ability to change pH in the presence of different biological systems (such as tumor tissues pH (6.6–6.8), lysosomes pH (4.0–5.0), endosomes pH (5.5–6.8)) [1,20]. Recently, gold (Au) nanorods with zwitterionic surfaces have been prepared and demonstrated to have enhanced circulation time and antifouling behavior. The surface charge (initially negative) changes to positive after exposure with the acidic environment of the tumor tissues, thus leading to better cell permeation and uptake [21,22]. Abraxane, the first US Food and Drug Administration (FDA)-approved protein (serum albumin)-based nanocarrier to be launched in the market in 2005 for delivering antitumor drug paclitaxel to breast cancer tissues utilizes the zwitterionic behavior of serum albumin [23].

The fate of the drug and its pharmacological effects largely depends on its "**administration route**," which includes oral, subcutaneous, intravenous, and intranasal [13,24]. In a study by Sonaje and group, a significant difference in the biodistribution and duration of glucose level reduction can be observed when NPs loaded with insulin are administered subcutaneously and orally. When the insulin-NPs are administered subcutaneously, they do not mimic the natural physiological pathway that insulin follows (when produced endogenously to reach liver), instead they enter the pathway of peripheral circulation. In contrast, orally administered insulin-NPs mimic the physiological pathway and reach the liver to maintain the blood glucose level [13,25]. Furthermore, several studies have been carried out with reference to the dependency of biodistribution and the elimination of NPs on the administration route. Recently, *in vivo* tumor uptake, biodistribution, and elimination of CDs injected via three different routes (intravenous, subcutaneous and intramuscular) were demonstrated by Huang and coworkers. They reported that blood clearance and renal accumulation of NPs followed the order intravenous > intramuscular > subcutaneous route with NPs injected subcutaneously showing the highest tumor uptake [1,13,24].

The "**specific geometry, conformation, and composition**" of NPs contribute to their *in vivo* activity and subsequent pharmacodynamics and toxicity. The composition of the NPs affects their biodegradability to a great extent. The usage of biodegradable polymers such as PEG, chitosan, and albumin improves their elimination process and reduces toxicity [1,13].

Therefore an improvement in pharmacokinetics leads to enhanced drug delivery efficiency. Besides PKs, there are several other advantages offered by nanotechnology [1,26].

(a) Increasing the solubility of drugs (especially hydrophobic) are otherwise insoluble in aqueous solution.
(b) Increasing half-life and bioavailability of the drugs.
(c) High loading capacity of drugs due to higher aspect ratio and therefore low administration dose.
(d) Increased targeted delivery of the drug molecules and reduced off-site side effects.
(e) Reduction in potential immunogenicity.
(f) Controlled drug release.

Over the past decade, rapid advances in the development of drug carriers using a variety of inorganic/organic NMs such as DNA, liposomes, metallic NPs, polymer conjugates, and peptides have been seen [1,2,27]. So far, over 350 therapeutic products based on nanotechnology have been approved by the FDA for medical use and many are in clinical trials and shall be approved soon [28].

3 NANOPHARMACOLOGY IN DISEASE TREATMENT

3.1 Cancer

Globally, cancer is a leading cause of death, especially in the developed countries. According to the recent statistics, it has been estimated that approximately 1,735,350 new cancer cases will be diagnosed and 609,640 cancer deaths will take place in 2018 (in the United States) [29]. The conventional treatment of cancer focuses mainly on chemotherapy and radiotherapy. The efficiency of chemotherapy is low due to certain limitations that include nonspecific biodistribution, rapid drug clearance, multidrug resistance, and occasionally toxicity (especially when drugs are administered at higher doses). Therefore extensive research is being carried out to understand the cellular and molecular mechanisms of cancer to design the appropriate treatment approach [27]. This has prompted the researchers to explore the development of nanoscale drug carriers with the ability to overcome the limitations of traditional tumor therapies. Due to some characteristic features and physiology of tumor cells, NPs can target them more actively and passively than other cells of the body [30]. Tumorous tissues have leaky blood vessels that preferentially allow only nanoscale particles to penetrate through them and accumulate, while normal tissues do not have leaky blood vessels and therefore do not permit the NPs to enter. This process is known as enhanced permeation retention (EPR) effect and it helps in targeted drug delivery to tumor tissues [1,30]. Fig. 7.2 depicts the mechanism behind the passive targeting of macromolecular drug-delivery nanocarriers to tumors. The nanocarriers circulate in the bloodstream, preferentially extravasate and accumulate in tumor tissues due to the EPR effect. The incorporated drug is then released (either by pH-mediated or endosome-mediated hydrolysis) in the tumor tissue [27].

Various nanocarriers including genes [31], soluble polymeric structures [32,33], polymeric micelles [34], NPs [35,36], liposomes [37], and nanocapsules [38] have been investigated to accumulate in tumor tissues via the EPR effect [1].

Doxil, a liposome incorporated with an antitumor drug DOX, was the first nanocarrier approved by FDA for tumor treatment [30]. Although liposomes encapsulated drugs have proved to improve their PKs and biodistribution, yet none of the approved liposomal carriers have shown potential to improve the overall survival (OS) as compared to the conventional drug [39]. However, recently, FDA has approved Vyxeos, the first nanomedicine for the treatment of high risk acute myeloid leukemia. Vyxeos, also known as CPX 351, is a liposomal formulation of two chemotherapeutic drugs

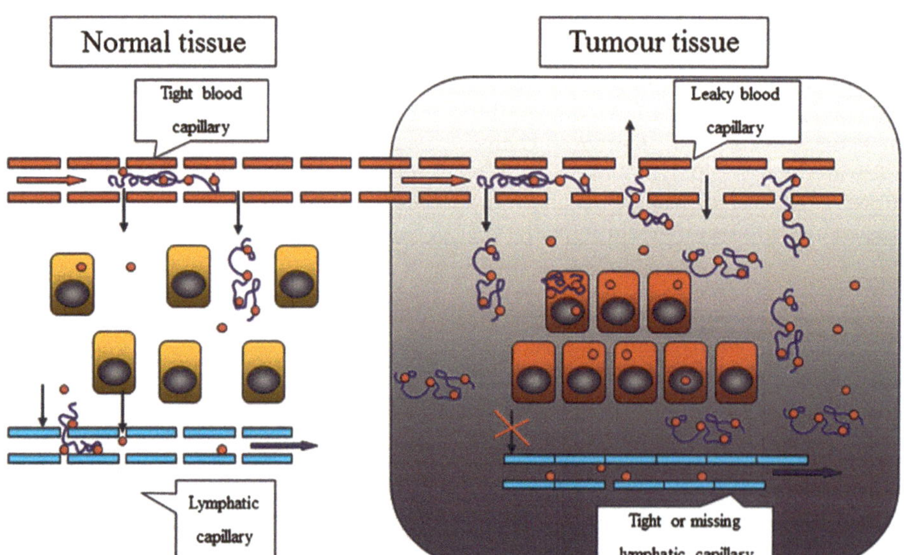

FIG. 7.2 Schematics of macromolecules accumulation in solid tumor tissue due to the EPR effect. (Red dots: Low Molecular Weight drug, blue line: macromolecule). (Reprinted with permission from Ulbrich K, Holá K, Šubr V, Bakandritsos A, Tuček J, Zbořil R. Targeted drug delivery with polymers and magnetic nanoparticles: covalent and noncovalent approaches, release control, and clinical studies. Chemical Reviews 2016;116(9): 5338e431. (Copyright (2016) American Chemical Society.)

cytarabine and daunorubicin and showed the marked improvement in OS from 5.95 months (in native chemotherapeutic drugs) to 9.56 (in Vyxeos) [40].

Along with lipids, several other nanocarriers have been approved by the FDA for cancer therapeutics. Some of them are polymeric micelles such as Genexol-PM2 [30], NK105 [41] and polymeric NP such as CRLX101 [42], AZD 2811 Accurin [43]. Inorganic NPs including gold nanoshells, hafnium oxide NPs, and iron oxide NPs are still being investigated for cancer treatment [30]. Iron oxide NP-based formulation known as NanoTherm has been accepted for marketing in European countries for the treatment of glioblastoma [44].

For many decades, nanomedicine has been widely accepted and reported for their use in chemotherapeutics delivery; more recently, it has been explored for the delivery of new antitumor agents such as small interfering RNA (siRNA) [45], molecularly targeted agents [43], antisense oligonucleotides [46], messenger RNA (mRNA) [47], and DNA inhibitor oligonucleotides [48]. Moreover, viral NPs such as lentivirus used for delivery of gene-based drugs for the treatment of cancer are under clinical trials [49]. Plant viruses such as tobacco mosaic virus [50] and potato virus X [51] are also being engineered and tested in animal models [30]. Plant viruses offer several advantages over other nanocarriers like noninfective toward mammals (due to their natural origin), easy large-scale production, and improved stability toward genetic or chemical modifications [50].

Lately, nanomedicine has shown immense potential in the area of cancer immunotherapy by replacing biologics such as monoclonal antibodies (mAb) with the NPs for addressing the issue of antidrug-antibodies (ADAs) formation leading to hypersensitivity reactions when they are administered into the body [30]. Kishimoto and group demonstrated the ability of tolerogenic NPs (synthesized using PLGA) loaded with drug rapamycin to prevent the formation of ADAs in nonhuman primates and mice. The human clinical trials for these NPs are currently under clinical trials [52].

3.2 Renal Diseases

Currently, the management and treatment of renal diseases present a huge burden, and this prompted the development of nanotechnology-based therapeutics for better treatment. Although the application of nanomedicine in renal therapy is still in its early developmental stage, still it has shown immense potential [15]. A deep understanding of the renal anatomy and designing of shape, size, structure, and surface chemistry of NPs is essential owing to the challenges that occur due to their engineering for renal diseases [53]. Most of the currently used small drugs accumulate in the liver or spleen or are cleared from the renal pathway even before achieving the desired therapeutic effect [15]. Therefore NPs designed to bind to, taken up, and retained by the different types of kidney cells or/and components of extracellular matrix (ECM) (such as **glomerular basement membrane (GBM)**) have been successfully exploited to improve therapeutic efficacy and reduce systemic toxicity [53,54]. Such NPs enhance targeted delivery of drug molecules [55], peptides and proteins [56], or nucleic acids [57] to the specific diseased renal cell. They can also deliver RNA interference molecules, but the efficiency of their treatment is still limited [58]. For example, ammonium functionalized CNTs (f-CNTs) were developed by Alidori and group to mediate delivery of small siRNA cargo selectively and efficiently to proximal tubule cells to treat acute kidney injury. The f-CNTs knocked down the expression of target genes Trp53, Mep1b, and Ctr1 more efficiently compared to free drugs [59].

It has been observed that the shape of the NPs greatly influences their localization in the renal cells. For instance, nonspherical NPs are cleared faster than the rod-shaped NPs [18]. Another important parameter for renal filtration is NPs charge, with anionic NPs showing better glomerular retention and minimized nonspecific uptake than cationic NPs. For example, fluorescence imaging of QDs (both anionic and cationic) demonstrated better retention of anionic QDs by mesangial cells, that is, for 30 days. This was due to the prevention of anionic QDs due to the negatively charged GBM [15]. Fig. 7.3 depicts the structure of a glomerular filtration barrier along with its different components.

Renal cell carcinoma is a lethal tumor and leaves patients with limited treatment options. Recently, several groups have demonstrated the potential of immunotherapy and targeted therapy for the treatment of metastatic renal cancer using nanomedicine [15]. Several targets including immunomodulatory mAb targeting programmed cell death protein 1 (PD-1) and cytotoxic T-lymphocyte-associated protein 4 (CTLA-4) have been approved and made available, yet their clinical value has not been shown to be comparable to the established treatments such as the mechanistic target of rapamycin or tyrosine inhibitors [59]. Nevertheless, nanotechnology-based carrier designing for targeted delivery of therapeutics and vaccines is under extensive research and development.

3.3 Cardiovascular Diseases

Despite several advances in the understanding and the subsequent treatments, cardiovascular diseases including heart failure remains the leading cause of death globally with approximately 50% of the patients diagnosed, dying within a short span of 5 years [60]. Although effective, cardiac transplants are very limited owing to the

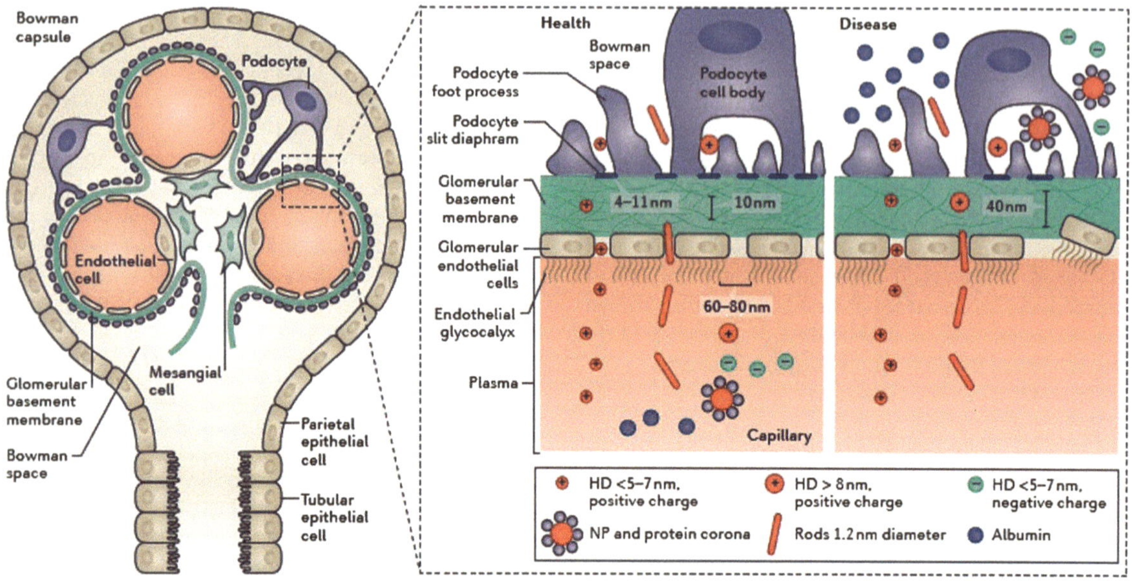

FIG. 7.3 Schematic representation of kidney glomerulus and glomerular basement membrane (GBM) in healthy and diseased conditions. (Reprinted with permission from Kamaly N, He JC, Ausiello DA, Farokhzad OC. Nanomedicines for renal disease: current status and future applications. Nature Reviews. Nephrology 2016;12(12):738e53. (Copyright (2016) American Chemical Society.)

availability of a compatible donor. Current therapies such as therapeutics, implant devices, and surgical reconstruction have enhanced the quality and life span for several patients, but systolic heart failure continues to be a progressive disease [61]. Hence, better therapeutic approaches are required. In this context, nanomedicine holds huge potential to improve the existing preventive, diagnostic, and therapeutic approaches for cardiac diseases [62,63].

Several nanoparticle systems are being engineered to prevent the stenosis and blocking of coronary arteries by breaking down the plaques. Nanodrugs can also treat the injured cells or tissues by reducing the effects of the injuries or regenerating the myocardium, by targeted and controlled delivery of biomolecules, cells, or paracrine factors [61].

Recently, as an alternative to therapeutic delivery, fabrication of acellular scaffolds mimicking the cellular and molecular patterns of heart tissue along with suitable physical–mechanical support have shown very promising outcomes for repairing tissue injury [61]. Moreover, nanoparticles can also reduce inflammation at the site of atherosclerosis by targeting and delivering therapeutic anti-inflammatory biomolecules [64–66]. Biodegradable polymeric nanomaterials including poly(D-lactic acid), PEG, and PLGA have been used to help resolve inflammation [66,67]. Anti-inflammatory biomolecules such as Ac2-26 peptide [67] and interleukin-10 cytokine [66] and drugs such as statin [68] have been documented to be released by the NPs to the site of atherosclerosis and plaques [61]. For instance, Kamaly and coworkers demonstrated *in vivo* delivery of anti-inflammatory biomolecule interleukin-10 to the atherosclerotic plaque using polymeric NPs (PLGA-PEG targeted with anticollagen IV) and a significant decrease in the necrotic core region in a mouse model [66].

Moreover, the targeted delivery of nanomedicine-based therapeutic stem cells can enhance therapeutic efficiency and reduce off-site drug accumulation [61]. In a study, dual antibody-conjugated iron oxide magnetic NPs were prepared to achieve targeted delivery, bioimaging (using fluorescence) and *in vivo* tissue repair. One antibody (anti-cd45) was specific to the therapeutic cells, and the other (anti-cd34) is specific to injured CMs (Fig. 7.4). This system demonstrated better accumulation of therapeutic cells in the injured heart region, reduced accumulation in the liver, and enhanced therapeutic efficiency of cells in tissue repair [69].

3.4 Ocular Diseases

Nanoscale drug-delivery systems (DDS) have enormous potential to overcome the limitations of conventional approaches that require the need for regular intravitreal dosing to ocular disorders [70]. For instance, Ozurdex is FDA-approved intravitreal implant containing drug dexamethasone (0.7 mg) loaded in a solid,

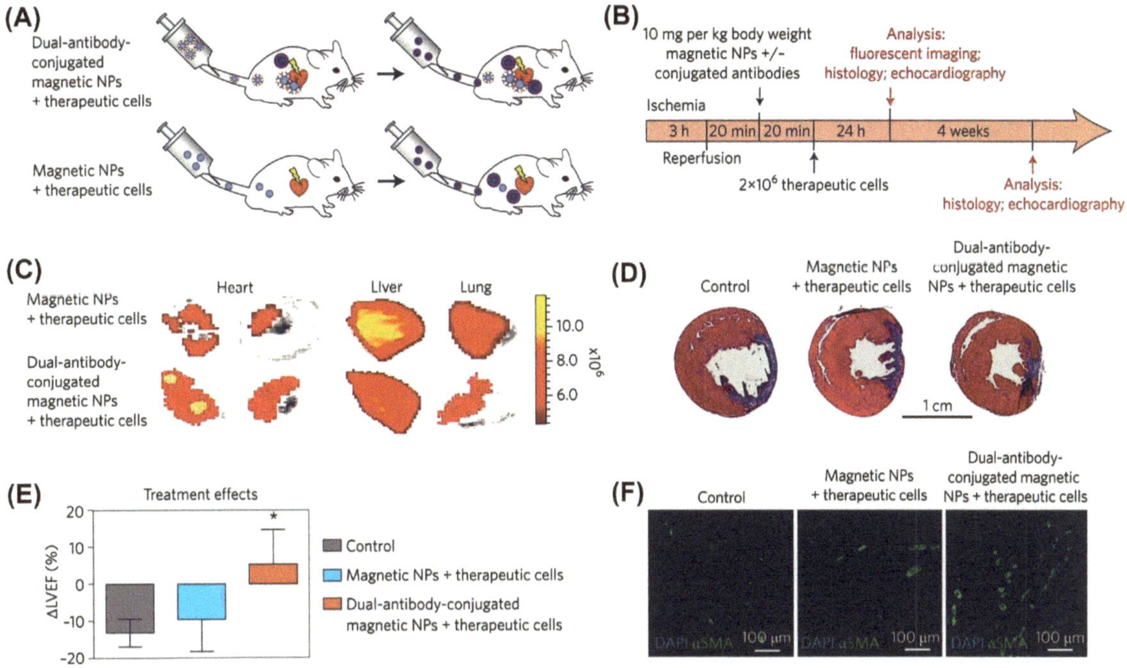

FIG. 7.4 Schematics of application of nanostructured cardiac biomaterial in repair or regeneration of the myocardial infarction (MI). (A) Different scenarios by which the engineered nanostructured scaffold can contribute to myocardial tissue repair following MI: (i) inhibition of adverse cardiac remodelling processes post-MI via physical mechanical support; (ii) targeted delivery and sustained release of anti-apoptotic factors; (iii) promoting angiogenesis (via delivery of angiogenic factors); and (iv) promoting proliferation of pre-existing native CMs (for example, through epicardial delivery of human FSTL1 peptide). (B) Schematic demonstration of cellular and molecular processes in healthy (top), post-MI (middle) and patch-treated (bottom) heart tissue. The bottom panel delineates the role of a nanostructured cardiac patch in regenerating the damaged myocardium by inducing cell cycle re-entry among pre-existing CMs. (Reprinted with permission from Lobatto ME, Fuster V, Fayad ZA, Mulder WJM. Perspectives and opportunities for nanomedicine in the management of atherosclerosis. Nature Reviews Drug Discovery 2011;10(11):835e52. (Copyright (2017) Nature Nanotechnology.)

biodegradable nanocarrier and used for the treatment of macular edema. It is directly injected into the vitreous and releases dexamethasone for the desired time, thus leading to excellent improvement in vision [71]. Although utilizing nanomedicine for the treatment of retinal disorders, special attention is needed to manipulate the NM's size and surface properties. It should not be too large to prevent efficient entry into the blood vessels, but it should not be so small that it cannot even penetrate the biological barriers restricting its flow to the retina [72]. Nanocarriers based on polyacrylic acid, chitosan, bovine serum albumin, and Eudragit RS have been used for ocular drug delivery [73]. Clinical trials utilizing these and other nanosystems are still underway.

3.5 Neurological Diseases

Our nervous system is very much prone to injuries and disorders such as trauma, tumor, or neurodegenerative diseases. Current therapeutic approaches include drugs, radiotherapy chemotherapy, and surgery that are only partially successful because the mortality rate and life quality even after the treatment remains unsatisfied. This is due to the lack of understanding of neurons, their circuits and therefore the deep investigation of the disease and its subsequent treatment cannot be carried out efficiently [1,74]. These challenges have encouraged researchers to develop strategies that can improve the knowledge of brain circuitry and therapeutic efficacy. In this regard, nanotechnology has shown an important path for deeper research in miniaturization and improvement in the efficiency of small artificial devices including nanodevices used for neural interfaces [75]. Nanotools such as nanowires have the potential to improve the fabrication and development of brain-inspired computing techniques that aid in the understanding of the operation of the neural system, communication of the neurons, and their organization into networks in different mental states [76,77]. Due to better solubility and ability to cross the blood–brain barrier, nanocarriers such as liposomes [78], micelles [79], dendrimers [80,81], or magnetic NPs [82] can be used to deliver drugs to the brain. Visudyne is one of a clinically approved liposomal nanocarrier for intravenous delivery of drug verteporfin to treat age-related macular degeneration, myopia (pathologic), and ocular histoplasmosis [83]. Additionally, nanomaterials based on polymers [84], CNTs [85] are being used as implants for patients with neurological disorders (such as paralysis, blindness, or epilepsy) due to their ability to enhance cell adhesion and sometimes cell growth [74].

Several groups have documented the use of nanotechnology-based treatment of Alzheimer's disease (AD), one of the most studied and complex neurodegenerative disorders [79,86–88]. Most of these therapies focus on the inhibition of β-amyloid and hyperphosphorylated tau aggregation. Recently, Chen and coworkers fabricated a novel multifunctional nanocomposite (CeNC/IONC/MSN-T807) with enhanced efficiency to bind and reduce aggregated hyperphosphorylated tau protein. The nanocomposite was constructed by modification of MSNP's surface with ultrasmall ceria nanocrystals (CeNCs) (antioxidants), iron oxide nanocrystals (IONCs) (MRI agents), and T807 (tau target), followed by loading with methylene blue (MB), an inhibitor of tau aggregation. They showed that the synthesized nanocomposites loaded with MB alleviated AD symptoms and rescued the deficits in memory of the challenged rat models [89].

4 FDA-APPROVED NANODRUGS

Since 1990, several nanodrugs have received FDA approval for use in medical applications, and many are currently in clinical trials [28,90]. Liposomes, polymers, and nanocrystals are the most frequently used nanocarriers for drugs along with micelles, inorganic NPs that are less frequently used. These are mainly administered intravenously or orally and sometimes transdermally. These nanodrugs have been approved for a wide variety of diseases such as cancer, severe combined immunodeficiency, liver malfunction, and so on with a maximum focus on various types of cancer [28,91]. A listing of FDA-approved nanodrugs is mentioned in Table 7.1.

5 POSSIBLE OUTCOMES OF NANOPHARMACOLOGY

Although recent technological innovations in the field of nanotechnology have led to a marked improvement in the medical status, it is accompanied by the potential health and environmental risks posed by the negative effects of the NPs [2]. Moreover, the scale-up of nanomedicine to the industrial level with reproducible and consistent efficiency is a time consuming, expensive, and complex process [1,27]. These concerns have somewhat limited their clinical translation. Therefore some

TABLE 7.1
FDA-Approved Nanopharmaceutical Formulations.

Name	Composition	Advantage	Targeted Disease	Year(s) Approved
POLYMER NANOPARTICLES				
Zilretta	Triamcinolone acetonide	Extended pain relief over 12 weeks	Osteoarthritis (OA) of the knee	2017
Rebinyn	Coagulation factor IX (recombinant)	Effective control in 95% of bleeding episodes; 98% of bleeds were treated with 1–2 infusions	Control and prevention of bleeding episodes and prevention of bleeding in the perioperative setting for hemophilia B patients	2017
ADYNOVATE (Baxalta)	Polymer–protein conjugate (PEGylated factor VIII)	Improved stability of protein through PEGylation	Hemophilia	2015
Plegridy® (Biogen)	Polymer–protein conjugate (PEGylated IFN beta-1a)	Improved stability of protein through PEGylation	Multiple sclerosis	2014

TABLE 7.1
FDA-Approved Nanopharmaceutical Formulations.—cont'd

Name	Composition	Advantage	Targeted Disease	Year(s) Approved
Cimzia®/ certolizumabpegol (UCB)	PEGylated antibody fragment (certolizumab)	Improved circulation time and greater stability *in vivo*.	Psoriatic arthritis Ankylosing Spondylitis Crohn's disease Rheumatoid arthritis	2013 2013 2008 2009
Mircera®/Methoxy polyethylene glycol-epoetin beta (Hoffman-La Roche)	Chemically synthesized ESA (erythropoiesis-stimulating agent)	Improved stability of aptamer as a result of PEGylation	Anemia associated with chronic kidney disease	2007
Macugen®/ Pegaptanib (Bausch & Lomb)	PEGylated anti-VEGF aptamer (vascular endothelial growth factor) aptamer	Improved stability of aptamer as a result of PEGylation	Macular degeneration, neovascular age-related	2004
Somavert®/ pegvisomant (Pfizer)	PEGylated HGH receptor antagonist	Improved stability of protein through PEGylation	Acromegaly	2003
Eligard® (Tolmar)	Leuprolide acetate and polymer (PLGH (poly (DL-lactide-co-glycolide)))	Controlled delivery of payload with longer circulation time	Prostate cancer	2002
Neulasta®/ pegfilgrastim (Amgen)	PEGylated GCSF protein	Improved stability of protein through PEGylation	Neutropenia, Chemotherapy induced	2002
Pegasys® (Genentech)	PEGylated IFN alpha-2a protein	Improved stability of protein through PEGylation	Hepatitis B; Hepatitis C	2002
PegIntron® (Merck)	PEGylated IFN alpha-2b protein	Improved stability of protein through PEGylation	Hepatitis C	2001
Renagel®[sevelamer hydrochloride]/ Renagel®[sevelamer carbonate] (Sanofi)	Poly(allylamine hydrochloride)	Increase circulation and therapeutic delivery	Chronic kidney disease	2000
Copaxone®/Glatopa (Teva)	Random copolymer of L-glutamate, L-alanine, L-lysine, and L-tyrosine	Large amino-acid based polymer with controlled molecular weight and clearance characteristics	Multiple sclerosis (MSs)	1996
Oncaspar®/ pegaspargase (Enzon Pharmaceuticals)	Polymer–protein conjugate (PEGylated L-asparaginase)	Improved stability of protein through PEGylation	Acute lymphoblastic leukemia	1994
Zinostatin	Polymer styrene maleic anhydride	Improve dispersion in the phase-contrast agent Lipiodol	Liver cancer	1994 (Japan)
Adagen®/ pegademase bovine (Sigma-Tau Pharmaceuticals)	PEGylated adenosine deaminase enzyme	Improved circulation time and decreased immunogenicity	Severe combined immunodeficiency disease (SCID)	1990

(continued)

TABLE 7.1
FDA-Approved Nanopharmaceutical Formulations.—cont'd

Name	Composition	Advantage	Targeted Disease	Year(s) Approved
LIPOSOME FORMULATIONS				
Onivyde® (Merrimack)	Liposomal Irinotecan	Long circulation time	Metastatic adenocarcinoma of the pancreas	2015
Abelcet® (Sigma-tau)	Liposomal amphotericin B lipid complex	Reduced toxicity	Fungal infections	2015
Onivyde® (Merrimack)	Liposomal Irinotecan	Increased delivery to tumor site; lower systemic toxicity arising from side effects	Pancreatic cancer	2015
Marqibo® (Onco TCS)	Liposomal vincristine	Increased delivery to tumor site; lower systemic toxicity Arising from side effects	Acute lymphoblastic leukemia	2012
Exparel® (Pacira)	Liposomal bupivacaine	Long circulation time	Pain management	2011
Doxil®/Caelyx™ (Janssen)	Liposomal doxorubicin	Improved delivery to the site of disease; decrease in systemic toxicity of the free drug.	Multiple myeloma Ovarian cancer Karposi's Sarcoma	2008 2005 1995
AmBisome® (Gilead Sciences)	Liposomal amphotericin B Liposomal morphine sulfate	Reduced nephrotoxicity Extended release	Analgesia (postoperative) Fungal/protozoal Infections	2004 1997
DepoDur® (Pacira Pharmaceuticals)				
Visudyne® (Bausch and Lomb)	Liposomal verteporfin	Increased delivery to the site of diseased vessels; photosensitive release	Macular degeneration, wet age-related; myopia; ocular Histoplasmosis	2002
Myocet®/Evacet®	Liposomal doxorubicin	Increased delivery to tumor site	Metastatic breast cancer	2000 (approved in Europe and Canada)
Curosurf®/Poractant alpha (Chieseifarmaceutici)	Liposome-proteins SP-B and SP-C	Increased delivery for smaller volume; reduced toxicity, increased delivery for smaller volume; reduced toxicity	Pulmonary surfactant for respiratory distress syndrome	1999

TABLE 7.1
FDA-Approved Nanopharmaceutical Formulations.—cont'd

Name	Composition	Advantage	Targeted Disease	Year(s) Approved
DepoCyt© (Sigma-tau)	Liposomal cytarabine	Increased delivery to tumor site; lower systemic toxicity arising from side effects	Lymphomatous meningitis	1996
DaunoXome® (Galen)	Liposomal daunorubicin	Increased delivery to tumor site; lower systemic toxicity arising from side effects	Karposi's sarcoma	1996
MICELLAR NANOPARTICLES				
Genexol-PM (Samayang)	Micellar paclitaxel	Improved solubility Reduced toxicity Improved efficacy Reduced hypersensitivity	Breast cancer, nonsmall-cell lung cancer	2007 (Korea, Europe)
Estrasorb™ (Novavax)	Micellar estradiol	Controlled delivery of therapeutic	Menopausal therapy	2003
PROTEIN NANOPARTICLES				
Abraxane®/ABI-007 (Celgene)	Albumin-bound paclitaxel nanoparticles	Improved solubility; improved delivery to tumor	Pancreatic cancer NSCLC Breast cancer	2013 2012 2005
Ontak® (Eisai Inc)	Engineered protein combining IL-2 and diphtheria toxin	Targeted T-cell specificity; lysosomal escape	Cutaneous T-Cell Lymphoma	1999
NANOCRYSTALS				
Ryanodex® (Eagle Pharmaceuticals)	Dantrolene sodium	Faster administration at higher doses	Malignant hypothermia	2014
Invega® Sustenna® (Janssen Pharms)	Paliperidone Palmitate	Allows slow release of injectable low solubility drug	Schizoaffective disorder Schizophrenia	2014 2009
EquivaBone® (Zimmer Biomet)	Hydroxyapatite	Mimics bone structure	Bone substitute	2009
Focalin XR® (Novartis)	Dexamethylphenidate HCl	Increased drug loading and bioavailability	Psychostimulant	2005
NanOss® (Rti surgical)	Hydroxyapatite	Mimics bone structure allowing cell adhesion and growth	Bone substitute	2005
Ostim® (HeraseusKulzer)	Hydroxyapatite	Mimics bone structure allowing cell adhesion and growth	Bone substitute	2004

(continued)

TABLE 7.1
FDA-Approved Nanopharmaceutical Formulations.—cont'd

Name	Composition	Advantage	Targeted Disease	Year(s) Approved
Tricor® (Lupin Atlantis)	Fenofibrate	Increases bioavailability, simplifies administration	Hyperlipidemia	2004
Emend® (Merck)	Aprepitant	Surface area allows faster absorption and increases the bioavailability	Antiemetic	2003
OsSatura® (IsoTis Orthobiologics)	Hydroxyapatite	Mimics bone structure allowing cell adhesion and growth	Bone substitute	2003
Vitoss® (Stryker)	Calcium phosphate	Mimics bone structure allowing cell adhesion and growth	Bone substitute	2003
Zanaflex® (Acorda)	Tizanidine HCl	Increased drug loading and bioavailability	Muscle relaxant	2002
Ritalin LA® (Novartis)	Methylphenidate HCl	Increased drug loading and bioavailability	Psychostimulant	2002
Avinza® (Pfizer)	Morphine sulfate	Increased drug loading and bioavailability; extended release	Psychostimulant	2002 (2015)
Megace ES® (Par Pharmaceuticals)	Megestrol acetate	Reduced dosing	Antianorexic	2001
Rapamune® (Wyeth Pharmaceuticals)	Sirolimus	Increased bioavailability	Immunosuppressant	2000
INORGANIC AND METALLIC NANOPARTICLES				
Nanotherm® (MagForce)	Iron oxide	Allows cell uptake and introduces superparamagnetism	Glioblastoma	2010
Feraheme™/ ferumoxytol (AMAG pharmaceuticals)	Ferumoxytol SPION with polyglucose sorbitol carboxymethylether	Magnetite suspension allows for prolonged steady release, decreasing number of doses	Deficiency anemia iron deficiency in chronic kidney disease (CKD)	2009
GastroMARK™; umirem® (AMAG pharmaceuticals)	SPION coated with silicon	Superparamagnetic character	Imaging agent	2001 (2009)
Venofer® (Luitpold Pharmaceuticals)	Iron sucrose	Allows increased dose	Iron deficiency in chronic kidney disease (CKD)	2000
Ferrlecit® (SanofiAvertis)	Sodium ferric gluconate	Allows increased dose	Iron deficiency in chronic kidney disease (CKD)	1999

TABLE 7.1
FDA-Approved Nanopharmaceutical Formulations.—cont'd

Name	Composition	Advantage	Targeted Disease	Year(s) Approved
Feridex®/Endorem® (AMAG pharmaceuticals)	SPION coated with dextran	Superparamagnetic character	Imaging agent	1996 (2008)
INFeD® (SanofiAvertis)	Iron dextran (low MW)	Allows increased dose	Iron deficiency in chronic kkidney dlsease (CKD)	1957
DexIron®/ Dexferrum® (SanofiAvertis)	Iron dextran (high MW)	Allows increased dose	Iron deficiency in chronic kidney disease (CKD)	1957

issues need to be taken into consideration to grab maximal benefits from nanopharmacological innovations. Firstly, a deep understanding of the effect of the physicochemical properties (such as size, surface chemistry, shape, hydrophobic, hydrophilicity, biodegradation, stability) on the *in vivo* biological response and fate is necessary [1,2]. Another major issue that should be taken care of is the formation of the protein corona. Protein corona is a layer of blood proteins (including albumin, fibrinogen, immunoglobulin, and lipoproteins) adsorbed by the injection of NPs in the blood that often alters the biokinetics of the NPs depending on the type and surface modification of the polymer used. This, in turn, may lead to undesirable accumulation of NPs at few sites [1,92,93]. Therefore these properties need to be utilized in a way that improves the therapeutic efficiency of the approach. Another important aspect of nanomedicine is the toxicity that limits its approval and use in medical applications. Increased exposure of humans to NPs via different routes (injection, ingestion, cutaneous uptake, or inhalation) and their interaction with the body on the molecular and cellular levels have encouraged the evaluation of safety and toxicity associated with these NMs [1,2,94]. This toxicological profile of the NMs on human health and society is addressed by a discipline in nanotechnology, known as "**Nanotoxicology**" [2]. Due to the large surface area-to-volume ratio, NPs are more effective in penetrating cells and tissues, often leading to the release of immune responses [95]. They may acquire access to the systemic circulation and subsequently translocate to the critically delicate organs mainly the heart, brain, spleen, bone marrow, and lymph. Once inhaled, NPs may enter the respiratory tract by-passing normal phagocytic pathways and lead to inflammatory responses [92]. Often, the cells in the reticuloendothelial systems take up NPs resulting in their accumulation in spleen and liver [1,96,97]. Hence, besides the efficiency of the NMs, it is also important to consider its safety and toxicity on the human health, taking an example of one of the outstanding candidates for biomedical application cobalt ferrite (CoFe) NPs. In spite of its remarkable features, extensive usage may result in its easy penetration in the human body by different pathways and subsequent induction of oxidative stress, cytotoxicity, inflammatory response, and aggregation of Co and Fe [98]. Other NMs such as CNTs [99], graphene [100], QDs [101], magnetic NPs, metal oxides, and MSNPs have been found to induce toxicity in various biological systems via different molecular mechanisms [102]. The toxicity problems related to QDs synthesized especially from heavy metals (like cadmium and mercury) come from their chemical composition [101]. These heavy metal ions may release into the blood circulation leading to their accumulation in various organs of the body namely spleen, kidney, and liver and lead to subsequent cell death [90,103]. The generation of reactive oxygen species and oxidative stress is the basic mechanism behind the cellular and genotoxicity. They severely affect the genetic makeup of the cell by inducing carcinogenesis, mutagenesis, or aging promoting diseases. Often, they lead to disruption in the regular cell signaling pathway and mobility [104]. Additionally, NPs in the range of 20–200 nm might not be excreted from the bloodstream through renal clearance or liver uptake, thus resulting in their long circulation period in the blood [1]. Therefore to develop safe medical devices based on nanotechnology, it is essential to assess the potential toxicological effects of these NMs on the biological systems.

6 CONCLUSION AND FUTURE PROSPECTIVE

New nanomaterials and their formulations are emerging for nanomedicine especially in the areas of diagnostics, bioimaging, drug delivery, antibacterial resistance, and cancer therapy. Recent studies have demonstrated the use of multiple NMs (such as NDs and proteins) or single nanomaterial functionalized with a variety of therapeutic molecules and contrast agents that can successfully image, target, and treat a wide variety of diseases with enhanced efficiency. Nanotechnology in cellular, preclinical, and clinical research has led to prominent advancements both at the fundamental and translational levels. However, most of these studies are related to tumor detection, diagnosis, and treatment and very less focus has been applied to other diseases. This bias emphasis on the tumor is expected to be considered in the upcoming decades with more focus extending to other challenging conditions including antibiotic resistance and engineering of artificial organs. Therefore it is expected that in the upcoming years, nanomedicine will reach every aspect of healthcare and NMs shall be able to amend the current therapeutic strategies significantly. Although nanomedicine has increased the expectations and hopes for better medical technologies, it has raised several scientific challenges and medical problems too. This is mainly due to insufficient knowledge about the NMs interaction with the biological systems. However, extensive research is being carried to address these issues for a better understanding of nanomaterial–living system interactions. This is propitious to reach the stage where real nanomedical solutions can emerge.

REFERENCES

[1] Pelaz B, Alexiou C, Alvarez-Puebla RA, Alves F, Andrews AM, Ashraf S, Balogh LP, Ballerini L, Bestetti A, Brendel C, et al. Diverse applications of nanomedicine. ACS Nano 2017;11(3):2313–81.

[2] Jain K, Mehra NK, Jain NK. Potentials and emerging trends in nanopharmacology. Current Opinion in Pharmacology 2014;15:97–106.

[3] Irby D, Du C, Li F. Lipid–drug conjugate for enhancing drug delivery. Molecular Pharmaceutics 2017;14(5):1325–38.

[4] Hua X-W, Bao Y-W, Wu F-G. Fluorescent carbon quantum dots with intrinsic nucleolus-targeting capability for nucleolus imaging and enhanced cytosolic and nuclear drug delivery. ACS Applied Materials and Interfaces 2018;10(13):10664–77.

[5] Wang Z, Colombi Ciacchi L, Wei G. Recent advances in the synthesis of graphene-based nanomaterials for controlled drug delivery. Applied Sciences 2017;7(11):1175.

[6] Rastogi V, Yadav P, Bhattacharya SS, Mishra AK, Verma N, Verma A, Pandit JK. Carbon nanotubes: an emerging drug carrier for targeting cancer cells. Journal of Drug Delivery 2014;2014:1–23.

[7] Matai I, Gopinath P. Hydrophobic myristic acid modified PAMAM dendrimers augment the delivery of tamoxifen to breast cancer cells. RSC Advances 2016;6(30):24808–19.

[8] Sapsford KE, Tyner KM, Dair BJ, Deschamps JR, Medintz IL. Analyzing nanomaterial bioconjugates: a review of current and emerging purification and characterization techniques. Analytical Chemistry 2011;83(12):4453–88.

[9] Gobin AM, Lee MH, Halas NJ, James WD, Drezek RA, West JL. Near-infrared resonant nanoshells for combined optical imaging and photothermal cancer therapy. Nano Letters 2007;7(7):1929–34.

[10] Altinoğlu EI, Russin TJ, Kaiser JM, Barth BM, Eklund PC, Kester M, Adair JH. Near-infrared emitting fluorophore-doped calcium phosphate nanoparticles for *in vivo* imaging of human breast cancer. ACS Nano 2008;2(10):2075–84.

[11] Tibbitt MW, Dahlman JE, Langer R. Emerging frontiers in drug delivery. Journal of the American Chemical Society 2016;138(3):704–17.

[12] Shi J, Votruba AR, Farokhzad OC, Langer R. NIH public access. Nano Letters 2005;257(5):2432–7.

[13] Kang H, Mintri S, Venugopal Menon A, Yeon Lee H, Soo Choi H, Kim Hak Soo Choi J. Pharmacokinetics, pharmacodynamics and toxicology of theranostic nanoparticles HHS public access. Nanoscale 2015;7(45):18848–62.

[14] Soo Choi H, Liu W, Misra P, Tanaka E, Zimmer JP, Itty Ipe B, Bawendi MG, Frangioni JV. Renal clearance of quantum dots. Nature Biotechnology 2007;25(10):1165–70.

[15] Kamaly N, He JC, Ausiello DA, Farokhzad OC. Nanomedicines for renal disease: current status and future applications. Nature Reviews. Nephrology 2016;12(12):738–53.

[16] Salatin S, Maleki Dizaj S, Yari Khosroushahi A. Effect of the surface modification, size, and shape on cellular uptake of nanoparticles. Cell Biology International 2015;39(8):881–90.

[17] Dasgupta S, Auth T, Gompper G. Shape and orientation matter for the cellular uptake of nonspherical particles. Nano Letters 2014;14(2):687–93.

[18] Zhao Y, Wang Y, Ran F, Cui Y, Liu C, Zhao Q, Gao Y, Wang D, Wang S. A comparison between sphere and rod nanoparticles regarding their *in vivo* biological behavior and pharmacokinetics. Scientific Reports 2017;7(1):4131.

[19] Suk JS, Xu Q, Kim N, Hanes J, Ensign LM. PEGylation as a strategy for improving nanoparticle-based drug and gene delivery. Advanced Drug Delivery Reviews 2016;99(Pt A):28–51.

[20] Yuan Y-Y, Mao C-Q, Du X-J, Du J-Z, Wang F, Wang J. Surface charge switchable nanoparticles based on zwitterionic polymer for enhanced drug delivery to tumor. Advanced Materials 2012;24(40):5476–80.

[21] Liu X, Li H, Chen Y, Jin Q, Ren K, Ji J. Mixed-charge nanoparticles for long circulation, low reticuloendothelial system clearance, and high tumor accumulation. Advanced Healthcare Materials 2014;3(9):1439–47.

[22] Liu X, Chen Y, Li H, Huang N, Jin Q, Ren K, Ji J. Enhanced retention and cellular uptake of nanoparticles

in tumors by controlling their aggregation behavior. ACS Nano 2013;7(7):6244–57.
[23] Gradishar WJ. Albumin-bound paclitaxel: a next-generation taxane. Expert Opinion on Pharmacotherapy 2006;7(8):1041–53.
[24] Huang X, Zhang F, Zhu L, Choi KY, Guo N, Guo J, Tackett K, Anilkumar P, Liu G, Quan Q, et al. Effect of injection routes on the biodistribution, clearance, and tumor uptake of carbon dots. ACS Nano 2013;7(7):5684–93.
[25] Shah RB, Patel M, Maahs DM, Shah VN. Insulin delivery methods: past, present and future. International Journal of Pharmaceutical Investigation 2016;6(1):1–9.
[26] Shi J, Votruba AR, Farokhzad OC, Langer R. Nanotechnology in drug delivery and tissue engineering: from discovery to applications. Nano Letters 2010;10(9):3223–30.
[27] Ulbrich K, Holá K, Šubr V, Bakandritsos A, Tuček J, Zbořil R. Targeted drug delivery with polymers and magnetic nanoparticles: covalent and noncovalent approaches, release control, and clinical studies. Chemical Reviews 2016;116(9):5338–431.
[28] Bobo D, Robinson KJ, Islam J, Thurecht KJ, Corrie SR. Nanoparticle-based medicines: a review of FDA-approved materials and clinical trials to date. Pharmaceutical Research 2016;33(10):2373–87.
[29] Siegel RL, Miller KD, Jemal A. Cancer statistics, 2018. CA: A Cancer Journal for Clinicians 2018;68(1):7–30.
[30] Shi J, Kantoff PW, Wooster R, Farokhzad OC. Cancer nanomedicine: progress, challenges and opportunities HHS public access. Nature Reviews Cancer 2017;17(1): 20–37.
[31] Xiao Z, Ji C, Shi J, Pridgen EM, Frieder J, Wu J, Farokhzad OC. DNA self-assembly of targeted near-infrared-responsive gold nanoparticles for cancer thermo-chemotherapy. Angewandte Chemie, International Edition in English 2012;51(47):11853–7.
[32] Dahlman JE, Barnes C, Khan O, Thiriot A, Jhunjunwala S, Shaw TE, Xing Y, Sager HB, Sahay G, Speciner L, et al. *In vivo* endothelial SiRNA delivery using polymeric nanoparticles with low molecular weight. Nature Nanotechnology 2014;9(8):648–55.
[33] Key J, Palange AL, Gentile F, Aryal S, Stigliano C, Di Mascolo D, De Rosa E, Cho M, Lee Y, Singh J, et al. Soft discoidal polymeric nanoconstructs resist macrophage uptake and enhance vascular targeting in tumors. ACS Nano 2015;9(12):11628–41.
[34] Duan X, Xiao J, Yin Q, Zhang Z, Yu H, Mao S, Li Y. Smart PH-sensitive and temporal-controlled polymeric micelles for effective combination therapy of doxorubicin and disulfiram. ACS Nano 2013;7(7):5858–69.
[35] Ediriwickrema A, Zhou J, Deng Y, Saltzman WM. Multilayered nanoparticles for combination gene and drug delivery to tumors. Biomaterials 2014;35(34):9343–54.
[36] Tang S, Yin Q, Su J, Sun H, Meng Q, Chen Y, Chen L, Huang Y, Gu W, Xu M, et al. Inhibition of metastasis and growth of breast cancer by PH-sensitive poly (β-Amino ester) nanoparticles Co-delivering two SiRNA and paclitaxel. Biomaterials 2015;48:1–15.
[37] Bigdeli A, Palchetti S, Pozzi D, Hormozi-Nezhad MR, Baldelli Bombelli F, Caracciolo G, Mahmoudi M. Exploring cellular interactions of liposomes using protein corona fingerprints and physicochemical properties. ACS Nano 2016;10(3):3723–37.
[38] Fang J-H, Lai Y-H, Chiu T-L, Chen Y-Y, Hu S-H, Chen S-Y. Magnetic core-shell nanocapsules with dual-targeting capabilities and Co-delivery of multiple drugs to treat brain gliomas. Advanced Healthcare Materials 2014; 3(8):1250–60.
[39] Petersen GH, Alzghari SK, Chee W, Sankari SS, La-Beck NM. Meta-analysis of clinical and preclinical studies comparing the anticancer efficacy of liposomal versus conventional non-liposomal doxorubicin. Journal of Controlled Release 2016;232:255–64.
[40] Chen EC, Fathi AT, Brunner AM. Reformulating acute myeloid leukemia: liposomal cytarabine and daunorubicin (CPX-351) as an emerging therapy for secondary AML. Onco Targets and Therapy 2018;11:3425–34.
[41] Kato K, Chin K, Yoshikawa T, Yamaguchi K, Tsuji Y, Esaki T, Sakai K, Kimura M, Hamaguchi T, Shimada Y, et al. Phase II study of NK105, a paclitaxel-incorporating micellar nanoparticle, for previously treated advanced or recurrent gastric cancer. Investigational New Drugs 2012;30(4):1621–7.
[42] Clark AJ, Wiley DT, Zuckerman JE, Webster P, Chao J, Lin J, Yen Y, Davis ME. CRLX101 nanoparticles localize in human tumors and not in adjacent, nonneoplastic tissue after intravenous dosing. Proceedings of the National Academy of Sciences of the United States of America 2016;113(14):3850–4.
[43] Ashton S, Song YH, Nolan J, Cadogan E, Murray J, Odedra R, Foster J, Hall PA, Low S, Taylor P, et al. Aurora kinase inhibitor nanoparticles target tumors with favorable therapeutic index *in vivo*. Science Translational Medicine 2016;8(325):325ra17.
[44] Maier-Hauff K, Ulrich F, Nestler D, Niehoff H, Wust P, Thiesen B, Orawa H, Budach V, Jordan A. Efficacy and safety of intratumoral thermotherapy using magnetic iron-oxide nanoparticles combined with external beam radiotherapy on patients with recurrent glioblastoma multiforme. Journal of Neuro-Oncology 2011;103(2): 317–24.
[45] Yang T, Li B, Qi S, Liu Y, Gai Y, Ye P, Yang G, Zhang W, Zhang P, He X, et al. Co-delivery of doxorubicin and Bmi1 SiRNA by folate receptor targeted liposomes exhibits enhanced anti-tumor effects *in vitro* and *in vivo*. Theranostics 2014;4(11):1096–111.
[46] Dritschilo A, Huang CH, Rudin CM, Marshall J, Collins B, Dul JL, Zhang C, Kumar D, Gokhale PC, Ahmad A, et al. Phase I study of liposome-encapsulated c-raf antisense oligodeoxyribonucleotide infusion in combination with radiation therapy in patients with advanced malignancies. Clinical Cancer Research 2006;12(4):1251–9.
[47] Islam MA, Reesor EKG, Xu Y, Zope HR, Zetter BR, Shi J. Biomaterials for MRNA delivery. Biomaterials Science 2015;3(12):1519–33.

[48] Tolcher AW, Rodrigueza WV, Rasco DW, Patnaik A, Papadopoulos KP, Amaya A, Moore TD, Gaylor SK, Bisgaier CL, Sooch MP, et al. A phase 1 study of the BCL2-targeted deoxyribonucleic acid inhibitor (DNAi) PNT2258 in patients with advanced solid tumors. Cancer Chemotherapy and Pharmacology 2014;73(2):363–71.
[49] Naldini L. Gene therapy returns to centre stage. Nature 2015;526(7573):351–60.
[50] Czapar AE, Zheng Y-R, Riddell IA, Shukla S, Awuah SG, Lippard SJ, Steinmetz NF. Tobacco mosaic virus delivery of phenanthriplatin for cancer therapy. ACS Nano 2016; 10(4):4119–26.
[51] Shukla S, DiFranco NA, Wen AM, Commandeur U, Steinmetz NF. To target or not to target: active vs. Passive tumor homing of filamentous nanoparticles based on potato virus X. Cellular and Molecular Bioengineering 2015;8(3):433–44.
[52] Kishimoto TK, Ferrari JD, LaMothe RA, Kolte PN, Griset AP, O'Neil C, Chan V, Browning E, Chalishazar A, Kuhlman W, et al. Improving the efficacy and safety of biologic drugs with tolerogenic nanoparticles. Nature Nanotechnology 2016;11(10):890–9.
[53] Blanco E, Shen H, Ferrari M. Principles of nanoparticle design for overcoming biological barriers to drug delivery. Nature Biotechnology 2015;33(9):941–51.
[54] Zuckerman JE, Davis ME. Targeting therapeutics to the glomerulus with nanoparticles. Advances in Chronic Kidney Disease 2013;20(6):500–7.
[55] Morrison EE, Bailey MA, Dear JW. Renal extracellular vesicles: from physiology to clinical application. The Journal of Physiology 2016;594(20):5735–48.
[56] Zhu X, Wu J, Shan W, Tao W, Zhao L, Lim J-M, D'Ortenzio M, Karnik R, Huang Y, Shi J, et al. Polymeric nanoparticles amenable to simultaneous installation of exterior targeting and interior therapeutic proteins. Angewandte Chemie, International Edition in English 2016;55(10):3309–12.
[57] Kim J, Kim J, Jeong C, Kim WJ. Synergistic nanomedicine by combined gene and photothermal therapy. Advanced Drug Delivery Reviews 2016;98:99–112.
[58] Tabernero J, Shapiro GI, LoRusso PM, Cervantes A, Schwartz GK, Weiss GJ, Paz-Ares L, Cho DC, Infante JR, Alsina M, et al. First-in-Humans trial of an RNA interference therapeutic targeting VEGF and KSP in cancer patients with liver involvement. Cancer Discovery 2013;3(4):406–17.
[59] Kasenda B, Larkin J, Gore M. Immunotherapies in early and advanced renal cell cancer. Progress in Tumor Research 2015;42:1–10.
[60] Ziaeian B, Fonarow GC. Epidemiology and aetiology of heart failure. Nature Reviews. Cardiology 2016;13(6): 368–78.
[61] Mahmoudi M, Yu M, Serpooshan V, Wu JC, Langer R, Lee RT, Karp JM, Farokhzad OC. Multiscale technologies for treatment of ischemic cardiomyopathy. Nature Nanotechnology 2017;12(9):845–55.
[62] Begum M, Sharma HK. Scope of nanomedicines against coronary artery disease: a review.
[63] Libby P, Ridker PM, Hansson GK. Progress and challenges in translating the biology of atherosclerosis. Nature 2011;473(7347):317–25.
[64] Lobatto ME, Fuster V, Fayad ZA, Mulder WJM. Perspectives and opportunities for nanomedicine in the management of atherosclerosis. Nature Reviews Drug Discovery 2011;10(11):835–52.
[65] Korin N, Kanapathipillai M, Matthews BD, Crescente M, Brill A, Mammoto T, Ghosh K, Jurek S, Bencherif SA, Bhatta D, et al. Shear-activated nanotherapeutics for drug targeting to obstructed blood vessels. Science 2012;337(6095):738–42.
[66] Kamaly N, Fredman G, Fojas JJR, Subramanian M, Choi WI, Zepeda K, Vilos C, Yu M, Gadde S, Wu J, et al. Targeted interleukin-10 nanotherapeutics developed with a microfluidic chip enhance resolution of inflammation in advanced atherosclerosis. ACS Nano 2016;10(5):5280–92.
[67] Fredman G, Kamaly N, Spolitu S, Milton J, Ghorpade D, Chiasson R, Kuriakose G, Perretti M, Farokhzad O, Tabas I. Targeted nanoparticles containing the proresolving peptide Ac2-26 protect against advanced atherosclerosis in hypercholesterolemic mice. Science Translational Medicine 2015;7(275). 275ra20.
[68] Duivenvoorden R, Tang J, Cormode DP, Mieszawska AJ, Izquierdo-Garcia D, Ozcan C, Otten MJ, Zaidi N, Lobatto ME, van Rijs SM, et al. A statin-loaded reconstituted high-density lipoprotein nanoparticle inhibits atherosclerotic plaque inflammation. Nature Communications 2014;5(1):3065.
[69] Cheng K, Shen D, Hensley MT, Middleton R, Sun B, Liu W, De Couto G, Marbán E. Magnetic antibody-linked nanomatchmakers for therapeutic cell targeting. Nature Communications 2014;5(1):4880.
[70] Dickmann L. Ocular therapeutics: drug delivery and pharmacology. Molecular Pharmaceutics 2016;13(9): 2875–6.
[71] Eris E, Erdogan G, Perente I, Demir G, Kepez Yildiz B, Aygit ED. A new side effect of intravitreal dexamethasone implant (Ozurdex®). Case Reports in Ophthalmological Medicine 2017;2017:6369085.
[72] Kim TW, Lindsey JD, Aihara M, Anthony TL, Weinreb RN. Intraocular distribution of 70-KDa dextran after subconjunctival injection in mice. Investigative Ophthalmology & Visual Science 2002;43(6):1809–16.
[73] Jiang S, Franco YL, Zhou Y, Chen J. Nanotechnology in retinal drug delivery. International Journal of Ophthalmology 2018;11(6):1038–44.
[74] Kumar A, Tan A, Wong J, Spagnoli JC, Lam J, Blevins BD, Natasha G, Thorne L, Ashkan K, Xie J, et al. Nanotechnology for neuroscience: promising approaches for diagnostics, therapeutics and brain activity mapping. Advanced Functional Materials 2017;27(39):1–30.
[75] Bekyarova E, Ni Y, Malarkey EB, Montana V, McWilliams JL, Haddon RC, Parpura V. Applications of carbon nanotubes in biotechnology and biomedicine. Journal of Biomedical Nanotechnology 2005;1(1):3–17.

[76] Khawaja AM. The legacy of nanotechnology: revolution and prospects in neurosurgery. International Journal of Surgery 2011;9(8):608–14.
[77] Mattei TA, Rehman AA. "Extremely minimally invasive": recent advances in nanotechnology research and future applications in neurosurgery. Neurosurgical Review 2015;38(1):27–37.
[78] Maurer N, Fenske DB, Cullis PR. Developments in liposomal drug delivery systems. Expert Opinion on Biological Therapy 2001;1(6):923–47.
[79] Wilson B, Samanta MK, Santhi K, Sampath Kumar KP, Ramasamy M, Suresh B. Significant delivery of tacrine into the brain using magnetic chitosan microparticles for treating Alzheimer's disease. Journal of Neuroscience Methods 2009;177(2):427–33.
[80] Kumar A, Chen F, Mozhi A, Zhang X, Zhao Y, Xue X, Hao Y, Zhang X, Wang PC, Liang X-J. Innovative pharmaceutical development based on unique properties of nanoscale delivery formulation. Nanoscale 2013; 5(18):8307.
[81] Wilhelm S, Tavares AJ, Dai Q, Ohta S, Audet J, Dvorak HF, Chan WCW. Analysis of nanoparticle delivery to tumours. Nature Reviews Materials 2016;1(5):16014.
[82] Veiseh O, Gunn JW, Zhang M. Design and fabrication of magnetic nanoparticles for targeted drug delivery and imaging. Advanced Drug Delivery Reviews 2010;62(3): 284–304.
[83] Khanbabaie R, Jahanshahi M. Revolutionary impact of nanodrug delivery on neuroscience. Current Neuropharmacology 2012;10(4):370–92.
[84] Gilmore JL, Yi X, Quan L, Kabanov AV. Novel nanomaterials for clinical neuroscience. Journal of Neuroimmune Pharmacology 2008;3(2):83–94.
[85] Lovat V, Pantarotto D, Lagostena L, Cacciari B, Grandolfo M, Righi M, Spalluto G, Prato M, Ballerini L. Carbon nanotube substrates boost neuronal electrical signaling. Nano Letters 2005;5(6):1107–10.
[86] Le Droumaguet B, Nicolas J, Brambilla D, Mura S, Maksimenko A, De Kimpe L, Salvati E, Zona C, Airoldi C, Canovi M, et al. Versatile and efficient targeting using a single nanoparticulate platform: application to cancer and Alzheimer's disease. ACS Nano 2012;6(7): 5866–79.
[87] Kwon HJ, Cha M-Y, Kim D, Kim DK, Soh M, Shin K, Hyeon T, Mook-Jung I. Mitochondria-targeting ceria nanoparticles as antioxidants for Alzheimer's disease. ACS Nano 2016;10(2):2860–70.
[88] Hajipour MJ, Santoso MR, Rezaee F, Aghaverdi H, Mahmoudi M, Perry G. Advances in Alzheimer's diagnosis and therapy: the implications of nanotechnology. Trends in Biotechnology 2017;35(10):937–53.
[89] Chen Q, Du Y, Zhang K, Liang Z, Li J, Yu H, Ren R, Feng J, Jin Z, Li F, et al. Tau-targeted multifunctional nanocomposite for combinational therapy of Alzheimer's disease. ACS Nano 2018;12(2):1321–38.
[90] Ventola CL. Progress in nanomedicine: approved and investigational nanodrugs. P & T: A Peer-Reviewed Journal for Formulary Management 2017;42(12):742–55.
[91] Hodson M, Izadi T, Lewinski N, McInnes B. Creation of an annotated library on FDA approved nanomedicines. Undergraduate Research Posters 2015.
[92] Brown DM, Donaldson K, Borm PJ, Schins RP, Dehnhardt M, Gilmour P, Jimenez LA, Stone V. Calcium and ROS-mediated activation of transcription factors and TNF-α cytokine gene expression in macrophages exposed to ultrafine particles. The American Journal of Physiology-Cell Physiology 2004;286(2):L344–53.
[93] Brown DM, Wilson MR, MacNee W, Stone V, Donaldson K. Size-dependent proinflammatory effects of ultrafine polystyrene particles: a role for surface area and oxidative stress in the enhanced activity of ultrafines. Toxicology and Applied Pharmacology 2001;175(3):191–9.
[94] Radomska A, Leszczyszyn J, Radomski M. The nanopharmacology and nanotoxicology of nanomaterials: new opportunities and challenges. Advances in Clinical and Experimental Medicine 2016;25(1):151–62.
[95] Sun X, Shi J, Fu X, Yang Y, Zhang H. Long-term *in vivo* biodistribution and toxicity study of functionalized near-infrared persistent luminescence nanoparticles. Scientific Reports 2018;8(1):1–11.
[96] Wang B, He X, Zhang Z, Zhao Y, Feng W. Metabolism of nanomaterials *in vivo*: blood circulation and organ clearance. Accounts of Chemical Research 2013;46(3): 761–9.
[97] Zhu M, Nie G, Meng H, Xia T, Nel A, Zhao Y. Physicochemical properties determine nanomaterial cellular uptake, transport, and fate. Accounts of Chemical Research 2013;46(3):622–31.
[98] Ahmad F, Zhou Y. Pitfalls and challenges in nanotoxicology: a case of cobalt ferrite ($CoFe_2O_4$) nanocomposites. Chemical Research in Toxicology 2017;30(2):492–507.
[99] Alshehri R, Ilyas AM, Hasan A, Arnaout A, Ahmed F, Memic A. Carbon nanotubes in biomedical applications: factors, mechanisms, and remedies of toxicity. Journal of Medicinal Chemistry 2016;59(18): 8149–67.
[100] Kryuchkova M, Fakhrullin R. Kaolin alleviates graphene oxide toxicity. Environmental Science & Technology Letters 2018;5(5):295–300.
[101] Contreras EQ, Cho M, Zhu H, Puppala HL, Escalera G, Zhong W, Colvin VL. Toxicity of quantum dots and cadmium salt to Caenorhabditis elegans after multigenerational exposure. Environmental Science & Technology 2013;47(2):1148–54.
[102] Ngamchuea K, Batchelor-McAuley C, Compton RG. The fate of silver nanoparticles in authentic human saliva. Nanotoxicology 2018;12(4):305–11.
[103] Medina C, Santos-Martinez MJ, Radomski A, Corrigan OI, Radomski MW. Nanoparticles: pharmacological and toxicological significance. British Journal of Pharmacology 2007;150(5):552–8.
[104] Xia Q, Hwang H-M, Ray PC, Yu H. Mechanisms of nanotoxicity: generation of reactive oxygen species. The Journal of Food and Drug Analysis 2014;22(1):64–75.

CHAPTER 8

Translational Nano-medicine Lab to Clinic

MOHAMMAD NADEEM KHAN, PHD

1 INTRODUCTION

As the technologies are proposed based on a clear understanding of a particular disease, diseases-specific focus is required for the development of novel pharmaceuticals. In addition, it will be essential to develop a case-by-case approach to clinical and ethical evaluation of every nanopharmaceutical. Nanotechnology is now widely recognized as a transformative technology platform to improve many life science innovations. In less than a decade, nanotechnology has established itself as a central feature intended to enhance product performance in many applications, including food technology, cosmetics, surfaces, new materials and medical products. Nanomedicines are also reported as nanodrugs, nanocarriers, nanoconstructs, nanoparticles, nanomaterials, or nanotherapeutics [1–3]. In general, the terms "nanomaterial" and "nanosystem" include a wide variety of objects with at least one dimension in the nanometric scale (typically 1–1000 nm) (Fig. 8.1). The term nanomedicine appeared in the 1990s, and since then, its interest has experienced an exponential increase within the scientific community, notably during the last 20 years, after the government of the United States announced that it was starting to fund nanomedicinal studies [4]. More recently, the European Technology platform on Nanomedicine (ETPN) defined the term "nanomedicine" as: "The application of nanotechnology to achieve breakthroughs in healthcare that exploits the improved and often novel physical, chemical and biological properties of materials at the nanometer scale" [5]. The European Medicines Agency (EMA), the EU agency responsible for the scientific evaluation and supervision of medicines, refers to nanomedicine as "a system designed for clinical applications that is composed of at least one nano scale component, resulting in specific advantageous characteristics, such as better targeting and bioavailability of therapeutics, new modes of therapeutic action, and nano structured surfaces/scaffolds for engineered tissues".

High priority should be given to increasing communication and exchange of information among academia, industry, and regulatory agencies encompassing all facets of this multidisciplinary approach [6]. These include pharmaceuticals and drug delivery products, medical imaging, the ranostics and diagnostics, implantable materials, sensors and devices, regenerative medicine, and combinations of technologies comprising several of these features into single platforms. Chemotherapy (chemo) cyclical treatment is widely used in many cancer types in order to eradicate primary cancer cells and metastasis. Often, more than a single chemotherapeutic agent with or without surgical intervention and/or radiation is more effective and might improve survival benefit compared to a monotherapy approach [1,2] more than 100 different types of chemo drugs are available for cancer treatment and can be divided according to pharmacological activity. For example, "alkylating-like" agents such as platinum-based drugs and cell-cycle nonspecific anthracyclines directly damage the DNA inside the cancer cells and prevent cancer cells from dividing. Nucleoside analogues, such as gemcitabine, 5-FU, and fludarabine, replace the natural building blocks of DNA and RNA and interfere with the cell metabolism and nucleic acid synthesis. Topoisomerase inhibitors such as irinotecan, topotecan, and SN-38 (the active metabolite of irinotecan) are extracted from natural sources. Others such as taxanes are isolated from plants (also known as plants alkaloids), and some are isolated from a soil-dwelling fungus such as *Streptomyces* (daunorubicin and doxorubicin). Administering bolus doses of cytotoxic chemo drugs causes damage not only to cancer cells but also to healthy tissues [7,8]. The effectiveness of the treatment is affected by a delicate balance between the ability to selectively kill the cancer cells and damage the healthy cells. Due to systemic, nonselective distribution, chemo treatment usually is much less

Intelligent Nanomaterials for Drug Delivery Applications. https://doi.org/10.1016/B978-0-12-817830-0.00008-4

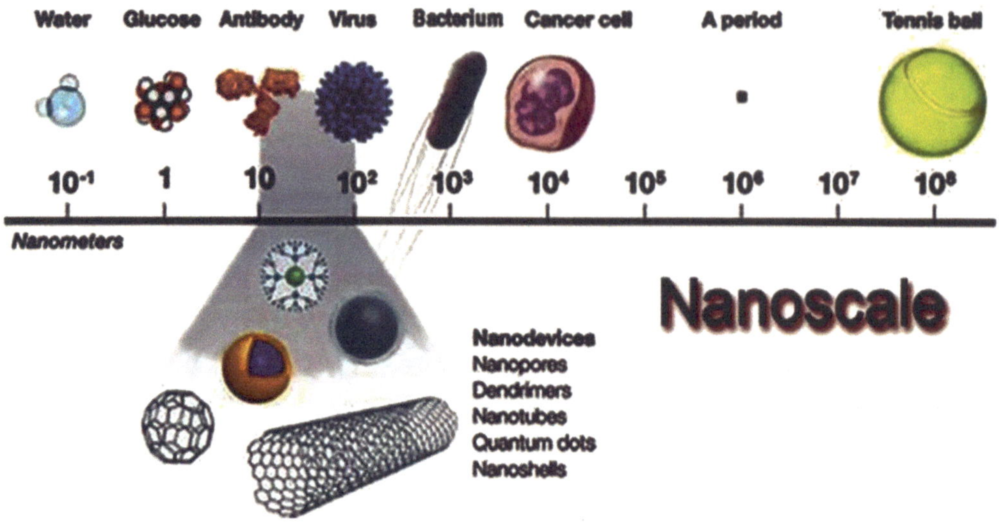

FIG. 8.1 Schematic representation of a scale bar to visualize the range of nano-materials and nano systems as compared with biological components.

effective against cancer cells since the therapeutic level reaching the target cells is low, and thus it may cause only a marginal improvement in patient overall survival rates. Other issues such as liver, lung, kidney, and immune toxicity and tumor drug resistance are major factors that need special attention in order to improve patient survival rate and quality of life. Therefore, more effective cancer therapies are needed both in term of new classes of cytotoxic drugs and in term of developing focused drug delivery vehicles to minimize the adverse effects and maximize the therapeutic effect, thus enhancing the survival of patients with metastatic or an advanced stage cancer disease [9]. In terms of education, a need for more formal interdisciplinary training courses in the area of nanomedicine was foreseen. At the undergraduate level these would cover the basic scientific disciplines including molecular biology, colloidal chemistry, cell physiology, surface chemistry, and membrane biophysics. In addition, new programmes at master's or early postgraduate level (preferably with combined medical and scientific training) are needed to support the rapidly developing field of nanomedicine. This will be essential to ensure a trained workforce to oversee clinical development of new nanomedicines. Timely exploitation of newly emerging nanomedicine technologies was seen as a key issue. This is an area of general weakness. There is a need to establish specific nanomedicine-related schemes to promote academic-commercial partnerships [10]. The growth of clusters or highly selected teams, chosen for personal excellence or track record, would ensure most rapid progress. To facilitate industrial development the establishment of more manufacturing sites with a Good Manufacturing Practice designation able to support small and medium enterprises would help in transferring projects more rapidly into clinical development. Establishment of good communication is a universal challenge for research and development at the interface of scientific disciplines and at the leading edge. There is a continuing need to promote more truly transdisciplinary conferences. Encouragement of goal-oriented research partnerships between large medical centers and university research groups is essential to progress the field. It was noted that scientifically qualified politicians are not common, and the regional, national, and European programmes seldom show alignment, suggesting less than optimal communication. Serious effort needs to be made by scientific opinion leaders in nanomedicine to ensure that politicians are well briefed. Not only will better diagnostics, treatments, and prevention for life-threatening and debilitating disease bring healthcare benefits to an aging population, research and development in nanomedicine also offers the potential of employment and economic benefits with a reduction of healthcare budgets. There the impact of nanotechnology in medicine can mainly be seen in diagnostic methods, drug-release techniques, and regenerative medicine. Diagnostic methods are essential for the early detection of diseases to enable their prompt treatment, minimizing possible damage to the rest of the organism [11]. The importance of imaging methods to diagnose, treat, and follow up cancer, cardiovascular, and neurological patients is well known. Diagnostic techniques based on the use of nanoparticles offer higher sensitivity and assist the early detection of disease, offering a better prognosis and greater possibilities of successful treatment [12]. Once the diagnosis is established, the fight against the disease begins, with medicaments playing a major role. There are numerous obstacles to the development of novel drugs against disease. Conventional drugs suffer from the major limitation of adverse effects, the result of the nonspecificity of their action, and from a lack of effectiveness due to improper or ineffective dosages, e.g., in cancer chemotherapy and antidiabetic therapy. Nanotechnology offers the possibility of designing novel drugs with greater cell specificity and new drug-release systems that act selectively on specific targets and protect the drug from degradation en route. This allows the administration of smaller but more effective doses, minimizing adverse effects. Nanotechnology can also be used to optimize drug formulations, increasing drug solubility, and alter the pharmacokinetics to sustain the release of the drug, thereby prolonging its bioavailability [13]. The diverse platforms of nanotechnology can be utilized to develop more sophisticated, cell-targeted therapies and to combine different drugs into a single nanotherapeutic agent for synergistic therapeutic benefits.

2 DETERRENT IN TARGETING, TRANSPORT, AND DELIVERY OF THERAPEUTICS

In drug therapeutics a perennial problem has always been that, due to challenges presented by its targeting, transport, and delivery, a pharmacologically active concentration in infective cells is often only achieved at the expense of what 'Couvrer' terms massive contamination of the rest of the body [14]. This toxicity can result in the use of suboptimal and/or intermittent dosing, to allow the body to rest, or in some cases to forgo therapeutic drugs altogether [15].Many traditional drugs have poor stability and aqueous solubility. Due to this limitation, many drugs, despite significant biological activity, are disregarded at early stages of drug screening in the laboratory. In addition, distribution of some drugs is too general; with only a small fraction of drugs reaching the pathological site, injected agents are often cleared by the monocytes and macrophages of the reticuloendothelial system (RES) [16,17]. To be successful, a therapeutically sufficient quantity of the drug, still in a viable state, must survive clearing and be delivered to different regions of tumors via blood vessels, cross the vessel wall, and then finally penetrate through the interstitial space to reach the target [18], where unpredictable blood flow and often abnormal vasculature in tumors, particularly in necrotic and semi-necrotic regions, can make accurate delivery even more difficult [19,20].Other than conventional therapeutic drugs, biological molecules, such as antibodies(protein) and nucleic acids, are being widely explored for the treatment of different diseases, including cancer. Nucleic acid drugs, such as aptamers, antisense DNA/RNA, and small-interfering RNA (siRNA), have shown great promise in the treatment of various diseases. However, these drugs are greatly limited by serum nucleases, opsonization and clearance by macrophages, and clearance by the renal system. Some of the nucleic acid therapeutics, such as stable nucleic-acid lipid particle (SNALP), have used nanocarriers to effectively overcome the abovementioned barriers and are being used in the treatment of cancer [21].

3 CURRENT TRENDS IN NANOTHERAPEUTICS: OVERCOMING THE DETERRENT DRUG DELIVERY SYSTEM

Promising results have emerged from combining nanoparticle-based optical contrast agents with existing optical imaging technologies [22]. Their programmable surface properties and potential for passive or active targeting make nanoparticles ideal for diagnostic imaging. The ability of nanoshells, constructed with a silica core and gold shell, to absorb specific wavelengths of light, has great potential for cancer imaging and therapeutic applications [23].

4 IMPROVING TARGETING, TRANSPORT, AND DELIVERY

Nanoparticles are increasingly utilized because of the multiple benefits that they offer [24].Nanodelivery systems can make the use of therapeutic drugs more safe and efficient by improving delivery, cell uptake, and targeting and have been shown to improve their pharmacokinetic profiles and enhance their targeting at the required site [25]. This success relies on two main factors; first is the EPR and second is the potential ability of nanodrug delivery systems to overcome the shortcomings of many therapeutic drugs including anticancer drugs [26].

1. The EPR, or enhanced permeation retention effect, exists because of two properties of tumors. Firstly, infective tissues have increased vasculature which allows the entry of macromolecules and colloidal

particles of diameter up to ~600 nm. Secondly, the lymphatic system is not effective in clearing the interstitial fluid from the tumor tissues [27]. Normal tissues other than the spleen, liver, and kidney are impermeable to molecules that are larger than 2 nm. Hence, nanoparticles can selectively target infective tissues reducing toxic side effects [28]. Together, the enhanced permeation and retention (EPR) properties of the tumor over the normal tissues cause the nanoparticle to have prolonged contact with the tumor cells. In addition, nanocarriers also release the drug slowly, ultimately resulting in reduced drug distribution and toxicity to normal tissues [29].

2. Once the nanoparticles reach the target tissue, cell surface receptors interact with ligand-coated nanoparticles leading to their uptake by endocytosis. Cellular uptake of uncoated nanoparticles is governed by their differences in size, shape, and charge. It is suggested that positively charged nanoparticles are taken up more readily due to electrostatic attraction [30]. Interaction with specific serum proteins results in the formation of a corona, promoting cell entry. Recent studies indicate that nonspherical molecules, such as rod-shaped structures, are internalized better than spherical structures [31]. Uptake of larger nanoparticles disrupts the membrane surface, thereby inducing cell death. Nonspecific uptake of the nanoparticles by the lung epithelial cells and red blood cells could cause toxic side effects.
3. Nanosized drug delivery systems can potentially overcome the shortcomings of many anticancer drugs, such as low aqueous solubility and stability and high nonspecific toxicity [32]. For example, paclitaxel nanoparticles stabilized with pluronic F-68 are stable for years, while the same drug in dissolved form degrades completely in less than 48 h [33], and magnetic nanoparticles (MNPs) are increasingly used because their targeting ability reduces systemic distribution of cytotoxic compounds in vivo and enhances uptake at the target site, resulting in effective treatment at lower doses [34].

5 BENCH TO BEDSIDE: TRANSLATIONAL PERSPECTIVES

Nanotechnology has raised as many questions as it has answered and spawned new and unpredicted fields [35]. A full road map to translation process outlines (Fig. 8.2) with its vast array of potential applications in so many fields of science and industry, it is a prime candidate for multidisciplinary collaboration, and the urgent need to see laboratory breakthroughs translated to clinical successes is increasingly recognized (Table 8.1). Biochemists are increasingly working with scientists from fields not usually associated with medicine, and Nanomedicine Development Centers are staffed by multidisciplinary research teams, including biologists, physicians, mathematicians, engineers, and computer scientists, whose first task has been to research the chemical and physical properties of cells and specific intracellular sites in infected cells [36,37]. Base line work of this sort is vital if clinicians are to have the knowledge to develop new therapies. Murday et al. claim that translational research has been "a powerful process that drives the clinical research engine", but feel that a stronger research infrastructure is needed in future to strengthen and accelerate this critical part of the clinical research enterprise [38]. Kawasaki and Player agree that scientists from all fields must make a strenuous effort to integrate and coordinate the research in an approach that might now be described as systems biology. They hold up the 2004 article "Electronic structure and bonding of Au on a SiO_2 cluster: a nano bullet for tumors" [39], produced by physicists, as a prime example of how research in other fields can advance nanomedicine [40].

Current challenges facing nanotechnology translation to nanomedicine hype, overused terminology, and inflated expectations for nanomedicine's capabilities aside, the practical realities for advancing in vivo applications of nanotechnology toward clinical use face a set of technical and scientific challenges [41]. Despite interesting science, translating in vitro technologies to applications must assert clear performance advantages to adoption of select nanocomponent or miniature designs. By contrast, all in vivo biomedical applications of nanoparticles require direct absorption, inhalation, ingestion, or injection into host tissues or circulation. Both require reagent stability, namely resistance to particle aggregation in complex biological media. Both require compatibility assessments as to acceptable toxicity profiles in their context of use: in vitro assays with biological components have vastly different requirements than nanomaterials in vivo. Possible dose-response and toxicity effects from these new nanomaterials on both isolated biological elements (e.g., enzymes, cells, proteins), tissues, and to human health is a critical prerequisite to translation. Currently, there is no successful predictive test strategy for in vitro-in vivo correlations to provide such risk assessments. Each nanomaterial must be evaluated separately under the conditions of use after validating material purity and particle surface analytics [42]. Full in vivo profiling would then follow with absorption,

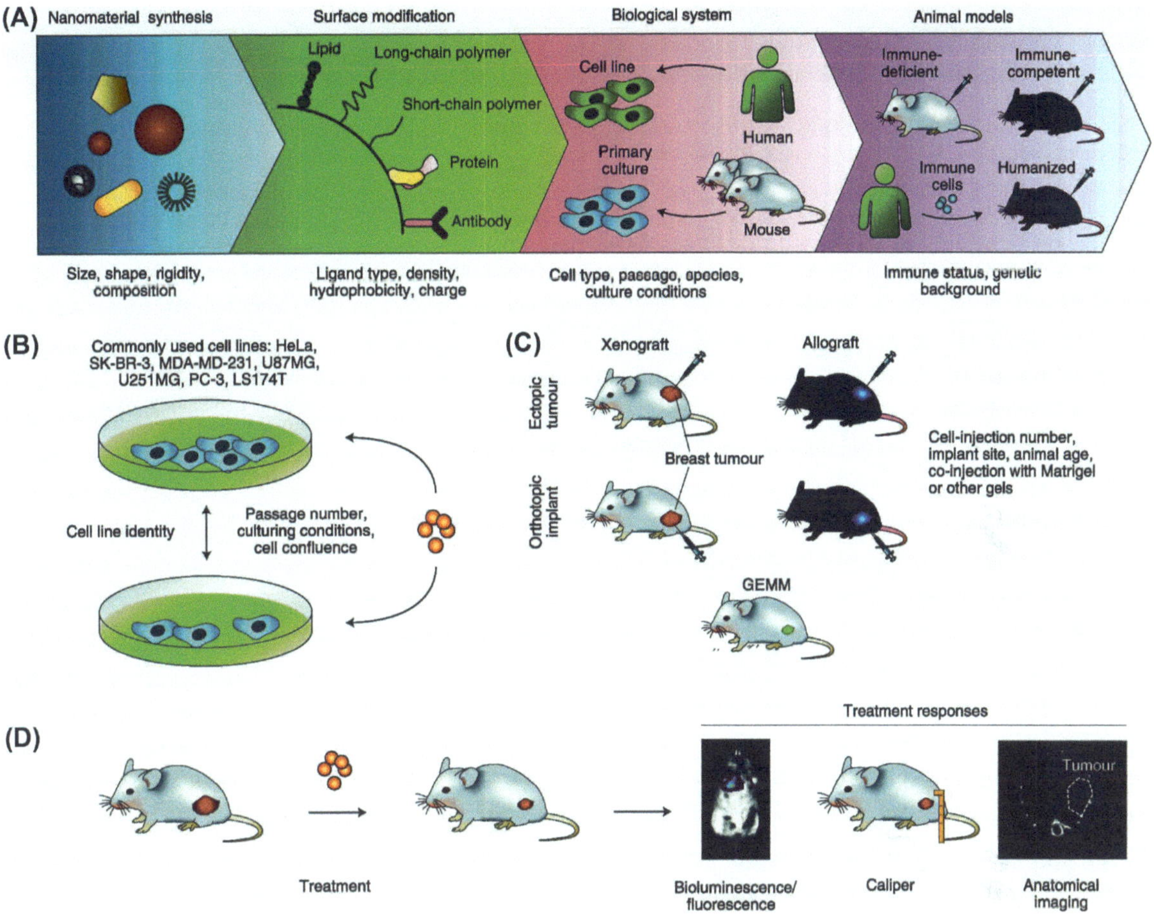

FIG. 8.2 Bridging the valley between developing nanomedicines at the lab scale to clinical translation.

distribution, metabolism, and excretion (ADME) tests and physicochemical and toxicological characterization, involving both in vitro tests and in vivo animal studies. Fig. 8.1 shows this conceptual design and performance evaluation. How these methods are performed is critical to understanding both risk and functional assessments of each nanomaterial's design. Few nanomaterials to date subscribe to a thorough, rigorous analytical protocol across in vitro and in vivo test beds [43]. The result is a frustrating and persistent gap between in vitro results and in vivo performance for poorly characterized nanomedicines that stymies ready translation [44].

- Oral nanomaterial bioavailability: Gastrointestinal barriers represent a formidable challenge for therapeutic fluxes of particles. Strategies to improve both passive and active transport of particles across the gut wall are an active focus [45]. Most published reports confirm that despite many surface and morphological changes, oral bioavailability of insoluble nanomaterials is consistent with that long known for oral penetration of microparticles: both are generally very poorly bioavailable [46]. While absolute efficiencies of gastric absorption and systemization reported for various nanoparticles in different oral preclinical models vary, their bioavailability is consistently less than 10% frequently much less than this. A recent highly accurate biodistribution study using radioactive gold nanoparticles showed less than 1% oral bioavailability from gastric instillation in rats, regardless of particle size or surface charge [47]. Hence, while nanoparticle uptake from the gut certainly has been shown that this particle flux and oral bioavailability can be general enough across a broad patient population, sufficient and sustained ability to produce reliable therapeutic dosing has not yet been shown.

TABLE 8.1
Considerations for the Translational Development of Nanomedicines.

Nanopharmaceutical Design (Key Considerations)	Preclinical Evaluation (Key Considerations)	Clinical Evaluation for Commercialization (Key Considerations)
• Route of administration • Reduce complexity in formulation design • Final dosage form for human use (biocompatibility and biodegradability) • Pharmaceutical stability (physical and chemical) • Current obstacles	• Need for validated and standardized assays for early detection of toxicity • Evaluation in appropriate animal models of disease • Adequate understanding of in vivo behavior, including cellular and molecular interactions	• simplification of development pathways from invention to commercialization to minimize time and expense • Evaluation of safety/toxicity in humans (acute and chronic) • Evaluation of therapeutic efficacy in patients • Optimal clinical trial design
• *E.g., Reproducibility, infrastructure, techniques, expertise, and cost* Quality control assays for characterization • *E.g., Size and polydispersity, morphology, charge, encapsulation, surface modifications, purity, and stability*	• *Pharmacokinetics (absorption, distribution, metabolism, and excretion)* • *Pharmacodynamics (intracellular trafficking, functionality, toxicity, and degradation)*	

- Circulating lifetimes of systemic nanoparticles: Most systemically injected nanomedicines have circulating half-lives of less than 24 h and many less than 6 h [48]. When injected intravenously, particles are cleared rapidly from systemic circulation, predominantly by the liver Kupffer cells and spleen (marginal zone and red pulp) macrophage populations [49]. Larger particles, 200 nm and above, are more efficient at activating the human complement system and are hence cleared faster from the blood by Kupffer cells than their smaller counterparts. Regardless of steric stabilization, charge, shape, or size, circulating particles accumulate rapidly in the liver and spleen are and followed by filtration through the kidneys and lung capillary beds. Particle filtration efficiencies in nontarget organs including the liver, spleen, and lungs is generally greater than 90% and often 95% [50]. This means that only 5% of the injected bioavailable dose produces the observed therapeutic outcome at the target site. Most studies emphasize this 1%–5% disease targeting while neglecting the important information provided from analyzing the 90%–99% untargeted nanomedicine dose. Enhancing drug circulating lifetimes (i.e., creating long-circulating vehicles) to improve accessibility to disease sites has been a research focus for improving nanomedicines for decades [51]. Improved lifetimes have been most frequently achieved using grafted polymeric steric stabilizers (commonly polyethylene glycols to date) [50]. These hydrophilic polymer grafts are thought to minimize nonspecific serum protein opsonization and corona formation, limit particle-cell and particle-particle aggregation, and inhibit scavenging uptake by the host mononuclear phagocyte system (MPS) [52]. This is shown in Fig. 8.3. Clearly, improved methods to provide simultaneous stabilization of particles in blood while improving circulation times and accessibility, either passively or actively to disease sites, are clearly required to improve nanomedicine dosing to target sites. Even small changes in the on-site to off-site particle distributions would have dramatic impacts on therapy versus adverse drug side effects (i.e., liver and kidney toxicity). One study reported attachment of polymeric nanoparticles to the surfaces of red blood cells, enhancing their circulating lifetimes as long as the attachment was stable [53]. It is unclear from such a "piggy-back circulation strategy" how these particles reach their intended target site without a programmed release mechanism. Other emerging strategies include surface grafting with self-

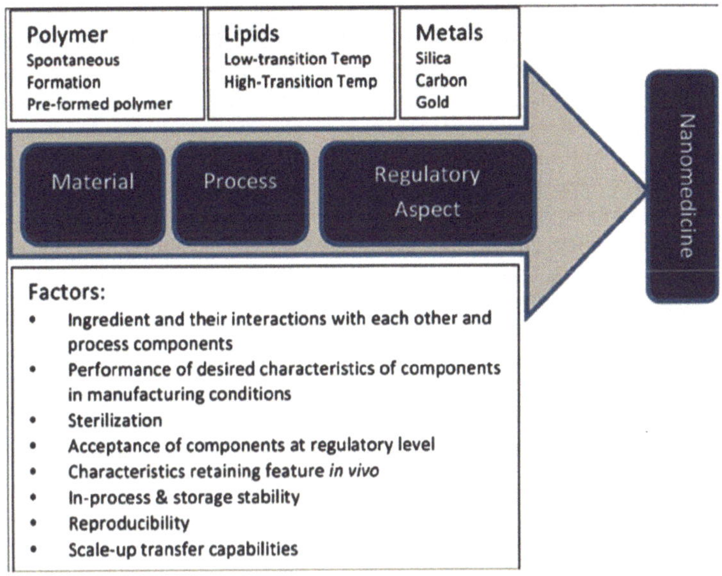

FIG. 8.3 Schematic representation of material and process parameters, and regulatory aspects in the development of a nanomedicine.

recognition peptides [52] and erythrocyte membrane surface mimics [54].

- Targeting nano medicines (Fact and Fiction): Most nanomedicines either approved for clinical use or in development are intended to become systemically bioavailable, that is, circulate widely within the vascular and lymphatic transport systems to find their intended sites of action. Lacking innate modes of self-propulsion to direct their movements, nanomedicine particles are held hostage to passive transport mechanisms within the systemic circulation: diffusion, partitioning, entrainments in blood flow (hydrodynamic phenomena). Their movement and transport in the body is therefore largely stochastic: particles can only move where these passive mechanisms take them. In addition, (nano)particle retention at any tissue site depends upon frequency of encounter with that site, hydraulic permeability and hydrodynamic transport via flow that facilitate particle exit from the circulatory flow fields, and two-body forces at that site. In a nonuniform fluid shear field, a particle experiences heterogeneous fluid velocities around it, and the resultant force from this nonuniform flow field causes the particle to move across the flow field. If the particle is not spherical, its center of mass and hydrodynamic centers are not coincident so the particle experiences imbalanced fluid forces and will move in a nonuniform manner. Many vascular conduits have laminar and even turbulent flow fields and are fenestrated, and extravascular, acellular extracellular matrix is a porous mesh of cross-linked macromolecules. Once transported by random flow and diffusive forces to a fenestrated vascular wall or to extracellular matrix, particles can become entrained in their porous network, removing them from flow and depositing them where intermolecular attractive forces determine their retention and residency. The interactions and relative contributions of these passive transport phenomena define passive targeting for nanomedicines: random particle encounters using stochastic processes to enable tissue site retention and particle removal from fluid flow [54]. Further addition of specific ligands to nanomaterials to improve intermolecular forces, engagement, and site-specific retention post transport is termed active targeting. But active targeting is short-range: that is, targeting ligands can only physically extend across space at a nanometer range, while the intermolecular forces that govern their avidity, affinity, and specificity with a target molecule are subnanometer in range. Particles with active targeting strategies must therefore use passive mechanisms to first provide requisite transport and entrainment close to their intended site of binding. Only then can active targeting ligand engagement facilitate retention at the specific target site, and only if intermolecular forces favor binding in the local environment during the residence time of the particle at that site. Both thermodynamics and kinetics play important roles in active targeting to tissue sites. It is important to note that both passive and active targeting processes with particles are saturable [55]. That is, the passive removal of suspended particles from flow depends on the initial removal characteristics of the filtering bed (i.e., fenestrated endothelium, extracellular matrix) and subsequent effects of particle deposition and retention within the initial outermost exposed porous tissue layer on subsequent particle encounters and retention efficiencies. Initial particle retention within exposed pores can cause particle-pore "jamming" and also alter the local hydrodynamic parameters, reducing both transport to and retention within the pores for further particle populations that might randomly encounter the site [56]. As more particle deposition occurs within tissue pores adjacent to the flow field, the relative porosity of the tissue bed is naturally changed, and with it the local hydrodynamics and the hydraulic sieving properties of the bed. Consequently, mean interstitial porous flow velocity near the tissue bed increases, reducing the probability of particle surface encounter, retention, and deposition for suspended particles. Additionally, higher interstitial pore velocities may induce a fraction of the deposited particles to detach and return to systemic flow. In addition to ubiquitous passive deposited particle saturation by jamming and resulting hydrodynamic transport alterations in the tissue bed directly adjacent to the vascular or lymphatic flow field [54], active targeting processes also saturate based on ligand-target binding. In contrast to what might be inferred from the abundant literature popularizing this strategy, active targeting is simply not a matter of pairing a high-affinity binding entity for a disease target onto a particle and injecting it. In practice, active targeting is not simply an exploitation of classic lock-key ligand-binding principles. Active targeting can in fact yield less site-specific nanomedicine accumulation than for passive analogs, and be plagued with the same nonspecific filtration issues [57]. Additionally, intrinsic tumor heterogeneity leads to heterogeneity in tumor site expression of specific receptors, both temporally and spatially in tissues of therapeutic interest, and makes reliable "active targeting" very difficult to establish. Targeting specific disease ligands often has failed to show much therapeutically relevant influence on nanomedicine tumor accumulations in animal models, despite showing increased agent internalization in animal models [58]. Reliable evidence showing substantial active targeting in vivo despite in vitro cell line support being presently scant [59]. Strategies for active targeting should consider the physiology of the local disease site, the context of site-specific delivery, and sometimes, the actual disadvantages of using high-affinity ligands for delivery in certain cases. In the context of solid tumor targeting, target-specific design parameters have been clearly described [60]. Nonetheless, recent clinical success for an actively targeted antibody therapy has reignited unreasonable expectations for general extensions for targeted nanoparticle approaches, despite the enormous differences in vehicle physical properties and transport mechanisms in tumor sites [47,48,58,60,61]. The EPR effect has been the basis for explaining passive targeting for macromolecules and particles to disease sites with pathology producing inflammation and loosening of tight junctions in the vascular endothelium. As first reported 40 years ago [62], micrometer-sized gaps between normally tight endothelial cells permit transport of large molecules and particles typically unable to traverse the vascular wall. This provides a passive route of therapeutic transport specifically enhanced in sites of inflammation, mostly observed in solid tumors to date, and recognized early as an avenue for enhancing drug delivery [63]. The EPR effect has been observed primarily in select preclinical animal disease models to date in solid tumor sites [64]. EPR-based tumor and tissue disease site leakiness is attributed to unnatural openings in the endothelium rather than fenestrations found in normal tissues. This local vascular wall hydraulic permeability and extravasation mechanism is slow and passive, certainly contributing to local flux, delivery, vehicle partitioning, and availability of therapeutics via this passive accumulation route. However, beyond claims to cures in animal disease models, the operable efficacy of the EPR effect in improving nanomedicine delivery and therapeutic outcomes in humans is not evident to date. Many reasons have been offered to explain the lack of translation of EPR-exploited therapeutic efficacy from animal models to humans [65]. Both physiological and transport properties of the disease site and the physical and chemical attributes of the nanomedicine construct amenable to exploiting the EPR effect must be considered. As these critical parameters are site-, disease-, and delivery vehicle–specific, there is "no-one-size-fits-all" nano medicine solution to address varieties of disease. Vehicle properties critical to EPR performance and passive targeting pathways include particle surface charge, steric stabilizers, particle size and shape, aggregation states, deaggregation and degradation rates, matrix transport rates, and ligand binding affinities if bearing active targeting components [66]. That these properties must be carefully and uniquely designed for each nanomedicine on a case-by-case therapeutic context seems increasingly important. Ultimately, the EPR effect may not be a generally exploitable transport phenomenon useful for all nanomedicines [67]. Claims to its specific utility in assessing nanomedicine efficacy require experimental validations of operable EPR mechanisms in producing therapy [68]. Despite many years of advocacy, research, and development, only a few nanomedicines are in clinical use, and these suffer from many of the challenges described above, including high MPS scavenging, short circulation lifetimes, low fractional dose to disease site, aggregation issues, and poor targeting [69]. Additionally, nanocarriers that display promising efficacy in cell culture models or in accepted animal disease models do not show this same success in clinical trials. This unsatisfactory outcome makes their commercial translation rather risky, costly, and a tenuous ambition for pharma companies. Despite strong, compelling, unmet clinical needs in clinical oncology, the risks associated with translating many nanomedicines remain unattractive for a large industrial sector. Reasons for lack of translation consistency from bench to bedside include: (1) a general distrust on the reliability of academic research and its quality and reproducibility [70]; (2) difficulties translating particle dosing determined for in vitro cell culture studies to in vivo models of tumors; (3) lack of validation of cancer cell lines, contaminated cell lines, irrelevant culture and assay conditions [71]; (4) lack of standards and standard methods to facilitate valid interstudy comparisons; and (5) glaring inequities between preclinical animal model results and clinical studies in humans [72]. In fact, only rodent models are used for in vivo tumor studies and most still lack validation. Most rodent tumors are generated ectopically with a large injection of humanized tumor cells and develop a large mass over relatively short periods of time compared to human tumor generation. This process produces tumors in rodents that if allometrically scaled would be kg size in humans, and with substantially different physiology. The current small rodent end result is not comparable with human tumor development. Yet, no large animal tumor models currently exist that might better replicate scaling, pathology, and physiology issues relevant to human conditions. Nanotechnology's link to developing nanomedicine as a most favored approach to improve cancer therapeutics is readily apparent [73]. The opening paragraphs of many publications using nanoparticles for cancer therapy mention the EPR effect as the justification for their approach. However, the EPR effect has only reliably been shown in ectopic tumor models in rodents without much human validation, and most of that using intrahepatic arterial (not systemic intravenous!) infusions of drugs. The current scientific community's extension of the EPR effects is largely from rodent tumor model data using unnaturally generated ectopic solid tumors where inflammation and pathology is dramatic and unrealistic compared to human tumors. The EPR effect is the exception rather than the rule for passive particle targeting and therefore should not be assumed to be operative in any experimental model. Additionally, extrapolation to human therapeutic potential is without much scientific or clinical support [74] and unjustified as a presumed mechanism until actually proven operative in treating human tumors [75]. From a nanomedicine formulation perspective, control of solid-state drug polymorphs may be more difficult with small crystalline particles or solid nanoparticle formulations that require complex, multistep chemical and formulation methods, quality control validation processes and extended storage. For example, active targeting often requires increasing compositional complexity The associated increases in yields, process chemistry scaling issues, unit costs, quality assurance/quality control (QA/QC) tasks, and regulatory obstacles are formidable enough to preclude most industrial interest in taking the formulation forward to human trials (Fig. 8.3). Nanomedicine's impact will be found in successful translation; one-off academic prototypes may find an audience in high-profile journals but their real test of value is in how many patients get treated. Such translation must anticipate the practical reality of economical therapeutic production, ready product validation, and quality systems management in an actual product life cycle [76]. Many nanomedicines do not yet subscribe to this reality. Nanoparticle transport and through-barrier movements have been reported both in vitro and in vivo for nanomedicine penetration through cell membranes and various tissue barriers (i.e., intestine, vascular wall, brain, stroma, etc.). Particles can penetrate biological barriers. However, the statistical argument aside, the actual critical therapeutic question is whether these particle fluxes are sufficient and reliable across many disease barriers, patient phenotypes, and diverse particle chemistries to produce therapeutically important effects without toxic side effects from mega-dosing. This difficult functional question remains to be answered. Additionally, when particle delivery to tumor sites is shown, quantifying fractional dose of these particles that actually penetrate into tumor matrix beyond the vascular wall is not well understood. The important therapeutic and mechanistic differences of directing nanomedicines "to a tumor" versus "into a tumor" has been raised to question the current operative philosophy and unjustified as a presumed mechanism until actually proven operative in treating human tumors [77].

5.1 Large-Scale Manufacturing

One of the important factors contributing to the slow pace in the clinical translation of NNMs is the structural and physicochemical complexity of the formulation itself. Platforms that require complex and/or laborious synthesis procedures generally have limited clinical translation potential, as they can be quite problematic to pharmaceutically manufacture on a large scale [78]. Pharmaceutical manufacturing development is centered on quality and cost. Quality includes the manufacturing process and stability of the formulation, with NNM manufacturing being challenged by potential issues related to: (1) poor quality control; (2) scalability complexities; (3) incomplete purification from contaminants, lack of clear regulatory guidelines specific for NNMs Complexity of NNM patents and IP Limited understanding of the biological interaction of NNM with the biological environment in the body of patients and starting materials (e.g., by-products); (4) high material and/or manufacturing costs; (5) low production yield; (6) insufficient batch-to-batch reproducibility, consistency, and storage stability of the final product (e.g., regarding size distribution, porosity, charge, and mass); (7) lack of infrastructure and/or in-house expertise; (8) chemical instability or denaturation of the encapsulated compound during the manufacturing process; and (9) scarcity of venture funds and pharmaceutical industry investment [79].

An essential requirement for clinical translation is to have access to a preparation method that allows the production of large scalable quantities of NNMs, which is also consistently manufactured at the same high level of quality and batch-to-batch reproducibility to set specifications [80]. The challenges arise when the NNM system becomes more complex. For example, surface modification with coatings, inclusion of multiple targeting components by encapsulating more than one therapeutic agent. Integration of multiple components into a single nanosized carrier requires multiple steps in the production process, which inevitably poses problems for large-scale good manufacturing (cGMP) production, increases the cost of production, and makes the quality assurance and quality control (QA and QC) evaluation of such products more difficult [81].Characteristics of the manufactured NNM need to be well defined and reproducibly generated to allow initiation of clinical translation. Chemistry, Manufacturing, and Controls (CMC) information is required for investigational new drugs (INDs) at each phase of investigation to ensure proper identity, strength or potency, quality, and purity of the drug substance and drug product (FDA, 2003). The type of information submitted will depend on the phase of the investigation. The extent of the human study, the duration of the investigation, the nature and source of the drug substance, and the drug product dosage form (FDA, 2003). The characterization and validation of more complex NNMs can be particularly challenging due to the sheer number of parameters to address (e.g., size distribution, morphology, charge, purity, drug encapsulation efficiency, coating efficiency, and density of conjugated ligand/s. Batch-to-batch variation of NNMs can potentially lead to significant changes to their physicochemical properties (e.g., polarity and size), pharmacokinetic parameters (i.e., absorption, distribution, metabolism, and excretion), and/or pharmacodynamic interactions (e.g., cellular interaction and activity). In addition, NNMs need to be stable after the manufacturing process, during long-term storage, and upon clinical administration (i.e., to avoid massive drug release or aggregation in the bloodstream en route to the site of action).

5.2 Intellectual Property

There is no shortage of excitement, exuberance, hype, and misinformation when it comes to anything "nano". Many claim it to be the next industrial revolution, some purport that it has already permeated virtually every sector of our economy, while others dismiss it as nothing new but a repackaging of old concepts and technologies with a new label. In reality, nanotechnology is the natural continuation of the miniaturization of materials and biomedicine. One can now say with relative certainty that research and development (R&D) is in full swing with novel nanomedical products starting to arrive, or at least inching their way, into the marketplace (Fig. 8.3). Everyone agrees that this decade has witnessed relatively more advances and product development in nanomedicine than years prior. Too often though, companies, academia, and policymakers exaggerate basic R&D in nanotechnology as potentially revolutionary advances and claim these early-stage discoveries as confirmation of downstream novel products and applications to come. Many have desperately tagged or thrown around the "nano" prefix to suit their purpose, whether it is for federal research funding, patent approval of supposedly novel technologies, raising venture capital funds, running for office, or seeking publication of their manuscript. All of this is happening while thousands of over-the-counter products containing silver nanoparticles, titanium dioxide, and carbon nanoparticles continue to stream into the marketplace without adequate safety testing, labeling, or regulatory review. Nanoproducts and applications do continue to evolve and play a pivotal role in various industry segments, spurring new directions in research, patents, commercialization, and technology transfer. In spite of anemic product development in the 1980s and 1990s, patent filings and patent grants have continued unabated [82]. It is no secret that nanopatents of dubious scope and breadth continue to be granted by patent offices around the world. In fact, since the early 1980s, "patent prospectors" have been on a global quest for "nano patent land grabs". How should regulators define nanotechnology in the context of pharma? What does nanoscale drug delivery mean? Are all engineered nanotherapeutics unique purely due to a specific size range? Is there a regulatory or legally plausible definition of nanotech from a pharma perspective with respect to size? Does a specific or unique size range of nanotherapeutics impart nanotoxicity? All these questions are related to the definition of the prefix "nano." In fact, this definitional issue, or lack thereof, continues to be one of the most significant problems shared by regulators, policymakers, drug companies, patent practitioners, and legal professionals. In particular, regulatory agencies and governmental entities such as the US Food and Drug Administration (FDA), The EMA, Environmental Protection Agency, Centers for Disease Control and Prevention, National Institute for Occupational Safety and Health, and the US Patent and Trademark Office are grappling with this critical issue. Clearly, the need for an internationally agreed definition for key terms like nanotechnology, nanoscience, nanomedicine, nanobiotechnology, nanodrug, nanotherapeutic, nanopharmaceutical, and nanomaterial has gained urgency. Similarly, we use these terms interchangeably: nano, nanotech, and nanotechnology. So, what does the prefix "nano" refer to? Any term with this prefix is broad in scope. Consider the widely used terms nanotechnologies, nanomedicine, and nanopharmaceutical, all of which are misnomers/misleading because they do not refer to a single technology. The terms nanotechnology and nanomedicine refer to interdisciplinary areas that draw from the interplay among numerous materials, products, and applications from several technical and scientific fields. In other words, "nano" is an umbrella term/prefix encompassing several technical/scientific fields, processes, and properties at the nano/micro scale. Clearly, there is confusion over the definition of nanotechnology. Moreover, due to a lack of any standard nomenclature, various inconsistent definitions have sprung up over the years. For instance, nanotechnology has been inaccurately defined by the US National Nanotechnology Initiative (NNI) since the 1990s as "the understanding and control of matter at the nanoscale, at dimensions between approximately 1 and 100 nanometers, where unique phenomena enable novel applications". Some definitions increase the upper limit to 200 nm or 300 nm or even 1000 nm. Some definitions omit a lower range. Others refer to sizes in one, two, or three dimensions while still others require a size plus special/unique property or vice versa. Given this backdrop, Dr. Bawa proposed the following definition of nanotechnology in 2007, one that is unconstrained by an arbitrary size limitation. "The design, characterization, production and application of structures, devices, and systems by controlled manipulation of size and shape at the nanometer scale (atomic, molecular, and macromolecular scale)that produces structures, devices and systems with at least one novel/superior characteristic or property" [83]. This definition has four key features. First, the definition recognizes that the properties and performance of the synthetic, engineered structures, devices, and systems are inherently rooted in their nanoscale dimensions. Second, the focus of this flexible definition is on "technology" that has commercial potential from a consumer perspective, not "nanoscience" or basic R&D in a laboratory setting that may lack commercial implication. Third, the structures, devices, and systems that result from or incorporate nano must be novel/superior compared to their bulk, conventional counterparts. Fourth, the concept of controlled manipulation (as compared to "self-assembly") is critical here. One of the major impacts of nanomedicine is taking place in the context of drug delivery. Novel nanodrugs and nanocarriers are being developed that address fundamental problems of traditional drugs, ranging from poor water solubility to unacceptable toxicity profiles, poor bioavailability, solubility issues, physical/

chemical instability, and a lack of target specificity. The first-generation nanodrugs mainly address single challenges such as targeted delivery while the second and third generations currently in development can offer two or more functions together (e.g., delivery and imaging) or overcome multiple physiological barriers to deliver their therapeutic payloads.

5.3 Baby Steps Lead to Regulatory Uncertainty: the FDA as an Example

The consequences of the overregulation of nanomedicine are potentially severe, ranging from unnecessary barriers to commercialization to inadvertent harm to the public and the environment. However, clearly there exists a need for creating regulatory guidelines that follow a science-based approach and are responsive to shifting risks. In other words, a fine balance needs to be struck, one that is flexible enough to be modified as technological innovation proceeds [84]. A balance is always desired between promoting the development of nanomedicine and exercising effective and timely oversight when warranted. After all, laws and regulations always try to keep up with scientific advances in emerging technologies and if they lag too far, as we believe in this case, that is when things become murky, commercialization stifles, public health is negatively impacted, and venture recedes. The "baby steps" that the FDA has undertaken over the past decade or so have generally contributed to regulatory uncertainty. As a result, the pharma and device industries, the drug and patent lawyers, business and venture communities, policymakers, and the public are left to dig up whatever information they can find on nanoregulation or on marketed products that may incorporate nanomaterials or involve nanotech. Currently, there are few reliable means to identify marketed "nano-containing" products, and consumers are unable to pinpoint which ones may be toxic. This is obviously not very comforting for the consumer, given that the FDA is an agency tasked with protecting public health [85]. Currently, major global regulatory systems, bodies, and regimes regarding nanomedicines are not fully mature, hampered in part by a lack of specific protocols for preclinical development and characterization. Additionally, in spite of numerous harmonization talks and meetings, there is lack of consensus on procedures, assays, and protocols to be employed during preclinical development and characterization of nanomedicines. On the other hand, there is a rise of diverse nano-specific regulatory arrangements and systems, contributing to a dense global nanotechnology regulatory landscape, full of gaps and devoid of central coordination. Moreover, governmental regulators often lack technical and scientific knowledge to support risk-based regulation, thereby leaving a significant regulatory void. Therefore, guidance is critically needed to provide clarity and legal certainty to all stakeholders: manufacturers, policymakers, healthcare providers, and the consumer. There are hundreds, if not thousands, of nanoproducts in the market for human use, but little is known of their health risks, safety data, or toxicity profiles. A one-size-fits-all approach to nanogovernance may not be ideal. It is well established that certain nanoproducts marketed for direct and indirect human consumption may be unsafe [31,32]. These products could present unexpected human toxicity effects due to various factors, such as an increased reactivity compared to their "bulk" counterparts and an increased potential to traverse biological barriers/membranes to reach or accumulate in tissues. Not all nanoscale materials are created equal: some may be toxic and their toxicities dependent upon various factors that are material-specific (charge, polarity, and chemical residues) and/or geometry specific (size, shape, and nanoscale features). As with other emerging technologies, advances in both basic and applied nanotechnology science may be unpredictable, rapid, and unevenly distributed across product applications and risk management tools. Therefore, the optimal regulatory approach is iterative, adaptive, and flexible. It is iterative by developing and delivering incremental components of a regulatory system, such as guidances specific to product areas, each as warranted and when ready. It is adaptive by providing a mechanism, within statutory constraints, to change the rules, presumptions, or pathways for these regulatory components, in light of new information gained from research or from experience in regulating earlier products. And it is flexible by using all available means, ranging from workshops to consultations to guidances to rules, in order to match the burden of regulation to its need. However, in spite of public and industry concerns, no clear guidelines or regulations have been proposed by or expected from the FDA. The "broadly inclusive initial approach" referenced above by the FDA Commissioner needs to be expanded into practical, science-based regulatory guidelines that can be depended upon by industry and consumers alike. Obviously, not every nanoproduct should to be regulated, but more is clearly expected from the FDA than guidance documents that are in draft format, broad lectures that fail to identify key regulatory issues, or policy papers that are often short on specifics. In this regard, although the FDA has made important strides, especially in the Hamburg era, numerous challenges continue to confront the agency as important unanswered questions linger regarding nanoregulation. Size changes within the nanoscale range, and the potential unpredictability arising there from are likely to add complexity to the FDA review process. The traditional product-by-product regulatory model that the FDA currently employs may not be effective for all nanoproducts because it may be difficult to put them into one of the available traditional classifications (i.e., drug, device, and biological or combination product). However, in many cases, the FDA may view nanoproducts as technologically overlapping key questions for regulators: balancing public health and encouraging nanomedicine.

- Many products that the FDA and EMA see during their review utilize nanotechnologies or contain nanomaterials; some lack disclosure of this while others identify them as such. Should these products be regulated? If so, how and to what degree?
- It is likely that some marketed nanoproducts (e.g., sunscreens containing zinc oxide and titanium dioxide) warrant safety labeling to alert the unsuspecting consumer. Are most nanomaterials used in nanoproducts inherently toxic?
- Are advances in nanoscience and nanomedicine proceeding too fast for meaningful review to take place? Can regulations truly tame the vastness encompassed by "nano"? For example, should a different set of principles apply for regulating cosmetics versus cancer nanomedicines?
- How much harmonization of drug regulation can reasonably be expected between the United States and the rest of the world?
- Should pharmaco economic data be required prior to commercialization of nanomedicines to demonstrate both social and economic added value in comparison to "conventional" established treatments?
- It appears that regulatory agencies are pushing industry to provide product-specific data for areas like cosmetics, where the FDA lacks statutory premarket review authority. Are such voluntary industry measures transparent and meaningful?
- Is the FDA's so-called "broadly inclusive approach" of considering whether products contain nanomaterials or involve nanotechnology sufficient?
- Can the safety and efficacy of complex follow-on nanotherapeutics (nanosimilar products) ever be assured without a full slate of clinical trials?
- Have the FDA and EMA kept pace with emerging advances in nanotech R&D with respect to predicting, defining, measuring, and monitoring potential "nanotoxicities"?
- Globally, should there be a wider coordinated effort on the part of regulatory agencies to review, amend, or create nanoregulations where appropriate and warranted? Who in addition to regulatory agencies like the FDA and EMA be given the key responsibility to regulate nanomedical products for human use?
- Can nanomedicine, as applied to public health, be solely regulated under existing regulations and laws?
- Are new regulations needed for all regulated products containing nanomaterials or involving nanotechnology or should they be limited to only a subset of products containing nanomaterials?
- Have delayed and uncoordinated efforts hurt venture and commercialization activities in the United States and Europe?
- The FDA has unofficially embraced the inaccurate definition of nanotechnology proposed by the NNI. What is the "official" position of the FDA regarding the definition of nanotechnology, nanoscale, nanotherapeutic, nanodrug, nanomaterial, and nanomedicine?

Industry and stakeholders fully understand that regulatory agencies cannot formulate generalized guidelines, assay protocols, or tests for all nanomedicines or nanoproducts. Nevertheless, regulatory agencies should provide selective guidelines, at least on a case-by-case basis (instead of nonbinding, unofficial "draft guidelines" or "position papers"). For industry to rely upon in determining whether their nanoformulations or nanodrug products might be subject to regulatory examination beyond what is typical for small-molecule drugs? (Miniaturization will blur distinctions between different categories) from a review perspective, and therefore consider them as highly integrated nanomedical combination products. These complexities are likely to pose additional challenges and review issues for the FDA [86]. The FDA's ability to regulate nanoproducts effectively will depend largely on the category into which the product seeking approval falls. However, as alluded to above, certain therapeutics are "combination products", which consist of two or more regulated components (drug, biological, or device) that are physically, chemically, or otherwise combined or mixed to produce a single entity [39,40]. In such cases, the FDA determines the "primary mode of action (PMOA) of the product, which is "the single mode of action of a combination product that provides the most important therapeutic action". This process is frequently imprecise, as it is not always possible to clearly elucidate a combination product's PMOA. Determining which framework will apply to any combination product is the task of the Office of Combination Products (OCP).Obviously, the

OCP will be the first office within the FDA to review many nanoproducts. The OCP makes its assignments on a case-by-case basis depending on the PMOA. But again, this process is frequently imprecise as it is not always possible to clearly elucidate a combination product's PMOA, often because at the time of an investigational application it is not clear which mode of action provides the most important therapeutic action, or the product has two different equally critical modes of action. It is quite possible that some nanoproducts will blur the distinction between mechanical and chemical action at the nanoscale or that they may be both therapeutic and diagnostic in operation.

- The FDA has unofficially embraced the inaccurate definition of nanotechnology proposed by the NNI. What is the "official" position of the FDA regarding the definition of nanotechnology, nanoscale, nanotherapeutic, nanodrug, nanomaterial, and nanomedicine?
- Industry and stakeholders fully understand that regulatory agencies cannot formulate generalized guidelines, assay protocols, or tests for all nanomedicines or nanoproducts. Nevertheless, regulatory agencies should provide selective guidelines, at least on a case-by-case basis (instead of nonbinding, unofficial "draft guidelines" or "position papers"), for industry to rely upon in determining whether their nanoformulations or nanodrug products might be subject to regulatory examination beyond what is typical for small-molecule drugs?

Another key issue is the limited information currently available correlating the physicochemical properties of nanoscale materials with risks, and lack of validated preclinical screens and animal models for the assessment of nanomaterials. In addition, the toxicity issues surrounding many nanoscale materials may not be fully apparent until they are widely used and their exposure fully felt by a diverse population. Hence, some sort of postmarket tracking or a surveillance system must be adopted or legislated to assist in product recalls. The above mentioned points are recommendations for regulatory agencies to consider as a regulatory framework for nanomedicine. Another critical issue is an important one: How does the FDA currently approve nanoscale therapeutics? Are unique nano-enabled properties assessed? Are these products reviewed on a case-by-case basis? It is clear that some nanomaterial-containing formulations ("nanoformulations") are indeed new chemical entities (NCEs). When warranted, nanoversions of active ingredients should be treated by the FDA as NCEs. This will ensure that drugs and biologicals that have been previously approved by the FDA but later modified as nanoversions will undergo a new and rigorous round of safety testing/clinical trials in order to obtain regulatory approval. The FDA on its website highlights certain critical issues in its nanoregulatory approach that, at face value, appear most appropriate [86]. However, underneath this veneer remains the real point for nanoregulation: "attributes" of materials at the nanoscale (nanocharacter) may require unique assays/testing and premarket approval beyond what the laws on the books can currently tackle to provide safe products to the consumer. There is urgency in this regard and clear guidelines are needed, not nonbinding draft reports that are meaningless from an industry perspective. Real progress in nanoregulation is urged, not mere regulatory discussions or a listing of what will be done in future. Nanoregulation with respect to pharmaceutical products, no matter how challenging or complex, needs to be addressed transparently with proper industry and public feedback. Regulation of nanomedicine must be conducted with the overarching principle that it is science based, not politically motivated or policy based. Sadly, the conclusions drawn a few years ago regarding the regulation of nanomedicine are no different from those faced by the industry and the public [56]. For now, nanoproducts submitted for FDA review will continue to be subjected to an uncertain regulatory pathway. This could negatively impact venture funding and stifle research and development in nanomedicine.

5.4 Safety and risk

- On a case-by-case basis and in conjunction with industry, identify unique safety issues associated with nanoparticles and nanomedical products. The FDA should meet its regulatory and statutory obligations by offering technical advice and guidance to industry beyond what its track record currently reflects.
- Actively seek product safety data from industry where FDA statutory authority exists for premarket review.
- Incentivize and encourage voluntary industry submissions of safety data on nanomaterials or products that incorporate nanotechnology prior to market launch, especially in cases (e.g., cosmetics) where the FDA lacks statutory premarket review authority.
- Correlate physiochemical properties with in vivo biological behavior and therapeutic outcome.
- Since there are few protocols to characterize nanomedicines at the physicochemical, biological, and

physiological levels, it is essential to develop a research strategy that involves adsorption, distribution, metabolism, and excretion (ADME) studies. A holistic approach to understanding ADME can be realized through the integration of mechanistic ADME data through the mathematical algorithms that underpin physiologically based pharmacokinetic (PBPK) modeling, routinely utilized to support regulatory submissions for conventional medicines in the United States by the FDA and in Europe by the EMA.

- Develop toxicology tests and conduct physicochemical characterization (PCC) studies for nanomaterials. Although complexity and diversity of nanomedicines and nanomaterials poses a problem, biocompatibility and immunotoxicity must be taken into consideration during preclinical assessment.
- Understand mass transport across biological membranes and body compartments as well as biodistribution profiles following administration via a specific delivery route.
- Develop standards that correlate the biodistribution of various nanomaterials with safety/efficacy by using parameters such as size, surface charge, stability, surface characteristics, solubility, crystallinity, and density.
- With industry input, create a comprehensive public databank relating to the biointeractions and toxicity profiles of engineered nanomaterials (ENMs).
- Adapt existing methodologies, as well as develop new paradigms for evaluating in vivo animal and clinical data pertaining to safety and efficacy of nanomedical products before and during the product life cycle.
- Develop guidance that provides specifics as to what kind of data is required at each step of the nanomedical translational process.
- Share data in a transparent and harmonized manner. Seek additional data on safety or effectiveness during premarket review process when warranted. The FDA's excessive reliance on publicly available or voluntarily submitted information, adverse-event reporting, and on postmarket surveillance activities may not be ideal in the case of ENMs for human use.

5.5 Standardization and Nomenclature

- Create reference classes for ENMs that are synthesized and characterized.
- Develop consensus-testing protocols to provide benchmarks for the creation of classes of nanoscale materials, both ENMs and native.
- Create uniform nomenclature for and/or working definitions of nanomaterials. Refine the current definitions of nanomaterial, nanotechnology, nanodrug, nanopharmaceutical, nanoscale, and nanomedicine for regulatory purposes.
- Further explore international regulatory harmonization efforts and formal treaties with relevant stakeholders.
- In addition to governmental bodies, involve various standard-setting organizations such as the ISO and ASTM International.
- Consult and collaborate with other federal agencies in a more effective, transparent, and science-based manner.

The FDA's current engagement in policy dialogue with other federal agencies (via the Emerging Technologies Interagency Policy Coordination Committee and other forums) has not produced any important guidelines for industry. The FDA should limit the number of nonbinding draft guidance and policy papers that it periodically issues.

5.6 Tools and Techniques

Assist in developing unique tools and techniques to characterize nanoscale materials. For instance:

- To develop imaging modalities for visualizing tissue biodistribution.
- To develop mathematical and computer models for risk/benefit analysis that can monitor quality, safety, and effectiveness vis-à-vis standard ENMs.

The end result of this could be a delay in or loss of commercialized nanoproducts. Whether the FDA eventually creates new regulations, tweaks existing ones, or establishes a new regulatory center to handle nanoproducts, for the time being it should at least look at nanoproducts on a case-by-case basis. The FDA should not attempt regulation of nanomedicine by applying existing statutes alone, especially where scientific evidence suggests otherwise. Incorporating nanomedicine regulation into the current regulatory scheme is unwise. Regulation of nanotech must balance innovation and R&D with the principle of ensuring maximum public health protection and safety. Regulatory oversight must evolve in concert with newer generations of nanomedical products [87]. It is hoped that the "baby steps" that the FDA has taken in the past decade regarding nanogovernance will translate into more meaningful, flexible, and science-based guidance in the near future. In the end, the long-term prognosis of nanomedicine will hinge on effective, valid nanogovernance requiring the full commitment of regulatory agencies such as the FDA, as well as the regulated community such as the manufacturing sector.

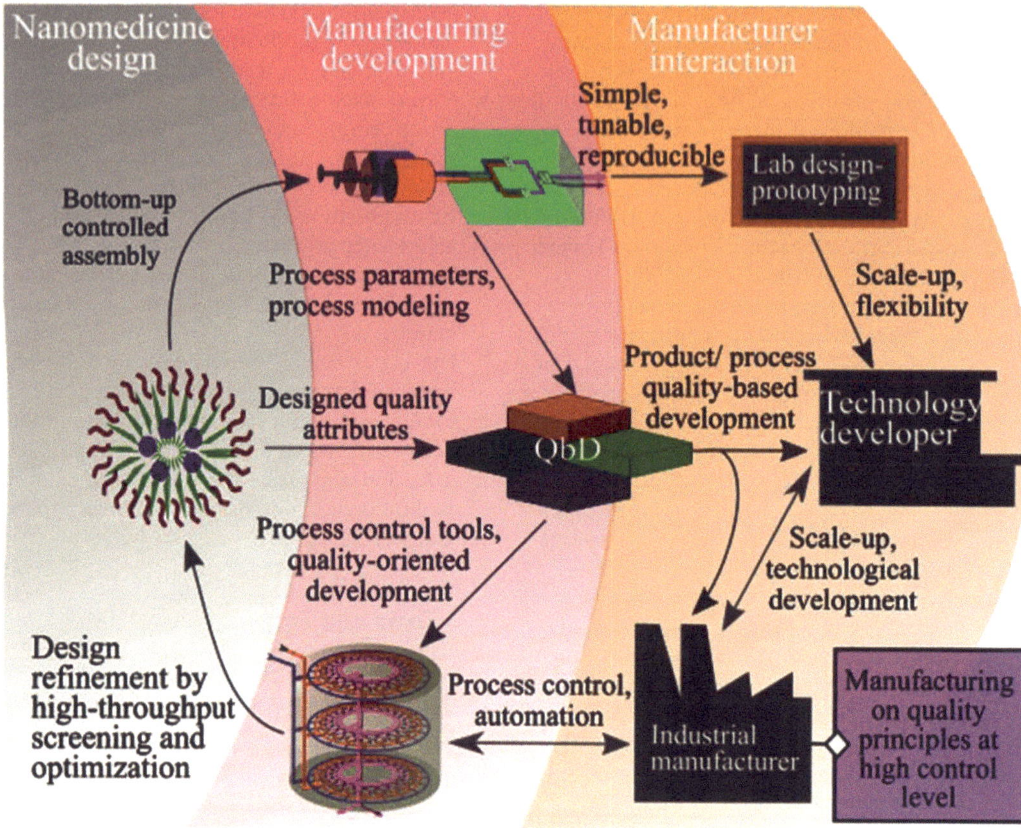

FIG. 8.4 Schematic representation of the stakeholders involved in the translation of nanomedicines from the lab to the market.

5.7 Approved Nanodrugs

Since 1995, 50 nanopharmaceuticals have received FDA approval and are currently available for clinical use [87]. Nanodrugs are typically administered orally or intravenously, and less frequently transdermally. Polymeric, liposomal, and nanocrystal formulations are heavily represented among approved nanodrugs [88] A diverse array of drug-delivery platforms based on other NPs has also been used in approved nanodrugs, including micelles and inorganic NPs (metals/metal oxides and other inorganic nanomaterials) [89]. Nanodrugs have been approved for a variety of indications, including cancer. To achieve this milestone, they should be provided with a wide knowledge from the point of view of the many disciplines involved in a nanomedical study (Fig. 8.4). Approvals of nanodrugs by the FDA peaked between 2001 and 2005, followed by a significant drop after 2006, possibly because of lower investment due to the 2008 financial crisis. A listing of approved nanodrugs appears in Table 8.2.

Most of the nanodrugs approved to date have demonstrated reduced toxicity rather than improved efficacy compared to conventional formulations. In fact, many nanodrugs have not survived clinical development because they were unable to demonstrate a significant improvement in efficacy and because improved toxicity could be achieved with other drugs or nanoformulations. However, nanoformulated versions of existing drugs that are undergoing clinical development have shown promising results with respect to improved efficacy and, therefore, appear likely to gain regulatory approval [90].

5.8 Nanodrugs in the Pipeline

New nanodrugs enter clinical investigation every year, but most are nanoformulations of previously approved drugs. As of October 2017, 56 clinical trials including the term "nano" were listed as "recruiting" or "active" on ClinicalTrials.gov [91]. There are also many nanodrugs in the very early stages of development; details

TABLE 8.2
List of Nanodrugs Approved for Commercial Applcation by FDA.

Trade name	Active Pharmaceutical Ingredient	Nanotechnology	Indication	Year Approved
Visudyne	Verteporfin	Liposome	Photodynamic therapy for age-related muscular degeneration	2000
Doxil/Caelyx	Doxorubicin	PEGylated liposome	Antineoplastic	1995
AmBisome	Amphotericin B	Liposome	Fungal infections	1990
Abelcet	Amphotericin B	Liposome	Fungal infections	1995
Definity	Octafluoropropane	Liposome	Vitrectomy	2001
Myocet	Doxorubicin	Liposome	Cancer	2001
DepoCyte	Cytarabine	Liposome	Lymphomatous meningitis	1999
DepoDur	Morphine	Liposome	opioidmedicines	2004
DaunoXome	Daunorubicin	Liposome	Antineoplastic	1996
Octocog alfa	Factor VIII	Liposome	hemophillia	2009
Abraxane	paclitaxel	Albumin bound nanoparticle	Metastatic breast cancer	2005
Rapamune	Rapamycin	Nanocrystal Elan	Immunosuppressant	2000
Emend	Aprepitant	Nanocrystal Elan	Antiemetic	2003
Tricor	Fenofibrate	Nanocrystal Elan	Hypercholesterolemia	2004
Megace ES	Megestrol	Nanocrystal Elan	Antianoretic	2005
Triglide	Fenofibrate	IDP-P Skyepharma nanocrystal	Hypercholesterolemia	2005
Mepact	Mifamurtide	Liposome	Antitumor, Immune modulator	2009
Amphotec	Amphotericin B	Micelle	Fungal infections	1996
Estrasorb	Estradiol	Micelle	Vasomotor symptoms associated with menopause	2003
Taxotere	docetaxel	Micelle	Antineoplastic	1996
Somatuline depot	lanreotide	Nanotube	Acromegaly	2007
Feraheme injection	Ferumoxytol	SPIO Polymer	Treatment of iron deficiency anemia in patient with chronic kidney disease	2009
Krystexxa/Pegloticase (Horizon)	Improved stability of protein through PEGylation; introduction of unique mammalian protein	Polymer-protein conjugate (PEGylated porcine-like uricase)	Chronic gout	2010
Marqibo (Onco TCS)	Increased delivery to tumor site; lower systemic toxicity arising from side effects	Liposomal Vincristine	Acute Lymphoblastic Leukemia	2012

Continued

TABLE 8.2
List of Nanodrugs Approved for Commercial Applcation by FDA.—cont'd

Trade name	Active Pharmaceutical Ingredient	Nanotechnology	Indication	Year Approved
Abraxane/ABI-007 (celgene)	Improved solubility; improved delivery to tumor	Albumin-bound paclitaxel nanoparticles	Breast cancer NSCLC Pancreatic cancer	2013
Invega Sustenna (Janssen Pharms)	Allows slow release of injectable low solubility drug	Paliperidone Palmitate	Schizophrenia Schizoaffective Disorder	2014
Adynovate (Baxalta)	Improved stability of protein through PEGylation	protein conjugate (PEGylated factor VIII)	Hemophilia	2015
Genexol-PM	Paclitaxel	PEG-PLA polymeric micelle	Breast cancer, Lung cancer, Ovarian cancer	2007
Lipo-Dox	Doxorubicin	Liposome	Kaposi's sarcoma, Breast cancer, Ovarian cancer	1998
NanoTherm	superparamagnetic iron oxide	Iron oxide nanoparticle	Thermal ablation glioblastoma	2010
Oncaspar	L-asparaginase	Polymer protein conjugate	Leukemia	2006
Onivyde	Irinotecan	Liposome	Pancreatic cancer	2015
Rebinyn	Coagulation factor IX (recombinant),	Coagulation factor IX (recombinant), glycopegylated	Hemophilia B	2018
Ryanodex	Dantrolene sodium	**Nanocrystal NPs**	Malignant hypothermia	2014

Source: Adapted from US-FDA and EMA.

are unknown because this information is proprietary. The number of nanodrugs that have received investigational new drug (IND) approval from the FDA to undergo clinical trials has steadily increased since 2007 [92]. The years from 2013 to 2015 had the highest number of nanoformulations entering clinical trials; this suggests an increase in the availability of FDA-approved nanodrugs in the future [93].

There are currently more anticancer and antimicrobial nanodrugs in clinical trials than any other drug classes. However, there are also many formulations being developed for other indications, including autoimmune conditions, anesthesia, metabolic disorders, ophthalmic conditions, neurological and psychiatric diseases, and others [94]. Of the products in development, the majority incorporate NPs that have already proven successful are liposomes and polymers. The remainder of investigational nanodrugs demonstrates a trend toward agents using micelles, as well as the introduction of formulations using dendrimers. Polymer-based nanoformulations are less prevalent in investigational nanodrugs than approved nanodrugs. Modern approaches in protein engineering, as well as advances in polymer and inorganic chemistry, have also resulted in an expansion of novel nanomaterials that blur the boundaries between the categories of traditional materials. This has allowed the initial goals for nanopharmaceuticals (improved PK, efficacy, and safety) to evolve into system designs that allow for more complex functions, such as controlled release and active targeting. Although the majority of FDA-approved nanodrugs rely on passive targeting via the EPR effect, some next-generation drugs in clinical trials employ active targeting approaches [95]. One example of a targeted nanodrug that is being investigated is SGT-53 (SynerGene Therapeutics), which contains an antitransferrin antibody fragment that binds with a transferring glycoprotein receptor on cancer cells. This agent is in phase 1 and 2 clinical trials for the treatment of solid tumors, glioblastoma, and metastatic pancreatic cancer [96].

6 CONCLUSIONS

Nanomedical innovations are in full swing at both academic and industrial sectors as they relate to research, patents, commercial opportunities, and technology transfer. However, in the nanomedicine arena, currently there are knowledge gaps relating to understanding the mechanisms that impact benefits for many technologies in terms of pharmacokinetics and ADME. Nanotechnology has dramatically changed approaches to drug development, vehicle designs, and combination or multiplexed capabilities on a single injected payload. Rational retrospective analysis of what has been produced and what has succeeded and what has failed could yield huge benefits, rather than just pushing forward with the same compromised design paradigm. Changes in both of these biological parameters are central to the achieved changes in efficacy but may also underpin new safety considerations that require additional data, assays, or research tools to clarify their mechanistic basis. Importantly, a fundamental basis for the application of nanodrugs is the improvement of efficacy through a reduction in toxicity. However, as for any drug, nano-enabled products for human use require a robust, evidence-based assessment that evolves as gaps in knowledge are addressed. Poor rates of technology translation from research to treatment regimens are beginning to draw ire from peers and payors. The nanomedicine community stands to lose credibility unless a rational road map forward is produced to make clinical impacts. Lack of significant therapeutic improvements in drug efficacy and difficulties and complexities of nanomedicine regulatory approval processes compared to the parent original drug form are identified as key factors slowing nanomedicine development. The current inability to accurately evaluate and screen newly designed systems for efficacy and toxicity with reliable, inexpensive in vitro methods does not provide the desired industrial goal to discern "drug candidate failure faster" in order to reduce costs and risks of further development. Lastly, limiting hype, exaggerated claims, and overstated experiment results are also desired goals to provide objectivity of the real status of nanomedicines. Blinded optimism lacks scientific credibility and undermines future work. Identifying and solving critical problems to translating nanomedicines is the challenge. Only when clean, reproducible data consistently elucidate the barriers, challenges, and successes can design, optimization, and more targeted development of improved nanomedicines proceed toward patient benefits. The guiding principle here should reflect a balance of innovation and R&D with public health protection.

REFERENCES

[1] Etheridge ML, Campbell SA, Erdman AG, Haynes CL, Wolf SM, McCullough J. The big picture on nanomedicine: the state ofinvestigational and approved nanomedicine products. Nanomedicine 2013;9:1–14.

[2] Bawa R. Nanoparticle-based therapeutics in humans: a survey. Nanotechnology Law and Business 2008;5:135–55.

[3] Ventola CL. The nanomedicine revolution. Part 2: current and future clinical applications. Pharmacy and Therapeutics 2012;37(10):582–91.

[4] Bawa R. Regulating nanomedicine—can the FDA handle it? Current Drug Delivery 2011;8:227–34.

[5] Vashist SK, Venkatesh AG, Mitsakakis K, Czilwik G, Roth G, Stetten F, Zengerle R. Nanotechnology-based biosensors and diagnostics: technology push versus industrial/healthcare requirements. BioNanoScience 2012;2:115–26.

[6] Robert J, Fisher TP. Form nanoparticles via controlled crystallization. Chemical Engineering Progress 2008:33–9.

[7] Tlmpe C. Drug solubilization strategies applying nanoparticulate formulation and solid dispersion approaches in drug development. Pharmacological Reviews 2010;13:12–21.

[8] Muller RH, Jacobs C, Kayser O. Nanosuspensions as particulate drug formulations in therapy. Rationale for development and what we can expect for the future. Advanced Drug Delivery Reviews 2001;47:3–19.

[9] Merisko-Liversidge EM, Liversidge GG. Drug nanoparticles: formulating poorly water-soluble compounds. Toxicologic Pathology 2008;36:43–8.

[10] Johnson KC. Comparison of methods for predicting dissolution andthe theoretical implications of particle-size-dependent solubility. Journal of Pharmaceutical Sciences 2012;101:681–9.

[11] Kaptay G. On the size and shape dependence of the solubility of nano-particles in solutions. International Journal of Pharmaceutics 2012;430:253–7.

[12] Merisko-Liversidge E, Liversidge GG. Nanosizing for oral and parenteral drug delivery: a perspective on formulating poorly-water soluble compounds using wet media milling technology. Advanced Drug Delivery Reviews 2011;63:427–40.

[13] Miele E, Spinelli GP, Miele E, Tomao F, Tomao S. Albumin-bound formulation of paclitaxel (Abraxane ABI-007) in the treatment of breast cancer. International Journal of Nanomedicine 2009;4:99–105.

[14] Raj S, Jose S, Sumod US, Sabitha M. Nanotechnology in cosmetics:Opportunities and challenges. Journal of Pharmacy and Bioallied Sciences 2012;4:186–93.
[15] Wiechers JW, Musee N. Engineered inorganic nanoparticles and cosmetics: facts, issues, knowledge gaps and challenges. Journal of Biomedical Nanotechnology 2010;6:408–31.
[16] Mu L, Sprando RL. Application of nanotechnology in cosmetics. Pharmaceutical Research 2010;27:1746–9.
[17] Nohynek GJ, Antignac E, Re T, Toutain H. Safety assessment of personal care products/cosmetics and their ingredients. Toxicology and Applied Pharmacology 2010;243:239–59.
[18] Ryman-Rasmussen JP, Riviere JE, Monteiro-Riviere NA. Penetration of intact skin by quantum dots with diverse physicochemical properties. Toxicological Sciences 2006;91:159–65.
[19] Rouse JG, Yang J, Ryman-Rasmussen JP, Barron AR, Monteiro-Riviere NA. Effects of mechanical flexion on the penetration of fullerene amino acid-derivatized peptide nanoparticles through skin. Nano Letters 2007;7:155–60.
[20] Corot C, Warlin D. Superparamagnetic iron oxide nanoparticles for MRI: contrast media pharmaceutical company R&D perspective. Wiley Interdisciplinary Reviews. Nanomedicine and Nanobiotechnology 2013;5(5):411–22.
[21] Lodhia J, Mandarano G, Ferris N, Eu P, Cowell S. Development and use of iron oxide nanoparticles (Part 1): synthesis of iron oxide nanoparticles for MRI. Biomedical Imaging and Intervention Journal 2010;6:e12.
[22] Huang CH, Tsourkas A. Gd-based macromolecules and nanoparticles as magnetic resonance contrast agents for molecular imaging. Current Topics in Medicinal Chemistry 2013;13:411–21.
[23] Schellenberger EA, Bogdanov Jr A, Hogemann D, Tait J, Weissleder R, Josephson L. Annexin V-CLIO: a nanoparticle for detecting apoptosis by MRI. Molecular Imaging 2002;1:102–7.
[24] Harisinghani MG, Barentsz J, Hahn PF, Deserno WM, Tabatabaei S, van de Kaa CH, de la Rosette J, Weissleder R. Noninvasive detection of clinically occult lymph-node metastases in prostate cancer. The New England Journal of Medicine 2003;348:2491–9.
[25] Fortina P, Kricka LJ. Nanotechnology: improving clinical testing? Clinical Chemistry 2010;56:1384–9.
[26] Dobson MG, Galvin P, Barton DE. Emerging technologies for point-of-care genetic testing. Expert Review of Molecular Diagnostics 2007;7:359–70.
[27] Agasti SS, Rana S, Park MH, Kim CK, You CC, Rotello VM. Nanoparticles for detection and diagnosis. Advanced Drug Delivery Reviews 2010;62:316–28.
[28] Sanvicens N, Marco MP. Multifunctional nanoparticles—properties and prospects for their use in human medicine. Trends in Biotechnology 2008;26:425–33.
[29] Dandy DS, Wu P, Grainger DW. Array feature size influences nucleic acid surface capture in DNA microarrays. Proceedings of National Academy of Sciences of the United States of America 2007;104:8223–8.
[30] Rao AN, Grainger DW. Biophysical properties of nucleic acids at surfaces relevant to microarray performance. Biomaterials Science 2014;2:436–71.
[31] Park K. Facing the truth about nanotechnology in drug delivery. ACS Nano 2013;7:7442–7.
[32] Loeve S, Vincent BB, Gazeau F. Nanomedicine metaphors: fromwar to care. Emergence of an oecological approach. Nano Today 2013;8(6):560–5.
[33] Grainger DW, Castner DG. Nano biomaterials and nano analysis: how to improve the nanoscience for biotechnology. Advanced Materials 2008;20:867–77.
[34] Crist RM, Grossman JH, Patri AK, Stern ST, Dobrovolskaia MA, Adiseshaiah PP, Clogston JD, McNeil SE. Common pitfalls in nanotechnology: lessons learned from NCI's nanotechnology characterization laboratory. Integrative Biology 2013;5:66–73.
[35] Richman EK, Hutchison JE. The nanomaterial characterization bottleneck. ACS Nano 2009;3:2441–6.
[36] Buzea C, Pacheco II, Robbie K. Nanomaterials and nanoparticles: sources and toxicity. Biointerphases 2007;2:MR17–71.
[37] Baer DR, Engelhard MH, Johnson GE, Laskin J, Lai J, Mueller K, Munusamy P, Thevuthasan S, Wang H, Washton N, Elder A, BaischB L, Karakoti A, Kuchibhatla SVNT, Moon D. Surface characterization of nanomaterials and nanoparticles: important needs and challenging opportunities. Journal of Vacuum Science and Technology A 2013;31:50820.
[38] Stirland DL, Nichols JW, Miura S, Bae YH. Mind the gap: a survey of how cancer drug carriers are susceptible to the gap between research and practice. Journal of Controlled Release 2013;172:1045–64.
[39] Begley CG, Ellis LM. Drug development: raise standards for preclinical cancer research. Nature 2012;483:531–3.
[40] Ensign LM, Cone R, Hanes J. Oral drug delivery with polymeric nanoparticles: the gastrointestinal mucus barriers. Advanced Drug Delivery Reviews 2012;64:557–70.
[41] Yamanaka YJ, Leong KW. Engineering strategies to enhance nanoparticle-mediated oral delivery. Journal of Biomaterials Science, Polymer Edition 2008;19:1549–70.
[42] Yuan H, Chen CY, Chai GH, Du YZ, Hu FQ. Improved transport and absorption through gastrointestinal tract by PEGylated solid lipid nanoparticles. Molecular Pharmaceutics 2013;10:1865–73.
[43] Hussain N, Jaitley V, Florence AT. Recent advances in the understanding of uptake of microparticulates across the gastrointestinal lymphatics. Advanced Drug Delivery Reviews 2001;50:107–42.
[44] Schleh C, Semmler-Behnke M, Lipka J, Wenk A, Hirn S, Schäffler M, Schmid G, Simon U, Kreyling WG. Size and surface charge of gold nanoparticles determine absorption across intestinal barriers and accumulation in secondary target organs after oral administration. Nanotoxicology 2012;6(1):36–46.
[45] Moghimi SM, Hunter AC, Murray JC. Long-circulating and target-specific nanoparticles: theory to practice. Pharmacological Reviews 2001;53:283–318.

[46] Moghimi SM, Hunter AC, Murray JC. Nanomedicine: currentstatus and future prospects. Federation of American Societies for Experimental Biology Journal 2005;19:311–30.
[47] Bae YH, Park K. Targeted drug delivery to tumors: myths, reality and possibility. Journal of Controlled Release 2011;153:198–205.
[48] Kwon IK, Lee SC, Han B, Park K. Analysis on the current status of targeted drug delivery to tumors. Journal of Controlled Release 2012;164:108–14.
[49] Bertrand N, Leroux JC. The journey of a drug-carrier in the body: An anatomo-physiological perspective. Journal of Controlled Release 2012;161:152–63.
[50] Harris JM, Chess RB. Effect of pegylation on pharmaceuticals. Nature Reviews Drug Discovery 2003; 2:214–21.
[51] Chambers E, Mitragotri S. Long circulating nanoparticles via adhesion on red blood cells: mechanism and extended circulation. Experimental Biology and Medicine 2007;232:958–66.
[52] Rodriguez PL, Harada T, Christian DA, Pantano DA, Tsai RK, Discher DE. Minimal "self" peptides that inhibit phagocytic clearance and enhance delivery of nanoparticles. Science 2013;339:971–5.
[53] Fang RH, Hu CM, Zhang L. Nanoparticles disguised as red blood cells to evade the immune system. Expert Opinion on Biological Therapy 2012;12:385–9.
[54] Florence AT. "Targeting" nanoparticles: the constraints of physical laws and physical barriers. Journal of Controlled Release 2012;164:115–24.
[55] Gronow JR. Mechanisms of particle movement in porous media. Clay Minerals 1986;21:753–67.
[56] Pirollo KF, Chang EH. Does a targeting ligand influence nanoparticle tumor localization or uptake? Trends in Biotechnology 2008;26:552–8.
[57] Kirpotin DB, Drummond DC, Shao Y, Shalaby MR, Hong K, Nielsen UB, Marks JD, Benz CC, Park JW. Antibody targeting of long-circulating lipidic nanoparticles does not increase tumor localization but does increase internalization in animal models. Cancer Research 2006;66:6732–40.
[58] Umemura K. Hybrids of nucleic acids and carbon nanotubes for nanobiotechnology. Nanomaterials 2015;5: 321–50.
[59] Nikalje A. Nanotechnology and its applications in medicine. Medicinal Chemistry 2015;5:081–9.
[60] Roco MC. The long view of nanotechnology development: the national nanotechnology initiative at10 years. Springer; 2011.
[61] Weissig V, Guzman-Villanueva D. Nanopharmaceuticals (part 2): products in the pipeline. International Journal of Nanomedicine 2015;10:1245.
[62] Li J, Wang Y, Liang R, et al. Recent advances in targeted nanoparticles drug delivery to melanoma. Nanomedicine: Nanotechnology, Biology and Medicine 2015; 11(3):769–94.
[63] Weissig V, Pettinger TK, Murdock N. Nanopharmaceuticals (part 1): products on the market. International Journal of Nanomedicine 2014;9:4357.
[64] Tomalia DA. Interview: an architectural journey: from trees, dendrons/dendrimers to nanomedicine. Nanomedicine 2012;7(7):953–6.
[65] Liu KC, Yeo Y. Extracellular stability of nanoparticulate drug carriers. Archives of Pharmacal Research 2014;37: 16–23.
[66] Gullotti E, Yeo Y. Beyond the imaging: limitations of cellular uptake study in the evaluation of nanoparticles. Jounal of Controlled Release 2012;164:170–6.
[67] Torchilin VP. Multifunctional, stimuli-sensitive nanoparticulate systems for drug delivery. Nature Reviews Drug Discovery 2014;13:813–27.
[68] le Masne de Chermont Q, Chaneac C, Seguin J, Pelle F, Maitrejean S, Jolivet JP, Gourier D, Bessodes M, Scherman D. Nanoprobes with near-infrared persistent luminescence for in vivo imaging. Proceedings of the National Academy of Sciences of the United States of America 2007;104:9266–71.
[69] Cho EJ, Holback H, Liu KC, Abouelmagd SA, Park J, Yeo Y. Nanoparticle characterization: state of the art,challenges, and emerging technologies. Molecular Pharmaceutics 2013;10:2093–110.
[70] Chen H, Kim S, He W, Wang H, Low PS, Park K, Cheng JX. Fast release of lipophilic agents from circulating PEG-PDLLA micelles revealed by in vivo forster resonance energy transfer imaging. Langmuir 2008;24:5213–7.
[71] Zervantonakis IK, Hughes-Alford SK, Charest JL, Condeelis JS, Gertler FB, Kamm RD. Three-dimensional microfluidic model for tumor cell intravasation and endothelial barrier function. Proceedings of the National Academy of Sciences of the United States of America 2012;109:13515–20.
[72] Hua S. Targeting sites of inflammation: intercellular adhesion molecule-1 as a target for novel inflammatory therapies. Frontiers in Pharmacology 2013;4:127. https://doi.org/10.3389/fphar.2013.00127.
[73] Korostynska O, Gil E, Arshak A, Arshak K. Review on state-of-the-art in polymer based pH sensors. Sensors 2007;7:3027–42.
[74] Laouini A, Limayem-Blouza I, Sfar S, Charcosset C, Fessi H, Jaafar-Maalej C. Preparation, characterization and applications of liposomes: state of the art. Journal of Colloid Science and Biotechnology 2012;1(2):147–68.
[75] Cohen ZR, Ramishetti S, Peshes-Yaloz N, Goldsmith M, Wohl A, Zibly Z, Peer D. Localized RNAi therapeutics of chemoresistant grade IV glioma using hyaluronan-grafted lipid based nanoparticles. ACS Nano 2015;9: 1581–91.
[76] Ramishetti S, Kedmi R, Goldsmith M, Leonard F, Sprague AG, Godin B, Gozin M, Cullis PR, Dykxhoorn DM, Peer D. Systemic gene silencing in primary T lymphocytes using targeted lipid nanoparticles. ACS Nano 2015;9:6706–16.
[77] Kim TY, Kim DW, Chung JY, et al. Phase I and pharmacokinetic study of Genexol-PM, a cremophor-free, polymeric micelle-formulated paclitaxel, in patients with advanced malignancies. Clinical Cancer Research 2004; 10:3708–16.

[78] Lee KS, Chung HC, Im SA, et al. Multicenter phase II trial of Genexol-PM, a Cremophor-free, polymeric micelle formulation of paclitaxel, in patients with metastatic breast cancer. Breast Cancer Research and Treatment 2008;108:241–50.
[79] Kim DW, Kim SY, Kim HK, et al. Multicenter phase II trial of Genexol-PM, a novel Cremophor-free, polymeric micelle formulation of paclitaxel, with cisplatin in patients with advanced non–small-cell lung cancer. Annals of Oncology 2007;18:2009–14.
[80] Bansal S, Bansal M, Kumria R. Nanocrystals: current strategies and trends. International Journal of Research in Pharmaceutical and Biomedical Sciences 2012;3:406–19.
[81] Gao L, Liu G, Ma J, et al. Application of drug nanocrystal technologies on oral drug delivery of poorly soluble drugs. Pharmaceutical Research 2013;30:307–24.
[82] Bruchez Jr M, Moronne M, Gin P, et al. Semiconductor nanocrystals as fluorescent biological labels. Science 1998;281:2013–6.
[83] Maier-Hauff K, Ulrich F, Nestler D, et al. Efficacy and safety of intratumoral thermotherapy using magnetic iron-oxide nanoparticles combined with external beam radiotherapy on patients with recurrent glioblastoma multiforme. Journal of Neuro-Oncology 2011;103(2):317–24.
[84] Libutti SK, Paciotti GF, Byrnes AA, et al. Phase I and pharmacokinetic studies of CYT-6091, a novel PEGylated colloidal gold-rhTNF nanomedicine. Clinical Cancer Research 2010;16:6139–49.
[85] Fleischman T. Cancer killers: C dots show ability to induce cell death in tumors. Cornell Chronicle 2016. Available at: http://news.cornell.edu/stories/2016/09/cancer-killers-c-dots-show-ability-induce-cell-death-tumors.
[86] Weissig V, Pettinger TK, Murdock N. Nanopharmaceuticals (part 1): products on the market. International Journal of Nanomedicine 2015;10:1245–57.
[87] Caster JM, Patel AN, Zhang T, Wang A. Investigational nanomedicines in 2016: a review of nanotherapeutics currently undergoing clinical trials. Wiley Interdisciplinary Reviews Nanomedicine and Nanobiotechnology 2017;9(1).
[88] Desai N, Trieu V, Yao ZW, et al. Increased antitumor activity, intratumor paclitaxel concentrations, and endothelial cell transport of Cremophor-free, albumin-bound paclitaxel, ABI-007, compared with Cremophor-based paclitaxel. Clinical Cancer Research 2006;12(4):1317–24.
[89] Fuentes AC, Szwed E, Spears CD, et al. Denileukin diftitox (Ontak) as maintenance therapy for peripheral T-cell lymphomas: three cases with sustained remission. Case Reports in Oncological Medicine 2015;2015:123756. https://doi.org/10.1155/2015/123756.
[90] Foss FM, Sjak-Shie N, Goy A, et al. A multicenter phase II trial to determine the safety and efficacy of combination therapy with denileukin diftitox and cyclophosphamide, doxorubicin, vincristine, and prednisone in untreated peripheral T-cell lymphoma: the CONCEPT study. Leukemia and Lymphoma 2013;54(7):1373–9.
[91] Foss F. Clinical experience with denileukin diftitox (Ontak). Seminars in Oncology 2006;33(1 Suppl. 3):S11–6.
[92] Svenson S, Tomalia DA. Dendrimers in biomedical applications— reflections on the field. Advanced Drug Delivery Reviews 2005;57:2106–29.
[93] O'Loughlin J, Millwood IY, McDonald HM, et al. Safety, tolerability, and pharmacokinetics of SPL7013 gel (VivaGel): a dose ranging, phase 1 study. Sexually Transmitted Diseases 2010;37:100–4.
[94] Price CF, Tyssen D, Sonza S, et al. SPL7013 Gel (VivaGel) retains potent HIV-1 and HSV-2 inhibitory activity following vaginal administration in humans. PLoS One 2011;6(9):e24095.
[95] Cohen CR, Brown J, Moscicki AB, et al. A phase 1 randomized placebo controlled trial of the safety of 3% SPL7013 Gel (VivaGel) in healthy young women administered twice daily for 14 days. PLoS One 2011;6(1):e1625.
[96] Ventola CL. The nanomedicine revolution: part 3: regulatory and safety challenges. P and T 2012;37(11):631–9.

CHAPTER 9

Nanotoxicology and Its Remediation

VINAY KUMAR, PHD • SIVARAMA KRISHNA LAKKABOYANA, PHD • NEHA SHARMA, MSC • SUBHRANGSU SUNDER MAITRA, PHD • MARLIA MOHD HANAFIAH, PHD

1 NANOTOXICITY

Rapid developments in the field of nanoparticles in recent years has led to an exponential rise in their applications. Manufactured nanoparticles come with unique physical and chemical properties and find their applications in the fields of the catalyst [1–3], biomedicine [3a], and antibacterial materials [4]. With the development of nanotechnology and its applications in the biomedical field, the toxic effects of nanomaterials have attracted much attention. With the development, a subfield termed as nanotoxicology has emerged. Nanotoxicology deals with the study of the interactions between nanomaterials and biological systems. It emphasizes on identifying the relationship between the physicochemical parameters such as size, shape, surface modification of nanoparticles, and the occurrence of toxic effects [5].

The toxicity of the nanoparticles is dependent on their structure, dose, and their physicochemical properties. These properties include size, surface charge, and surface area [6]. Changes in the nanoparticle physicochemical properties will affect their translocation and distribution to the target organs [7]. Studies have demonstrated a direct correlation between the level of reactive oxygen species (ROS) generation and the size distribution of nanoparticles. One example of the size distribution is presented in a study where silver nanoparticles with a size in the range of 10 nm had higher liver distribution and more serious hepatobiliary toxicity as compared to the larger size range (40 and 100 nm) [8,9]. The study has reported that the pharmacokinetics is determined by the surface charge rather than the size of the nanoparticle [10]. The reason for the smaller nanoparticles increased dissolution as compared to the larger particles of the equivalent mass was their higher number. Due to the higher number, the toxicity imposed was higher. In a study, the surface modification of nevirapine nanosuspensions with serum albumin, polyethylene glycol, and polysaccharide enhanced their accumulation in organs such as liver, brain, and spleen, which is correlated to their toxicity in these organs [11]. Ionic forms of rather than the particulate forms accumulate in the lungs, liver, and kidneys following oral and intravenous administration of zinc oxide nanoparticles [10]. Fig. 9.1 presents the different nanoparticles types, experimental models used in the studies, and toxic effects of the nanoparticles.

The route for entry of nanoparticles includes endocytosis, which is facilitated by binding to the membrane receptors. The efficiency of endocytosis depends on the size of nanoparticles, and smaller nanoparticles have more probability of internalization and hence cause more cellular toxicities [12,13].

The studies on animal models have shown that nanoparticles have the ability to accumulate in different organs and organ systems. There are various mechanisms responsible for the cellular toxicity of nanoparticles. These include activation of inflammatory responses (activation of proinflammatory transcription factor NF-κB, nuclear translocation of NF-κB, and transcription of proinflammatory genes: TNF-α, IL-6, and IL-8), generation of ROS (cell injury, apoptosis, and altered DNA methylation) and epigenetic modifications, histone posttranslational modifications and histone hypoacetylation, and reduced gene transcription [14].

Nanoparticle-induced ROS production is mediated by mitochondrial dysfunction [13]. Mechanisms for nanoparticle-mediated oxidative stress include protein oxidation and increased release of nitric acid and glutamic acid [15,16]. Moreover, oxidative stress caused by nanoparticles may occur via the p38-Nrf-2 signaling pathway [17]. The studies have reported that the nanoparticles have the ability to increase the levels of inflammatory cytokines [18–20]. In the macrophage-rich organs such as spleen and liver, nanoparticles are usually taken up by the macrophages

Intelligent Nanomaterials for Drug Delivery Applications. https://doi.org/10.1016/B978-0-12-817830-0.00009-6

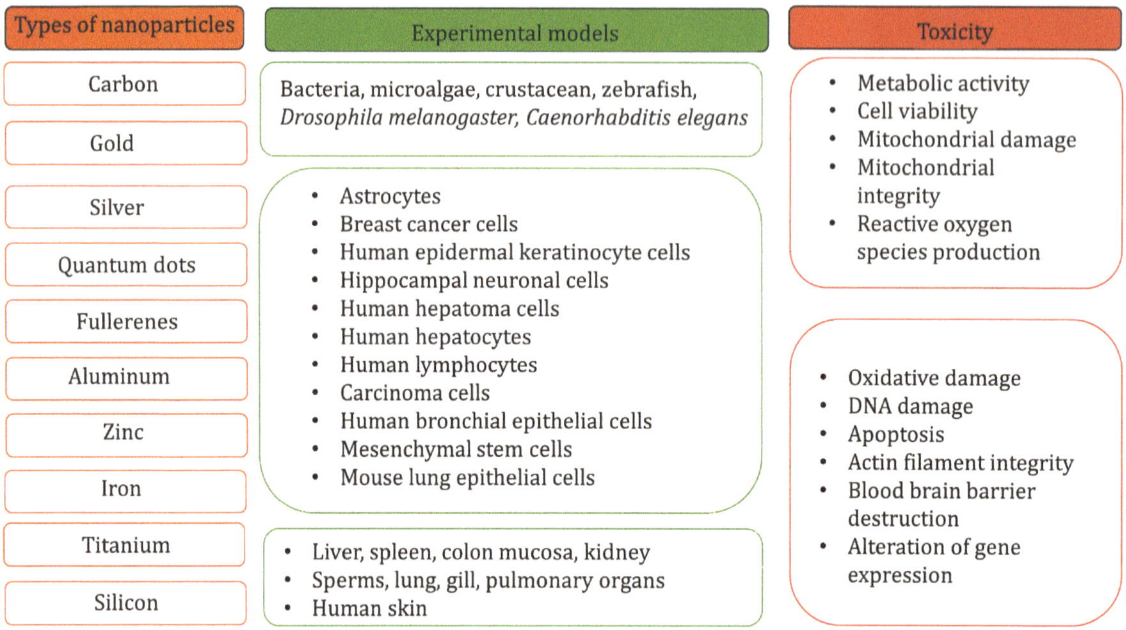

FIG. 9.1 Various types of nanoparticles synthesized, animal models used and their toxic effects.

that induce the release of the cytokines. Nanoparticles first bind to specific macrophage receptors [21]. Later on, nanoparticles are internalized and translocated inside the macrophages via endocytosis or micropinocytosis. Nanoparticles are known to induce epigenetic modifications in addition to inflammation and oxidative stress. Exposure to nanoparticles in vitro studies has demonstrated changes in DNA methylation, chromatin remodeling, histone posttranslational modifications, and RNA methylation [22,23].

2 GOLD NANOPARTICLES' TOXICITY

Due to the popularity of gold nanoparticles, they have been used in various biological systems such as NIR-responsive drug-delivery systems based on photothermal-chemotherapy [24,25]. Other applications include effective chemical drug delivery [26], NIR-responsive controlled release [27], promising agents in bioimaging [28], theranostic applications [29,30], and biosensing and as immunotherapy carriers [31]. With extensive applications of gold nanoparticles in the biomedical field, a strong interest has been exerted to explore their in vitro and in vivo adverse effects [32–34]. The cytotoxicity of gold nanoparticles depends on the particle size [35,36]. Uptake of gold nanoparticles is mediated by nonspecific adsorption of serum proteins onto the gold surface and they internalized by receptor-mediated endocytosis pathway [37,38]. When gold nanoparticles of three different sizes viz. 1.5, 4, and 14 nm were applied to the human embryonic stem cells, only 1.5 nm particles showed toxicity to the cells while the other two nanoparticles exhibited almost no toxic effects [39]. The amount of nanoparticle also plays an important role in the nanotoxicity. It has been observed that the increased amount of gold nanoparticles uptake may account for the toxicity and the immunological responses [40]. In another study, when human cerebral microvascular endothelial cells and human dermal microvascular endothelial cells were incubated with gold nanoparticles of size 10, 11, and 25 nm, no toxicity was observed rather a dose-dependent release of lactate dehydrogenase was seen after incubation with the particles [41]. When PC-3 and MCF-7 cells were treated with 2-nm gold nanoparticles, toxicity was observed and assessed by MTT and LDH assays. When 17-nm citrate-coated gold particles were used to treat the A549 cells, toxicity was observed and measured by MTT assay, LDH assay, and ATP depletion measurements [42].

Surface modification of the gold nanoparticles is another factor that affects the nanotoxicity. The application of these nanoparticles for therapeutic or diagnostic purposes often involves the functionalization of nanomaterials. The functionalization of nanoparticles improves their biocompatibility and achieves specific purposes such as drug delivery. Surface hydrophobicity of the gold nanoparticles was studied where a series of gold nanoparticles were synthesized and modified with quaternary ammonium functionalities of different lengths. An increase in both cellular toxicity and ROS production increased with increasing alkyl chain length. In addition, DNA damage decreased with increasing particle hydrophobicity [43]. The influence of PEG shells on the interaction of gold nanoparticles with the cells indicated that thinner and more hydrophilic coatings together with the partial functionalization with quaternary ammonium cations resulted in a more efficient cellular uptake. This relates to significant effects on structural and functional cell parameters [44].

3 SILVER NANOPARTICLES' TOXICITY

In spite of the wide applications of silver nanoparticles, very few studies have been conducted on their impact on human health and the environment. Regardless of a number of publications describing toxicity, the toxicological mechanism is still unclear. In general, in vitro assays, silver nanoparticles are highly toxic at 5–10 μg/mL and sizes from 10 to 100 nm, and they are known for the disruption of mitochondrial function [45,46]. A study reported that cytotoxicity of silver nanoparticles is mainly induced through the mitochondrial pathway by reducing glutathione (GSH), ROS-responsive genes, and high lipid peroxidation, and it causes apoptosis, DNA damage, and necrosis (Andrea Haase et al., 2012). A few in vivo studies reported that silver nanoparticles cause adverse effects on reproduction, morphological deformities, and malformations in nonmammalian animal models, in addition to the in vitro effects [47]. It was reported in a recent study that the cytotoxicity of silver nanoparticles is the result of the minimum release of silver ions [48–51]. There are various mechanisms by which silver nanoparticles are internalized by the cell. These include diffusion, phagocytosis, and endocytosis [52]. In human macrophages, silver nanoparticles can enter the cells in phagocytic and nonphagocytic methods [53]. The uptake range of the nanoparticles depends on the particle type and particle size [54–56]. Both active and passive transport are involved in flip-flop mechanism or direct penetration for the uptake of silver nanoparticles [53]. Silver nanoparticles uptake in the HT22 cell lines was continued even after 96 h of removal of nanoparticles [57]. In vitro studies have confirmed that cellular and biochemical alterations in the cells are caused by ROS-mediated toxicity [58]. The probable mechanism of silver nanoparticles' toxicity is considered to be oxidative stress. Extensive ROS can collapse the antioxidant defense system that leads to damage to DNA, proteins, and lipids [59]. Mitochondria releases ROS leading to disruption of ATP synthesis, oxidative stress, DNA damage, and apoptosis [60]. Due to an increase in ROS levels, the GSH level decreases dramatically and thereby leads to an increase in the LDH in the medium and induces apoptosis [61]. In neuronal cells, the increased levels of ROS lead to oxidative stress that might cause calcium dysregulation or neurodegeneration in the neuronal cell [62–64]. In human cell lines, oxidative stress resulting from silver nanoparticles can damage DNA, oxidative defense capacity of the cell, and finally apoptosis [58,61,65–68]. The major cytotoxic effect of silver nanoparticles is apoptosis-mediated cell death [69].

The cytotoxicity of silver nanoparticles is mediated by various pathways. Silver nanoparticles' toxicity is triggered by an increase in ROS generation [70]. Silver nanoparticles in vitro instillation could generate overproduction of intracellular ROS leading to activation of cell death-regulating pathways such as p53, AKT, and MAPK signaling apoptotic pathways [71]. Downregulation of AKT takes place after the overproduction of ROS that increases the expression of proapoptotic kinase p38. The decrease in PARP (poly ADP ribose polymerase) expression results in a significant increase in H2X, caspase-3, p-p53, and total p53 expressions [71]. Therefore silver nanoparticles can induce apoptosis followed by the p53 signaling pathway Fig. 9.2. Mitochondria are considered an important center for apoptosis signals.

Silver nanoparticles' toxicity to mitochondrial membrane permeability can cause loss of mitochondrial integrity that may regulate the c-Jun N-terminal kinase (JNK)-mediated caspase-dependent apoptosis [72]. Thus silver nanoparticles can induce apoptosis by caspase-dependent pathway via mitochondria mediated by JNK. Silver nanoparticles can also induce epigenetic dysregulation, which may have long-term effects on gene expression reprogramming [57]. Silver nanoparticles could affect inducing DNA methylation and cell cycle following p53 or p21 pathway [57]. Fig. 9.3 presents the possible uptake process and mechanism of cytotoxicity induced by silver nanoparticles in different cell lines.

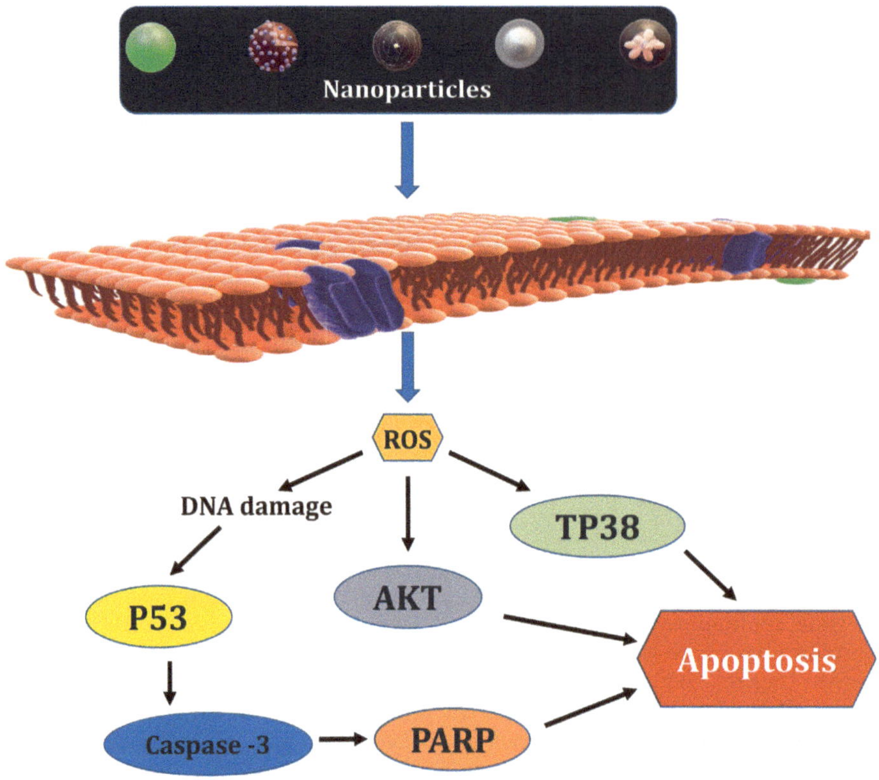

FIG. 9.2 Nanoparticles toxicity after penetration to the cells and the apoptosis pathway mediated by the JNK pathway.

A study conducted on the liver shows that silver nanoparticles are responsible for dose-dependent changes in the level of cholesterol and alkaline phosphatase. The study also demonstrated that nanoparticles administered intravenously or intraperitoneally tend to accumulate in the liver [73]. Another study also reported the accumulation of silver nanoparticles in the liver, and it induced activation of hepatic macrophages and was associated with liver damage [74]. Silver nanoparticles are known to cause tubular dilatation, inflammation, and mineralization in rat kidneys and increased proteinuria [75]. A study reported the gender associated accumulation of silver nanoparticles in kidneys which were twofold higher in female rat kidneys compared to males [73]. Dermal administration of silver nanoparticles resulted in an increase in the number of Langerhans cells and reduced thickness of the epidermis and capillary layer [76]. It was suggested that the permeation of the nanoparticles into the outermost stratum layers is dependent on the hydrophilic pathways including hair follicles, sweat pores, physiological properties of nanoparticles, and intracellular spaces between the corneocytes [77]. A study reported that oral administration of silver nanoparticles can induce changes in the cerebral myelin integrity in mice [78]. The result was supported by the changes in the expression of several myelin-specific proteins that include MAG (myelin-associated glycoprotein), CNP (2′,3′-cyclic nucleotide phosphodiesterase), and MOG (myelin/oligodendrocyte glycoprotein). In addition, silver nanoparticles exposure was associated with an increase in body weight and temperature [78].

4 ZINC NANOPARTICLES' TOXICITY

Zinc oxide (ZnO) nanoparticles have received popularity recently in medicine, cosmetics, chemical fiber, and other industries due to their enhanced redox capacity and large surface area [79,80]. However, zinc oxide

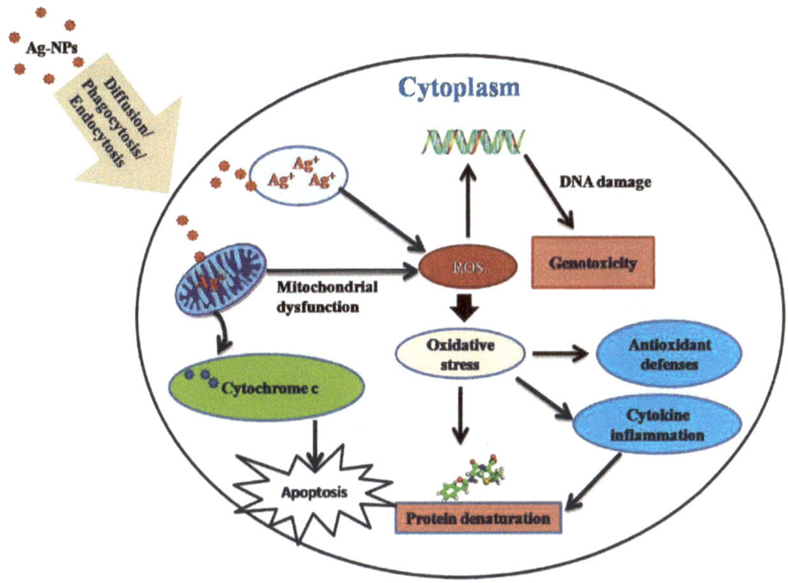

FIG. 9.3 The possible mechanisms of cytotoxicity induced by silver-nanoparticles in different cell lines.

nanoparticles are released in the environment during their transport, production, and disposal [81,82]. Toxicity studies on the *Escherichia coli* and *Staphylococcus aureus* demonstrated that *E. coli* was completely inhibited at ZnO concentration more than 3.4 mM and *Staphylococcus aureus* was inhibited at more than 1 mM [83]. ZnO nanoparticles can cause damage to the liver, spleen, and pancreas and increased blood viscosity in mice [15,16]. ZnO nanoparticles can also cause toxicity damage to the respiratory tract and digestive tract. Small ZnO nanoparticles may enter the nucleus through the nuclear pore or internalize through the cell wall in cell division [84,85]. Indirectly they may promote oxidative stress and cellular inflammatory response leading to the DNA damage [86]. In a study after exposure of ZnO for 21 days, the concentration of Zn^{2+} increased by 13.3%, 11.0%, 15.0%, and 20.2% in gill, liver, brain, and skin of *Danio rerio* at 5000 mg/L, respectively [87]. The extracellular Zn^{2+} can pass through the cell membrane and enters the cytoplasm. Zn^{2+} can induce ROS production by interfering with the electron transport chain once it binds to the intracellular mitochondria. This causes mitochondrial dysfunction and activation of the apoptotic pathway [88]. If the detoxification mechanism fails, oxidative stress can cause the dissolution of the mitochondria leading to more severe autophagy [89]. It was observed that the toxicity of the ZnO nanoparticles was concentration dependent. When *Caenorhabditis elegans* was treated with 1.6 and 4.1 mg/L of ZnO nanoparticles, it resulted in body length decrease by 32.7% and 41.0%, the eggs inside the body decrease by 77.1% and 100%, and the offspring per worm decrease by 89.4% and 100% after 5 days, respectively [62]. It was found that at high concentration ZnO nanoparticles provide high levels of zinc ions. When absorbed by the lysosomes, these ions can be released rapidly into the cells and production of free radicals or directly bind to the intracellular proteins, endoplasmic reticulum, lysosomes, and nucleic acid. It can result in disorganizing the cellular zinc homeostasis, interfere with physiological activity, inhibit enzyme activity, and cause cell damage [90]. Zn^{2+} can accelerate the expression of proteins (BiP, p-PERK, etc.) and genes (xbp-1s, chop, etc.) related to endoplasmic reticulum stress. This leads to the rapid activation of the apoptotic pathway associated with endoplasmic reticulum [91].

The studies demonstrated that the effects of the ZnO nanoparticles on microbial growth were specific. For example, the 24-h LC_{20} was higher for *Daphnia Magna* (2.45 mg/L) and lower for *Thamnocephalus platyurus* (0.12 mg/L) [92]. ROS can induce oxidative stress that can lead to cell damage including protein denaturation and DNA damage. ROS generation can be attributed to the electronic properties of the particle surface and can be promoted by reacting with intracellular components [90]. In general, there is a dynamic balance between the generation and scavenging of ROS in the cells [93]. Restoration of intracellular ROS can be performed by electron transport chain and antioxidant enzymes such as peroxidase, catalase, and superoxide dismutase [62]. The major mechanisms for the destruction of cellular functions and alteration of cell membrane properties are lipid peroxidation [94]. This damage can be shielded by antioxidant enzymes [95]. The toxicity of the ZnO nanoparticles can be affected by light conditions due to its photocatalytic activity. Due to this property, ZnO can promote ROS production under irradiation [96]. ZnO nanoparticles' toxicity was enhanced in *Candida albicans* under natural light irradiation as compared to darkness [97]. In a similar study, greater mortality was observed in natural sunlight as compared to under ambient laboratory light in *C. elegans* [37,38]. ZnO nanorods can be used as a photosensitizer to produce stronger toxicity by destroying the membrane integrity [98]. Zinc nanoparticles are easier to attach and adsorb on the surface of cells due to their large surface area. Therefore, these nanoparticles may block the protein transport pathway which results in the destruction of cell membrane permeability [99]. The shape of the ZnO nanoparticle determines the uptake as well as the toxicity of the nanoparticles. For example, the rod-shaped ZnO nanoparticles were more toxic than the sphere-shaped nanoparticles. The higher toxicity of the rod-shaped nanoparticles can be attributed to the large surface area and higher van der Waals force in the rod-shaped nanoparticles [100]. The toxicities of ZnO nanoparticles to *Vibrio fischeri* for 30 min NOEC, EC50, and EC20 were 1.0, 1.0, and 1.8 mg/L for bulk and 0.75, 0.9, and 1.9 mg/L for nano ZnO particles [92]. In study 30 min EC50 for nano ZnO was 4.8 mg/L and for bulk ZnO was 4.3 mg/L [101]. Growth inhibition study of ZnO nanoparticles to *Saccharomyces cerevisiae* found that the EC50 values calculated by cell viability for nano and bulk ZnO were 131 and 158 mg/L after 24 h exposure [102]. The studies demonstrated that the toxicity depends on the dissolved zinc ions in addition to the particle size. This may be due to the absence of a supramolecular transmission mechanism in bacteria to prevent ZnO nanoparticles from invading [103]. The exposure medium and exposure method also have an important role in the toxicity of ZnO nanoparticles. For example, ZnO nanoparticles in microplate format (30 min EC50 of 3.8 mg/L) were more toxic to *V. fischeri* than that in cuvette format (30 min EC50 of 4.8 mg/L) [101].

Another major factor controlling the direct contact between nanoparticles and cells is the surface potential of ZnO. The study demonstrated that higher is the surface cation charge of ZnO nanoparticle, the lower is the cytotoxicity [104]. The negatively charged surface of the bacteria can change the viscosity of the cell membrane, affect the transport and exchange of cellular substances, damage the specific ion pump, and further inhibit the growth of bacteria [99]. Due to the nonconductivity of the prokaryotic cell walls toward the active entry of ZnO nanoparticles, the particles can enter the cell by disrupting the cell membrane [105]. In contrast to the prokaryotic cells, the eukaryotic cells can uptake the nanoparticles by pinocytosis followed by mixing with organelles and degradation of the cell components [105].

5 TITANIUM OXIDE NANOPARTICLES' TOXICITY

The release of titanium oxide nanoparticles into the environment due to their huge production and uncontrolled disposal causes toxicity to the organisms [106,107]. Various studies have reported the toxic effects of titanium nanoparticles on bacteria [92], plants [108], algae [109], and animal cells [110]. In microorganism, titanium oxide nanoparticles' toxicity has been reported on *Escherichia* [111–119], *Bacillus subtilis* [111,113,115–117], *Vibrio fischeri* [92], *Pseudomonas aeruginosa*, *Pseudomonas stutzeri* CFY1 [114–117,120], *Staphylococcus aureus* [116,117], *Klebsiella pneumonia* [117], *Saccharomyces cerevisiae* [102,114], *Aeromonas hydrophila* [114], *Streptococcus pyogenes*, *Streptococcus pneumonia* [112,116], *Enterococcus faecalis* [114], and *Proteus vulgaris* [116]. Titanium oxide nanoparticles' toxicity studies were also conducted on invertebrates as well as vertebrates. Invertebrate *Daphnia Magna* [92,121], zebrafish (*Danio rerio*) [122,123], *Thamnocephalus platyurus* [92], *Caenorhabditis elegans* [62,63,124], *Tetrahymena thermophila* [125], *Eisenia fetida* [126], *Haliotis diversicolor* supertexta [127], *Xenopus laevis* embryos [128], and *Drosophila melanogaster* [129,130] were used. In vertebrates, rainbow trout (*Oncorhynchus mykiss*) [131], European carp (*Cyprinus carpio*) [132], mice [133], gold fish (*Carassius auratus*) [134], Japanese rich fish (*Oryzias latipes*) [135], fathead minnow [136], mice [137], and rat primary hippocampal neurons [138] were used. Long-term intranasal administration of titanium dioxide nanoparticles may cause hemorrhage in mice brain [17]. The studies have suggested that titanium dioxide nanoparticles can activate inflammatory signaling pathways including p38 MAP kinase, c-Src, and NF-κB [21].

6 CARBON NANOMATERIALS' TOXICITY

The rapid development of nanotechnology has produced a variety of carbon nanomaterials. These include single-walled carbon nanotubes (SWCNTs), multi-walled carbon nanotubes (MWCNTs), carbon nanofibers, fullerene (C60), and graphene. These are the most widely used carbon nanomaterials in the world [139]. SWCNTs and MWCNTs are the major carbon nanomaterials, and they have wide applications. New and recent studies have reported that carbon nanomaterials have adverse effects on organisms as well as the environment [140–143].

A number of studies have reported the toxicity of the carbon nanomaterials toward animals. There are various routes to enter the carbon nanomaterials in the body. These include inhalation, oral entry, intravenous injection, and intraperitoneal injection. Due to their ability to cross the biological membranes, they can affect the physiology of any cell type [144]. SWCNTs have been reported to cause notable cytotoxicity in mammalian cells, resulting in the induction of mitochondrial damage, oxidative stress, rupture of cell membranes, interference with intracellular metabolic routes, and cell death [145]. A study has reported that SWCNTs had greater cytotoxicity than MWCNTs [146]. The study has shown that MWCNTs could accumulate on the surface and in macrophages intracellularly. Another studied the pulmonary exposure of commercial MWCNTs. The results demonstrated that MWCNTs were deposited in the lungs of female mice by intratracheal instillation [147]. A study conducted on Swiss-Webster mice by treatment of functionalized SWCNTs resulted in increased ROS, decreased mitotic index, and increased DNA damage [148].

The treatment of SWCNTs and GNPs on *Mus musculus* resulted in mitochondrial damage, chronic inflammation, and apoptosis. The nanoparticles accumulated in the spleen, lungs, and kidney of mice [149,150]. The study conducted on RAW 264.7 macrophages by SWCNTs, O-MWCNTs, and pristine MWCNTs resulted in cell death and apoptosis. The study demonstrated that SWCNTs had greater cytotoxicity than MWCNTs [151]. The treatment of murine macrophages by MWCNTs, COOH-CNTs, raw MWCNTs, and MWCNTs-PEG resulted in changing the oxidative potential of macrophages. The COOH-MWCNTs produced a large number of ROS, activating MAPK and NF-B signal pathways and interrupting the synthesis of ATP [47]. In a study, *Rattus norvegicus* was treated with short SWCNTs and short MWCNTs. The study resulted in persistent pneumonia as compared to short-fiber MWCNTs. However, short fiber MWCNTs caused greater inflammation than short fiber SWCNTs [152]. In another study where *Rattus norvegicus* was treated with MWCNTs, the study resulted in penetration of lung blood-gas barrier and MWCNTs were accumulated in the liver and spleen [153]. When the bovine embryo was treated with COOH-MWCNTs, it resulted in altered gene expression [154]. The treatment of OH-MWCNTs on *Daphnia magna* resulted in increased adsorption capacity or arsenic [2,3]. The study conducted on *Caenorhabditis elegans* by the treatment of pristine MWCNTs and OH-MWCNTs resulted in the induction of the stress response in the organism [155]. When the *Paracyclopina nana* was treated with CNTs, it resulted in increased rates of cell death at high concentrations [156].

In microorganisms, when *Rhizoctonia solani* was treated with nitrogen-doped CNTs, it inhibited fungal growth [157]. When microflora was treated with MWCNTs, it reduced the degradation process and lead to microbial cell damage [158]. In another study where *Actinomycetes* and fungi were treated with MWCNTs and SWCNTs, it resulted in altered soil microbial community distribution ad PAH degradation [159].

7 NANOTOXICITY REMEDIATION

As far as the toxicity of nanoparticles is concerned, very little information is available regarding the nanotoxicity remediation. This is mainly due to incomplete or limited available information about the nontoxicity. One of the efforts to understand the toxicity imposed by the nanoparticles was demonstrated by the Research Team for Nano-Associated Safety Assessment (RT-NASA) [160]. It was organized in three parts where each part focused on different contents of research directions. Part I focused on the efficient management and international cooperation on nanosafety studies. Part II focused on the toxicity evaluation on the target organs such as the assessment of immunotoxicity, genotoxicity, and skin penetration; whereas, Part III evaluated the toxicokinetics of nanoparticles with newly developed techniques for toxicokinetic analyses and methods for estimating nanotoxicity. The study was carried out in six steps: need assessment, physicochemical property, toxicity evaluation, toxicokinetics, peer review, and risk communication. The need assessment was carried out by analyzing consumer responses based on age, gender, educational level, and household income [160]. In a similar effort, a working system was developed by the US Environmental Protection Agency (EPA), the Risk Science Institute developed working systems and the International Life Sciences Institute Research Foundation where experts from the academia and government worked together to conduct a risk assessment and toxicity screening associated with nanoparticles [161]. Moreover, the US and European experts have been focusing on mapping the potential risks of nanoparticles [162–164]. In addition, it has been suggested that there should be an exchange of ideas between chemists, toxicologists, material scientists, and medical doctors via effective networks. This will enhance the understanding of the area of nanotoxicology and its related risk to develop safe nanotechnology [165].

There are several approaches that can be used to overcome the toxic effects of nanoparticles. In a study, ZnO nanoparticles were doped with iron, which resulted in decreased dissolution and thereby reduced toxicity in the zebrafish embryos and rodent lung [166]. The study demonstrated alleviation in the lethal and sublethal toxicological responses after iron doping. Iron doping slows the rate of particle dissolution by changing the particle matrix. In another study, pH-responsive biodegradable nanoparticles were synthesized [167]. The designed nanoparticles were used for codelivery of Hydroxytyrosol and Doxorubicin hydrochloride in HT-29 colon cancer cells. The study resulted in the generation of homogenous spherical nanoparticles with a size of around 18 nm. The study demonstrated that the uptake of DOX@HT was more than 94%.

One of the methods to reduce the nanoparticles' toxicity is by safe design approach. It can include the modification of physicochemical properties. The approaches include the modification of core–shell structure [168,169], geometric control [170], surface coating [171,172], and doping [168]. Surface coating is the most effective and simple modification method. It can be used for different kinds of nanoparticles with similar surface properties. A study demonstrated that the toxicity of carbon nanotubes can be reduced by using pluronic molecules [170].

Another method to reduce the toxicity of the nanoparticles is the synthesis of nanoparticles using phytochemicals. Most of the nanoparticles are synthesized using toxic chemical reagents. Phytochemicals are extracted from plants such as vegetables and fruits. In water phytochemicals acts as stabilizers to convert metal ions into metal nanoparticles [173]. Phytochemicals can be used to produce nanoparticles under green conditions and non-toxic conditions for medical applications. Phytochemicals such as tannins and flavonoid were used to produce silver nanoparticles with a size range 8–10 nm [174]. These particles had shown antibacterial activity. Silver and gold nanoparticles were synthesized using phenolic compounds, phytosterols and saponins with a size range 15–20 nm. The synthesized nanoparticles demonstrated antimicrobial activities [175]. Gold nanoparticles were synthesized using gellan gum [176], catechin, trans-cinnaminc acid, trans-cinnamaldehyde [28], isoflavone and soy protein [177], arabic gum [178], catechin [178], protocatechuate and gallic acid [179]. The synthesized gold nanoparticles were used as anti-oxidant agents [179], X-ray contrast agents [180,28] and drug delivery systems [176]. Iron oxide nanoparticles with a size of 14 nm were synthesized using arabic gum and were used as MRI contrast agents [178]. Selenium nanoparticles in a size range 50–150 nm were synthesized using flavonoids and galactomannan and were used as anticancer agent [40]. In a study, the gold nanoparticles showed very good cell viability [181]. Gold nanoparticles were recently used as X-ray contrast agent and computer tomography agent [180,28]. A study evaluated the efficiency of the gold nanoparticles in imaging for cancer diagnosis [182,183]. The cellular uptake and cytotoxicity of magnetic nanoparticles synthesized using gum arabic was evaluated by Zhang et al. [178]. Antioxidant activity and cytotoxic effects of protocatchuic acid-isoflavones-gold and gallic acid-isoflavones-gold nanoparticles were evaluated on L-929 cells [179]. The study demonstrated that both the nanoparticles had excellent cell viability.

8 CONCLUSION

Although nanotechnology has revolutionized medical science by providing novel methods for the treatment of major diseases. However, overproduction, use, disposal, and transport of nanomaterials have created a negative impact on both the organisms and the environment. Recently, several studies have reported the toxicity imposed by the nanoparticles. However, these studies are not enough to make an accurate assessment of the toxicity caused. This is mainly because of the diverse nature and distribution of the nanoparticles. Due to insufficient knowledge about the toxicity of the nanomaterials, it is very difficult to discover methods to reduce or remediate the toxicity of these nanoparticles. In conclusion, there is an immediate requirement for the accurate assessment and evaluation of toxicity caused by nanomaterials so that some methods can be developed to reduce the toxicity.

REFERENCES

[1] Khan A, Wang J, Li J, Wang X, Chen Z, Alsaedi A, et al. The role of graphene oxide and graphene oxide-based nanomaterials in the removal of pharmaceuticals from aqueous media: a review. Environmental Science and Pollution Research 2017;24(9):7938–58.

[2] Wang X, Qu R, Allam AA, Ajarem J, Wei Z, Wang Z. Impact of carbon nanotubes on the toxicity of inorganic arsenic [AS(III) and AS(V)] to Daphnia magna: the role of certain arsenic species. Environmental Toxicology and Chemistry 2016;35(7):1852–9. https://doi.org/10.1002/etc.3340.

[3] Wang X, Yu S, Jin J, Wang H, Alharbi NS, Alsaedi A, et al. Application of graphene oxides and graphene oxide-based nanomaterials in radionuclide removal from aqueous solutions 石墨烯和石墨烯基纳米材料对放射性核素去除应用研究进展. Science Bulletin 2016;61(20):1583–93.

[3a] Guzman D, Katherine A, Finnegan MP, Banfield JF. Influence of surface potential on aggregation and transport of titania nanoparticles. Environmental science & technology 2006;40(24):7688–93.

[4] Sayılkan F, Asiltürk M, Kiraz N, Burunkaya E, Arpaç E, Sayılkan H. Photocatalytic antibacterial performance of Sn^{4+}-doped TiO_2 thin films on glass substrate. Journal of Hazardous Materials 2009;162(2–3):1309–16.

[5] Akter M, Sikder MT, Rahman MM, Ullah AKMA, Hossain KFB, Banik S, et al. A systematic review on silver nanoparticles-induced cytotoxicity: physicochemical properties and perspectives. Journal of Advanced Research 2018;9:1–16. https://doi.org/10.1016/j.jare.2017.10.008.

[6] Wu T, Tang M. Review of the effects of manufactured nanoparticles on mammalian target organs. Journal of Applied Toxicology 2018;38(1):25–40.

[7] Kreyling WG, Semmler-Behnke M, Seitz J, Scymczak W, Wenk A, Mayer P, et al. Size dependence of the translocation of inhaled iridium and carbon nanoparticle aggregates from the lung of rats to the blood and secondary target organs. Inhalation Toxicology 2009;21(Suppl. 1):55–60.

[8] Recordati C, De Maglie M, Bianchessi S, Argentiere S, Cella C, Mattiello S, et al. Tissue distribution and acute toxicity of silver after single intravenous administration in mice: nano-specific and size-dependent effects. Particle and Fibre Toxicology 2015;13(1):12.

[9] Wu T, Tang M. Toxicity of quantum dots on respiratory system. Inhalation Toxicology 2014;26(2):128–39.

[10] Paek H-J, Lee Y-J, Chung H-E, Yoo N-H, Lee J-A, Kim M-K, et al. Modulation of the pharmacokinetics of zinc oxide nanoparticles and their fates in vivo. Nanoscale 2013;5(23):11416–27.

[11] Shegokar R, Singh KK. Surface modified nevirapine nanosuspensions for viral reservoir targeting: in vitro and in vivo evaluation. International Journal of Pharmaceutics 2011;421(2):341–52.

[12] Shang L, Nienhaus K, Nienhaus GU. Engineered nanoparticles interacting with cells: size matters. Journal of Nanobiotechnology 2014;12(1):5.

[13] Teodoro JS, Silva R, Varela AT, Duarte FV, Rolo AP, Hussain S, Palmeira CM. Low-dose, subchronic exposure to silver nanoparticles causes mitochondrial alterations in Sprague–Dawley rats. Nanomedicine 2016;11(11):1359–75.

[14] Missaoui WN, Arnold RD, Cummings BS. Toxicological status of nanoparticles: what we know and what we don't know. Chemico-Biological Interactions 2018;295 [Elsevier].

[15] Wang B, Feng W, Wang M, Wang T, Gu Y, Zhu M, et al. Acute toxicological impact of nano- and submicro-scaled zinc oxide powder on healthy adult mice. Journal of Nanoparticle Research 2008;10(2):263–76. https://doi.org/10.1007/s11051-007-9245-3.

[16] Wang J, Chen C, Liu Y, Jiao F, Li W, Lao F, et al. Potential neurological lesion after nasal instillation of TiO_2 nanoparticles in the anatase and rutile crystal phases. Toxicology Letters 2008;183(1–3):72–80.

[17] Ze Y, Zheng L, Zhao X, Gui S, Sang X, Su J, et al. Molecular mechanism of titanium dioxide nanoparticles-induced oxidative injury in the brain of mice. Chemosphere 2013;92(9):1183–9.

[18] Manshian BB, Poelmans J, Saini S, Pokhrel S, Grez JJ, Himmelreich U, et al. Nanoparticle-induced inflammation can increase tumor malignancy. Acta Biomaterialia 2018;68:99–112.

[19] Pandey RK, Prajapati VK. Molecular and immunological toxic effects of nanoparticles. International Journal of Biological Macromolecules 2017.

[20] Roy R, Kumar S, Tripathi A, Das M, Dwivedi PD. Interactive threats of nanoparticles to the biological system. Immunology Letters 2014;158(1–2):79–87.

[21] Moon C, Park H-J, Choi Y-H, Park E-M, Castranova V, Kang JL. Pulmonary inflammation after intraperitoneal administration of ultrafine titanium dioxide (TiO_2) at rest or in lungs primed with lipopolysaccharide. Journal of Toxicology and Environmental Health, Part A 2010;73(5–6):396–409.

[22] Shyamasundar S, Ng CT, Lanry Yung LY, Dheen ST, Bay BH. Epigenetic mechanisms in nanomaterial-induced toxicity. Epigenomics 2015;7(3):395–411.

[23] Smolkova B, El Yamani N, Collins AR, Gutleb AC, Dusinska M. Nanoparticles in food. Epigenetic changes induced by nanomaterials and possible impact on health. Food and Chemical Toxicology 2015;77:64–73.

[24] Liao J, Li W, Peng J, Yang Q, Li H, Wei Y, et al. Combined cancer photothermal-chemotherapy based on doxorubicin/gold nanorod-loaded polymersomes. Theranostics 2015;5(4):345–56. https://doi.org/10.7150/thno.10731.

[25] Peng J, Qi T, Liao J, Chu B, Yang Q, Qu Y, et al. Mesoporous magnetic gold "nanoclusters" as theranostic carrier for chemo-photothermal co-therapy of breast

cancer. Theranostics 2014;4(7):678–92. https://doi.org/10.7150/thno.7869.

[26] Wu J, Liu B, Wu H, Wu Y, Zhang W, Zhao S, et al. A gold nanoparticle platform for the delivery of functional TGF-beta1 siRNA into cancer cells. Journal of Biomedical Nanotechnology 2016;12(4):800–10.

[27] Yavuz MS, Cheng Y, Chen J, Cobley CM, Zhang Q, Rycenga M, et al. Gold nanocages covered by smart polymers for controlled release with near-infrared light. Nature Materials 2009;8(12):935–9. https://doi.org/10.1038/nmat2564.

[28] Kim D, Park S, Lee JH, Jeong YY, Jon S. Antibiofouling polymer-coated gold nanoparticles as a contrast agent for in vivo X-ray computed tomography imaging. Journal of the American Chemical Society 2007;129(24):7661–5. https://doi.org/10.1021/ja071471p.

[29] Cheng B, He H, Huang T, Berr SS, He J, Fan D, et al. Gold nanosphere gated mesoporous silica nanoparticle responsive to near-infrared light and redox potential as a theranostic platform for cancer therapy. Journal of Biomedical Nanotechnology 2016;12(3):435–49. https://doi.org/10.1166/jbn.2016.2195.

[30] Sun L, Joh DY, Al-Zaki A, Stangl M, Murty S, Davis JJ, et al. Theranostic application of mixed gold and superparamagnetic iron oxide nanoparticle micelles in glioblastoma multiforme. Journal of Biomedical Nanotechnology 2016;12(2):347–56.

[31] Xu L, Liu Y, Chen Z, Li W, Liu Y, Wang L, et al. Surface-engineered gold nanorods: promising DNA vaccine adjuvant for HIV-1 treatment. Nano Letters 2012;12(4):2003–12. https://doi.org/10.1021/nl300027p.

[32] Fadeel B, Garcia-Bennett AE. Better safe than sorry: understanding the toxicological properties of inorganic nanoparticles manufactured for biomedical applications. Advanced Drug Delivery Reviews 2010;62(3):362–74. https://doi.org/10.1016/j.addr.2009.11.008.

[33] Kagan VE, Bayir H, Shvedova AA. Nanomedicine and nanotoxicology: two sides of the same coin. Nanomedicine 2005;1(4):313–6. https://doi.org/10.1016/j.nano.2005.10.003.

[34] Kumar V, Sharma N, Maitra S. In vitro and in vivo toxicity assessment of nanoparticles. International Nano Letters 2017;7:1–14.

[35] Ahn S, Seo E, Kim KH, Lee SJ. Physical property control on the cellular uptake pathway and spatial distribution of nanoparticles in cells. Journal of Biomedical Nanotechnology 2015;11(6):1051–70.

[36] Jiang W, Kim BY, Rutka JT, Chan WC. Nanoparticle-mediated cellular response is size-dependent. Nature Nanotechnology 2008;3:145. https://doi.org/10.1038/nnano.2008.30. https://www.nature.com/articles/nnano.2008.30#supplementary-information.

[37] Ma H, Kabengi NJ, Bertsch PM, Unrine JM, Glenn TC, Williams PL. Comparative phototoxicity of nanoparticulate and bulk ZnO to a free-living nematode *Caenorhabditis elegans*: the importance of illumination mode and primary particle size. Environmental Pollution 2011;159(6):1473–80. https://doi.org/10.1016/j.envpol.2011.03.013.

[38] Ma X, Wu Y, Jin S, Tian Y, Zhang X, Zhao Y, et al. Gold nanoparticles induce autophagosome accumulation through size-dependent nanoparticle uptake and lysosome impairment. ACS Nano 2011;5(11):8629–39. https://doi.org/10.1021/nn202155y.

[39] Senut MC, Zhang Y, Liu F, Sen A, Ruden DM, Mao G. Size-dependent toxicity of gold nanoparticles on human embryonic stem cells and their neural derivatives. Small 2016;12(5):631–46. https://doi.org/10.1002/smll.201502346.

[40] Yen HJ, Hsu SH, Tsai CL. Cytotoxicity and immunological response of gold and silver nanoparticles of different sizes. Small 2009;5(13):1553–61. https://doi.org/10.1002/smll.200900126.

[41] Freese C, Uboldi C, Gibson MI, Unger RE, Weksler BB, Romero IA, et al. Uptake and cytotoxicity of citrate-coated gold nanospheres: comparative studies on human endothelial and epithelial cells. Particle and Fibre Toxicology 2012;9:23. https://doi.org/10.1186/1743-8977-9-23.

[42] Choi SY, Jeong S, Jang SH, Park J, Park JH, Ock KS, et al. In vitro toxicity of serum protein-adsorbed citrate-reduced gold nanoparticles in human lung adenocarcinoma cells. Toxicology in Vitro 2012;26(2):229–37. https://doi.org/10.1016/j.tiv.2011.11.016.

[43] Chompoosor A, Saha K, Ghosh PS, Macarthy DJ, Miranda OR, Zhu ZJ, et al. The role of surface functionality on acute cytotoxicity, ROS generation and DNA damage by cationic gold nanoparticles. Small 2010;6(20):2246–9. https://doi.org/10.1002/smll.201000463.

[44] Del Pino P, Yang F, Pelaz B, Zhang Q, Kantner K, Hartmann R, et al. Basic physicochemical properties of polyethylene glycol coated gold nanoparticles that determine their interaction with cells. Angewandte Chemie International Edition in English 2016;55(18):5483–7. https://doi.org/10.1002/anie.201511733.

[45] Ahamed M, Karns M, Goodson M, Rowe J, Hussain SM, Schlager JJ, Hong Y. DNA damage response to different surface chemistry of silver nanoparticles in mammalian cells. Toxicology and Applied Pharmacology 2008;233(3):404–10. https://doi.org/10.1016/j.taap.2008.09.015.

[46] Arora S, Jain J, Rajwade JM, Paknikar KM. Interactions of silver nanoparticles with primary mouse fibroblasts and liver cells. Toxicology and Applied Pharmacology 2009;236(3):310–8. https://doi.org/10.1016/j.taap.2009.02.020.

[47] Zhang XF, Gurunathan S, Kim JH. Effects of silver nanoparticles on neonatal testis development in mice. International Journal of Nanomedicine 2015;10:6243–56. https://doi.org/10.2147/IJN.S90733.

[48] Beer C, Foldbjerg R, Hayashi Y, Sutherland DS, Autrup H. Toxicity of silver nanoparticles—nanoparticle or silver ion? Toxicology Letters 2012;208(3):286–92. https://doi.org/10.1016/j.toxlet.2011.11.002.

[49] Gorth DJ, Rand DM, Webster TJ. Silver nanoparticle toxicity in Drosophila: size does matter. International Journal of Nanomedicine 2011;6:343–50.

[50] Kruszewski M, Brzoska K, Brunborg G, Asare N, Dobrzyńska M, Dušinská M, et al. Chapter five – toxicity of silver nanomaterials in higher eukaryotes. In: Fishbein JC, editor. Advances in molecular toxicology, vol. 5. Elsevier; 2011. p. 179–218.
[51] van der Zande M, Vandebriel RJ, Van Doren E, Kramer E, Herrera Rivera Z, Serrano-Rojero CS, et al. Distribution, elimination, and toxicity of silver nanoparticles and silver ions in rats after 28-day oral exposure. ACS Nano 2012;6(8):7427–42. https://doi.org/10.1021/nn302649p.
[52] Murugan K, Choonara YE, Kumar P, Bijukumar D, du Toit LC, Pillay V. Parameters and characteristics governing cellular internalization and trans-barrier trafficking of nanostructures. International Journal of Nanomedicine 2015;10:2191–206. https://doi.org/10.2147/IJN.S75615.
[53] Haase A, Tentschert J, Jungnickel H, Graf P, Mantion A, Draude F, et al. Toxicity of silver nanoparticles in human macrophages: uptake, intracellular distribution and cellular responses. Journal of Physics: Conference Series 2011;304(1):012030.
[54] AshaRani PV, Low Kah Mun G, Hande MP, Valiyaveettil S. Cytotoxicity and genotoxicity of silver nanoparticles in human cells. ACS Nano 2009;3(2): 279–90. https://doi.org/10.1021/nn800596w.
[55] Jiang X, Foldbjerg R, Miclaus T, Wang L, Singh R, Hayashi Y, et al. Multi-platform genotoxicity analysis of silver nanoparticles in the model cell line CHO-K1. Toxicology Letters 2013;222(1):55–63. https://doi.org/10.1016/j.toxlet.2013.07.011.
[56] Sahu SC, Zheng J, Graham L, Chen L, Ihrie J, Yourick JJ, Sprando RL. Comparative cytotoxicity of nanosilver in human liver HepG2 and colon $Caco_2$ cells in culture. Journal of Applied Toxicology 2014;34(11):1155–66. https://doi.org/10.1002/jat.2994.
[57] Mytych J, Zebrowski J, Lewinska A, Wnuk M. Prolonged effects of silver nanoparticles on p53/p21 pathway-mediated proliferation, DNA damage response, and methylation parameters in HT22 hippocampal neuronal cells. Mol Neurobiol 2016:1–16.
[58] van Aerle R, Lange A, Moorhouse A, Paszkiewicz K, Ball K, Johnston BD, et al. Molecular mechanisms of toxicity of silver nanoparticles in zebrafish embryos. Environmental Science and Technology 2013;47(14): 8005–14. https://doi.org/10.1021/es401758d.
[59] Sriram MI, Kalishwaralal K, Barathmanikanth S, Gurunathani S. Size-based cytotoxicity of silver nanoparticles in bovine retinal endothelial cells. Nanoscience Methods 2012;1(1):56–77.
[60] Gurunathan S, Park JH, Han JW, Kim JH. Comparative assessment of the apoptotic potential of silver nanoparticles synthesized by Bacillus tequilensis and Calocybe indica in MDA-MB-231 human breast cancer cells: targeting p53 for anticancer therapy. International Journal of Nanomedicine 2015;10:4203–22. https://doi.org/10.2147/ijn.s83953.
[61] He D, Dorantes-Aranda JJ, Waite TD. Silver nanoparticle-algae interactions: oxidative dissolution, reactive oxygen species generation and synergistic toxic effects. Environmental Science and Technology 2012; 46(16):8731–8. https://doi.org/10.1021/es300588a.
[62] Wang H, Wick RL, Xing B. Toxicity of nanoparticulate and bulk ZnO, Al_2O_3 and TiO_2 to the nematode *Caenorhabditis elegans*. Environmental Pollution 2009;157(4): 1171–7. https://doi.org/10.1016/j.envpol.2008.11.004.
[63] Wang J, Rahman MF, Duhart HM, Newport GD, Patterson TA, Murdock RC, et al. Expression changes of dopaminergic system-related genes in PC12 cells induced by manganese, silver, or copper nanoparticles. NeuroToxicology 2009;30(6):926–33. https://doi.org/10.1016/j.neuro.2009.09.005.
[64] Ziemińska E, Stafiej A, Struzyńska L. The role of the glutamatergic NMDA receptor in nanosilver-evoked neurotoxicity in primary cultures of cerebellar granule cells. Toxicology 2014;315(1):38–48. https://doi.org/10.1016/j.tox.2013.11.008.
[65] Ahmadi F. Impact of different levels of silver nanoparticles (Ag-NPs) on performance, oxidative enzymes and blood parameters in broiler chicks. Pakistan Veterinary Journal 2012;32(3):325–8.
[66] Awasthi AK, Kishore K, Bisht GS, Awasthi S. In vitro antibacterial and antifungal activity of Carissa opaca Stapf ex Haines. International Journal of Current Pharmaceutical Research 2013;5:15–8.
[67] Cheng X, Zhang W, Ji Y, Meng J, Guo H, Liu J, et al. Revealing silver cytotoxicity using Au nanorods/Ag shell nanostructures: disrupting cell membrane and causing apoptosis through oxidative damage. RSC Advances 2013;3(7):2296–305. https://doi.org/10.1039/c2ra23131j.
[68] Haase A, Rott S, Mantion A, Graf P, Plendl J, Thünemann AF, et al. Effects of silver nanoparticles on primary mixed neural cell cultures: uptake, oxidative stress and acute calcium responses. Toxicological Sciences 2012;126(2):457–68. https://doi.org/10.1093/toxsci/kfs003.
[69] De Gusseme B, Hennebel T, Christiaens E, Saveyn H, Verbeken K, Fitts JP, et al. Virus disinfection in water by biogenic silver immobilized in polyvinylidene fluoride membranes. Water Research 2011;45(4):1856–64. https://doi.org/10.1016/j.watres.2010.11.046.
[70] Foldbjerg R, Olesen P, Hougaard M, Dang DA, Hoffmann HJ, Autrup H. PVP-coated silver nanoparticles and silver ions induce reactive oxygen species, apoptosis and necrosis in THP-1 monocytes. Toxicology Letters 2009;190(2):156–62. https://doi.org/10.1016/j.toxlet.2009.07.009.
[71] Li Y, Guo M, Lin Z, Zhao M, Xiao M, Wang C, et al. Polyethylenimine-functionalized silver nanoparticle-based co-delivery of paclitaxel to induce HepG2 cell apoptosis. International Journal of Nanomedicine 2016; 11:6693–702. https://doi.org/10.2147/IJN.S122666.

[72] Piao MJ, Kang KA, Lee IK, Kim HS, Kim S, Choi JY, et al. Silver nanoparticles induce oxidative cell damage in human liver cells through inhibition of reduced glutathione and induction of mitochondria-involved apoptosis. Toxicology Letters 2011;201(1):92–100. https://doi.org/10.1016/j.toxlet.2010.12.010.

[73] Kim YS, Song MY, Park JD, Song KS, Ryu HR, Chung YH, et al. Subchronic oral toxicity of silver nanoparticles. Particle and Fibre Toxicology 2010;7:20. https://doi.org/10.1186/1743-8977-7-20.

[74] Bartneck M, Ritz T, Keul HA, Wambach M, Bornemann J, Gbureck U, et al. Peptide-functionalized gold nanorods increase liver injury in hepatitis. ACS Nano 2012;6(10):8767–77. https://doi.org/10.1021/nn302502u.

[75] Sung JH, Ji JH, Park JD, Yoon JU, Kim DS, Jeon KS, et al. Subchronic inhalation toxicity of silver nanoparticles. Toxicological Sciences 2009;108(2):452–61. https://doi.org/10.1093/toxsci/kfn246.

[76] Korani M, Rezayat SM, Gilani K, Arbabi Bidgoli S, Adeli S. Acute and subchronic dermal toxicity of nanosilver in guinea pig. International Journal of Nanomedicine 2011;6:855–62. https://doi.org/10.2147/ijn.s17065.

[77] Tang L, Zhang C, Song G, Jin X, Xu Z. In vivo skin penetration and metabolic path of quantum dots. Science China Life Sciences 2013;56(2):181–8. https://doi.org/10.1007/s11427-012-4404-x.

[78] Dabrowska-Bouta B, Zieba M, Orzelska-Gorka J, Skalska J, Sulkowski G, Frontczak-Baniewicz M, et al. Influence of a low dose of silver nanoparticles on cerebral myelin and behavior of adult rats. Toxicology 2016; 363–364:29–36. https://doi.org/10.1016/j.tox.2016.07.007.

[79] Djurišić AB, Chen X, Leung YH, Man Ching Ng A. ZnO nanostructures: growth, properties and applications. Journal of Materials Chemistry 2012;22(14):6526–35. https://doi.org/10.1039/C2JM15548F.

[80] Serpone N, Dondi D, Albini A. Inorganic and organic UV filters: their role and efficacy in sunscreens and suncare products. Inorganica Chimica Acta 2007;360(3):794–802. https://doi.org/10.1016/j.ica.2005.12.057.

[81] Dimkpa CO, McLean JE, Britt DW, Anderson AJ. Antifungal activity of ZnO nanoparticles and their interactive effect with a biocontrol bacterium on growth antagonism of the plant pathogen Fusarium graminearum. Biometals 2013;26(6):913–24. https://doi.org/10.1007/s10534-013-9667-6.

[82] Westerhoff P, Nowack B. Searching for global descriptors of engineered nanomaterial fate and transport in the environment. Accounts of Chemical Research 2013; 46(3):844–53. https://doi.org/10.1021/ar300030n.

[83] Reddy KM, Feris K, Bell J, Wingett DG, Hanley C, Punnoose A. Selective toxicity of zinc oxide nanoparticles to prokaryotic and eukaryotic systems. Applied Physics Letters 2007;90(213902). https://doi.org/10.1063/1.2742324. 213902-213901–213902-213903.

[84] Magdolenova Z, Collins A, Kumar A, Dhawan A, Stone V, Dusinska M. Mechanisms of genotoxicity. A review of in vitro and in vivo studies with engineered nanoparticles. Nanotoxicology 2014;8(3):233–78. https://doi.org/10.3109/17435390.2013.773464.

[85] Navarro E, Piccapietra F, Wagner B, Marconi F, Kaegi R, Odzak N, et al. Toxicity of silver nanoparticles to Chlamydomonas reinhardtii. Environmental Science and Technology 2008;42(23):8959–64. https://doi.org/10.1021/es801785m.

[86] Golbamaki N, Rasulev B, Cassano A, Marchese Robinson RL, Benfenati E, Leszczynski J, Cronin MTD. Genotoxicity of metal oxide nanomaterials: review of recent data and discussion of possible mechanisms. Nanoscale 2015;7(6):2154–98. https://doi.org/10.1039/c4nr06670g.

[87] Johnston BD, Scown TM, Moger J, Cumberland SA, Baalousha M, Linge K, et al. Bioavailability of nanoscale metal oxides TiO_2, CeO_2, and ZnO to fish. Environmental Science and Technology 2010;44(3):1144–51. https://doi.org/10.1021/es901971a.

[88] Kao YY, Chen YC, Cheng TJ, Chiung YM, Liu PS. Zinc oxide nanoparticles interfere with zinc ion homeostasis to cause cytotoxicity. Toxicological Sciences 2012;125(2):462–72. https://doi.org/10.1093/toxsci/kfr319.

[89] Bacchetta R, Maran B, Marelli M, Santo N, Tremolada P. Role of soluble zinc in ZnO nanoparticle cytotoxicity in Daphnia magna: a morphological approach. Environmental Research 2016;148:376–85. https://doi.org/10.1016/j.envres.2016.04.028.

[90] Xia T, Kovochich M, Liong M, Mädler L, Gilbert B, Shi H, et al. Comparison of the mechanism of toxicity of zinc oxide and cerium oxide nanoparticles based on dissolution and oxidative stress properties. ACS Nano 2008;2(10):2121–34. https://doi.org/10.1021/nn800511k.

[91] Chen R, Huo L, Shi X, Bai R, Zhang Z, Zhao Y, et al. Endoplasmic reticulum stress induced by zinc oxide nanoparticles is an earlier biomarker for nanotoxicological evaluation. ACS Nano 2014;8(3):2562–74. https://doi.org/10.1021/nn406184r.

[92] Heinlaan M, Ivask A, Blinova I, Dubourguier H-C, Kahru A. Toxicity of nanosized and bulk ZnO, CuO and TiO_2 to bacteria Vibrio fischeri and crustaceans Daphnia magna and Thamnocephalus platyurus. Chemosphere 2008;71(7):1308–16.

[93] Hou J, Wang X, Hayat T, Wang X. Ecotoxicological effects and mechanism of CuO nanoparticles to individual organisms. Environmental Pollution 2017;221:209–17. https://doi.org/10.1016/j.envpol.2016.11.066.

[94] Rikans LE, Hornbrook KR. Lipid peroxidation, antioxidant protection and aging. Biochimica et Biophysica Acta 1997;31:116–27.

[95] Feris K, Otto C, Tinker J, Wingett D, Punnoose A, Thurber A, et al. Electrostatic interactions affect nanoparticle-mediated toxicity to gram-negative bacterium pseudomonas aeruginosa PAO1. Langmuir 2010; 26(6):4429–36. https://doi.org/10.1021/la903491z.

[96] Diamond SA, Peterson GS, Tietge JE, Ankley GT. Assessment of the risk of solar ultraviolet radiation to amphibians. III. Prediction of impacts in selected northern

midwestern wetlands. Environmental Science and Technology 2002;36(13):2866–74. https://doi.org/10.1021/es011197d.

[97] Lipovsky A, Nitzan Y, Gedanken A, Lubart R. Antifungal activity of ZnO nanoparticles—the role of ROS mediated cell injury. Nanotechnology 2011;22(10):105101. https://doi.org/10.1088/0957-4484/22/10/105101.

[98] Saito TK, Seki M, Tabata H. Self-organized ZnO nanorod with photooxidative cell membrane perforation enables large-scale cell manipulation. Analytical and Bioanalytical Chemistry 2008,391(7):2513–9. https://doi.org/10.1007/s00216-008-2226-2.

[99] Kumar A, Pandey AK, Singh SS, Shanker R, Dhawan A. Engineered ZnO and TiO_2 nanoparticles induce oxidative stress and DNA damage leading to reduced viability of *Escherichia coli*. Free Radical Biology and Medicine 2011;51(10):1872–81. https://doi.org/10.1016/j.freeradbiomed.2011.08.025.

[100] Yang ST, Liu JH, Wang J, Yuan Y, Cao A, Wang H, et al. Cytotoxicity of zinc oxide nanoparticles: importance of microenvironment. Journal of Nanoscience and Nanotechnology 2010;10(12):8638–45. https://doi.org/10.1166/jnn.2010.2491.

[101] Mortimer M, Kasemets K, Heinlaan M, Kurvet I, Kahru A. High throughput kinetic Vibrio fischeri bioluminescence inhibition assay for study of toxic effects of nanoparticles. Toxicology in Vitro 2008;22(5):1412–7. https://doi.org/10.1016/j.tiv.2008.02.011.

[102] Kasemets K, Ivask A, Dubourguier H-C, Kahru A. Toxicity of nanoparticles of ZnO, CuO and TiO_2 to yeast *Saccharomyces cerevisiae*. Toxicology in Vitro 2009;23(6):1116–22.

[103] Moorer WR, Genet JM. Antibacterial activity of gutta-percha cones attributed to the zinc oxide component. Oral Surgery, Oral Medicine, Oral Pathology 1982;53(5):508–17. https://doi.org/10.1016/0030-4220(82)90468-6.

[104] Hu X, Cook S, Wang P, Hwang HM. In vitro evaluation of cytotoxicity of engineered metal oxide nanoparticles. The Science of the Total Environment 2009;407(8):3070–2. https://doi.org/10.1016/j.scitotenv.2009.01.033.

[105] Neal AL. What can be inferred from bacterium-nanoparticle interactions about the potential consequences of environmental exposure to nanoparticles? Ecotoxicology 2008;17(5):362–71. https://doi.org/10.1007/s10646-008-0217-x.

[106] Aruoja V, Dubourguier H-C, Kasemets K, Kahru A. Toxicity of nanoparticles of CuO, ZnO and TiO_2 to microalgae Pseudokirchneriella subcapitata. The Science of the Total Environment 2009;407(4):1461–8.

[107] Minetto D, Ghirardini AV, Libralato G. Saltwater ecotox icology of Ag, Au, CuO, TiO_2, ZnO and C60 engineered nanoparticles: an overview. Environment International 2016;92:189–201.

[108] Atha DH, Wang H, Petersen EJ, Cleveland D, Holbrook RD, Jaruga P, et al. Copper oxide nanoparticle mediated DNA damage in terrestrial plant models. Environmental Science and Technology 2012;46(3):1819–27.

[109] Sadiq IM, Dalai S, Chandrasekaran N, Mukherjee A. Ecotoxicity study of titania (TiO_2) NPs on two microalgae species: *Scenedesmus* sp. and *Chlorella* sp. Ecotoxicology and Environmental Safety 2011;74(5):1180–7.

[110] Karlsson HL, Cronholm P, Gustafsson J, Moller L. Copper oxide nanoparticles are highly toxic: a comparison between metal oxide nanoparticles and carbon nanotubes. Chemical Research in Toxicology 2008;21(9):1726–32.

[111] Adams LK, Lyon DY, Alvarez PJ. Comparative ecotoxicity of nanoscale TiO_2, SiO_2, and ZnO water suspensions. Water Research 2006;40(19):3527–32.

[112] Ahmad R, Mohsin M, Ahmad T, Sardar M. Alpha amylase assisted synthesis of TiO_2 nanoparticles: structural characterization and application as antibacterial agents. Journal of Hazardous Materials 2015;283:171–7.

[113] Erdem A, Metzler D, Cha DK, Huang C. The short-term toxic effects of TiO_2 nanoparticles toward bacteria through viability, cellular respiration, and lipid peroxidation. Environmental Science and Pollution Research 2015;22(22):17917–24.

[114] Jayaseelan C, Rahuman AA, Roopan SM, Kirthi AV, Venkatesan J, Kim S-K, et al. Biological approach to synthesize TiO_2 nanoparticles using Aeromonas hydrophila and its antibacterial activity. Spectrochimica Acta Part A: Molecular and Biomolecular Spectroscopy 2013;107:82–9.

[115] Jiang W, Mashayekhi H, Xing B. Bacterial toxicity comparison between nano-and micro-scaled oxide particles. Environmental Pollution 2009;157(5):1619–25.

[116] Priyanka KP, Sukirtha TH, Balakrishna KM, Varghese T. Microbicidal activity of TiO_2 nanoparticles synthesised by sol–gel method. IET Nanobiotechnology 2016;10(2):81–6.

[117] Rajakumar G, Rahuman AA, Roopan SM, Khanna VG, Elango G, Kamaraj C, et al. Fungus-mediated biosynthesis and characterization of TiO_2 nanoparticles and their activity against pathogenic bacteria. Spectrochimica Acta Part A: Molecular and Biomolecular Spectroscopy 2012;91:23–9.

[118] Tong T, Binh CT, Kelly JJ, Gaillard J-F, Gray KA. Cytotoxicity of commercial nano-TiO_2 to *Escherichia coli* assessed by high-throughput screening: effects of environmental factors. Water Research 2013;47(7):2352–62.

[119] Vijay M, Selvarajan V, Ananthapadmanabhan P, Sreekumar K, Stengl V, Bondioli F. Bactericidal effects of reactive thermal plasma synthesized titanium dioxide photocatalysts. In: Paper presented at the journal of physics: conference series; 2010.

[120] Li J, Wang X, Zhao G, Chen C, Chai Z, Alsaedi A, et al. Metal–organic framework-based materials: superior adsorbents for the capture of toxic and radioactive metal ions. Chemical Society Reviews 2018;47(7):2322–56.

[121] Wiench K, Wohlleben W, Hisgen V, Radke K, Salinas E, Zok S, Landsiedel R. Acute and chronic effects of nano-

and non-nano-scale TiO_2 and ZnO particles on mobility and reproduction of the freshwater invertebrate Daphnia magna. Chemosphere 2009;76(10):1356–65.
[122] Boran H, Boyle D, Altinok I, Patsiou D, Henry TB. Aqueous Hg^{2+} associates with TiO_2 nanoparticles according to particle size, changes particle agglomeration, and becomes less bioavailable to zebrafish. Aquatic Toxicology 2016;174:242–6.
[123] Chen J, Dong X, Xin Y, Zhao M. Effects of titanium dioxide nano-particles on growth and some histological parameters of zebrafish (*Danio rerio*) after a long-term exposure. Aquatic Toxicology 2011;101(3–4):493–9.
[124] Roh J-Y, Park Y-K, Park K, Choi J. Ecotoxicological investigation of CeO_2 and TiO_2 nanoparticles on the soil nematode *Caenorhabditis elegans* using gene expression, growth, fertility, and survival as endpoints. Environmental Toxicology and Pharmacology 2010;29(2): 167–72.
[125] Rajapakse K, Drobne D, Valant J, Vodovnik M, Levart A, Marinsek-Logar R. Acclimation of Tetrahymena thermophila to bulk and nano-TiO_2 particles by changes in membrane fatty acids saturation. Journal of Hazardous Materials 2012;221–222:199–205. https://doi.org/10.1016/j.jhazmat.2012.04.029.
[126] Cañas JE, Qi B, Li S, Maul JD, Cox SB, Das S, Green MJ. Acute and reproductive toxicity of nano-sized metal oxides (ZnO and TiO_2) to earthworms (*Eisenia fetida*). Journal of Environmental Monitoring 2011;13(12): 3351–7. https://doi.org/10.1039/C1EM10497G.
[127] Zhu X, Zhou J, Cai Z. The toxicity and oxidative stress of TiO_2 nanoparticles in marine abalone (Haliotis diversicolor supertexta). Marine Pollution Bulletin 2011;63(5): 334–8. https://doi.org/10.1016/j.marpolbul.2011.03.006.
[128] Birhanli A, Emre FB, Sayilkan F, Güngördü A. Effect of nanosized TiO_2 particles on the development of *Xenopus laevis* embryos. Turkish Journal of Biology 2014;38(2): 283–8. https://doi.org/10.3906/biy-1307-37.
[129] Carmona ER, Escobar B, Vales G, Marcos R. Genotoxic testing of titanium dioxide anatase nanoparticles using the wing-spot test and the comet assay in *Drosophila*. Mutation Research: Genetic Toxicology and Environmental Mutagenesis 2015;778:12–21. https://doi.org/10.1016/j.mrgentox.2014.12.004.
[130] Sabat D, Patnaik A, Ekka B, Dash P, Mishra M. Investigation of titania nanoparticles on behaviour and mechanosensory organ of *Drosophila melanogaster*. Physiology & Behavior 2016;167:76–85. https://doi.org/10.1016/j.physbeh.2016.08.032.
[131] Federici G, Shaw BJ, Handy RD. Toxicity of titanium dioxide nanoparticles to rainbow trout (*Oncorhynchus mykiss*): gill injury, oxidative stress, and other physiological effects. Aquatic Toxicology 2007;84(4):415–30. https://doi.org/10.1016/j.aquatox.2007.07.009.
[132] Hao L, Wang Z, Xing B. Effect of sub-acute exposure to TiO_2 nanoparticles on oxidative stress and histopathological changes in Juvenile Carp (*Cyprinus carpio*). Journal of Environmental Sciences 2009;21(10):1459–66. https://doi.org/10.1016/S1001-0742(08)62440-7.
[133] Wang J, Zhou G, Chen C, Yu H, Wang T, Ma Y, et al. Acute toxicity and biodistribution of different sized titanium dioxide particles in mice after oral administration. Toxicology Letters 2007;168(2):176–85. https://doi.org/10.1016/j.toxlet.2006.12.001.
[134] Ates M, Demir V, Adiguzel R, Arslan Z. Bioaccumulation, subacute toxicity, and tissue distribution of engineered titanium dioxide nanoparticles in goldfish (*Carassius auratus*). Journal of Nanomaterials 2013;2013:6. https://doi.org/10.1155/2013/460518.
[135] Paterson G, Ataria JM, Hoque ME, Burns DC, Metcalfe CD. The toxicity of titanium dioxide nanopowder to early life stages of the Japanese medaka (*Oryzias latipes*). Chemosphere 2011;82(7):1002–9. https://doi.org/10.1016/j.chemosphere.2010.10.068.
[136] Jovanović B, Anastasova L, Rowe EW, Zhang Y, Clapp AR, Palić D. Effects of nanosized titanium dioxide on innate immune system of fathead minnow (*Pimephales promelas* Rafinesque, 1820). Ecotoxicology and Environmental Safety 2011;74(4):675–83. https://doi.org/10.1016/j.ecoenv.2010.10.017.
[137] Azim SA, Darwish HA, Rizk MZ, Ali SA, Kadry MO. Amelioration of titanium dioxide nanoparticles-induced liver injury in mice: possible role of some antioxidants. Experimental and Toxicologic Pathology 2015;67(4): 305–14. https://doi.org/10.1016/j.etp.2015.02.001.
[138] Hong F, Sheng L, Ze Y, Hong J, Zhou Y, Wang L, et al. Suppression of neurite outgrowth of primary cultured hippocampal neurons is involved in impairment of glutamate metabolism and NMDA receptor function caused by nanoparticulate TiO_2. Biomaterials 2015;53:76–85. https://doi.org/10.1016/j.biomaterials.2015.02.067.
[139] Ghorbanpour M, Hadian J. Multi-walled carbon nanotubes stimulate callus induction, secondary metabolites biosynthesis and antioxidant capacity in medicinal plant Satureja khuzestanica grown in vitro. Carbon 2015;94:749–59. https://doi.org/10.1016/j.carbon.2015.07.056.
[140] Chen M, Qin X, Li J, Zeng G. Probing molecular basis of single-walled carbon nanotube degradation and nondegradation by enzymes based on manganese peroxidase and lignin peroxidase. RSC Advances 2016;6(5): 3592–9. https://doi.org/10.1039/c5ra21814d.
[141] Chen M, Qin X, Zeng G. Single-walled carbon nanotube release affects the microbial enzyme-catalyzed oxidation processes of organic pollutants and lignin model compounds in nature. Chemosphere 2016;163:217–26. https://doi.org/10.1016/j.chemosphere.2016.08.031.
[142] Chen M, Zeng G, Xu P, Zhang Y, Jiang D, Zhou S. Understanding enzymatic degradation of single-walled carbon nanotubes triggered by functionalization using molecular dynamics simulation. Environmental Sciences: Nano 2017;4(3):720–7. https://doi.org/10.1039/c7en00050b.
[143] Chen M, Zhou S, Zeng G, Zhang C, Xu P. Putting carbon nanomaterials on the carbon cycle map. Nano Today 2018;20:7–9. https://doi.org/10.1016/j.nantod.2018.02.001.

[144] Ema M, Gamo M, Honda K. A review of toxicity studies on graphene-based nanomaterials in laboratory animals. Regulatory Toxicology and Pharmacology 2017;85:7–24. https://doi.org/10.1016/j.yrtph.2017.01.011.
[145] Kumarathasan P, Breznan D, Das D, Salam MA, Siddiqui Y, Mackinnon-Roy C, et al. Cytotoxicity of carbon nanotube variants: a comparative in vitro exposure study with A549 epithelial and J774 macrophage cells. Nanotoxicology 2015;9(2):148–61. https://doi.org/10.3109/17435390.2014.902519.
[146] Khaliullin TO, Fatkhutdinova LM, Zalyalov RR, Kisin ER, Murray AR, Shvedova AA. In vitro toxic effects of different types of carbon nanotubes. IOP Conference Series: Materials Science and Engineering 2015;98(1):012021.
[147] Poulsen SS, Knudsen KB, Jackson P, Weydahl IE, Saber AT, Wallin H, Vogel U. Multi-walled carbon nanotube-physicochemical properties predict the systemic acute phase response following pulmonary exposure in mice. PLoS One 2017;12(4):e0174167. https://doi.org/10.1371/journal.pone.0174167.
[148] Patlolla AK, Patra PK, Flountan M, Tchounwou PB. Cytogenetic evaluation of functionalized single-walled carbon nanotube in mice bone marrow cells. Environmental Toxicology 2016;31(9):1091–102. https://doi.org/10.1002/tox.22118.
[149] Park EJ, Lee GH, Han BS, Lee BS, Lee S, Cho MH, et al. Toxic response of graphene nanoplatelets in vivo and in vitro. Archives of Toxicology 2015;89(9):1557–68. https://doi.org/10.1007/s00204-014-1303-x.
[150] Principi E, Girardello R, Bruno A, Manni I, Gini E, Pagani A, et al. Systemic distribution of single-walled carbon nanotubes in a novel model: alteration of biochemical parameters, metabolic functions, liver accumulation, and inflammation in vivo. International Journal of Nanomedicine 2016;11:4299–316. https://doi.org/10.2147/ijn.s109950.
[151] Luo M, Chen P, Wang J, Deng X, Dong L, Wu M, Shen X. The cytotoxicity of oxidized multi-walled carbon nanotubes on macrophages. Science China Chemistry 2016;59(7):918–26. https://doi.org/10.1007/s11426-016-5595-y.
[152] Fujita K, Fukuda M, Endoh S, Maru J, Kato H, Nakamura A, et al. Pulmonary and pleural inflammation after intratracheal instillation of short single-walled and multi-walled carbon nanotubes. Toxicology Letters 2016;257:23–37. https://doi.org/10.1016/j.toxlet.2016.05.025.
[153] Chen X, Zhouhua W, Jie Z, Xinlu F, Jinqiang L, Yuwen Q, Zhiying H. Renal interstitial fibrosis induced by high-dose mesoporous silica nanoparticles via the NF-κB signaling pathway. International Journal of Nanomedicine 2015;10:1.
[154] Munk M, Camargo LSA, Quintão CCR, Silva SR, Souza ED, Raposo NRB, et al. Biocompatibility assessment of fibrous nanomaterials in mammalian embryos. Nanomedicine: Nanotechnology, Biology and Medicine 2016;12(5):1151–9. https://doi.org/10.1016/j.nano.2016.01.006.
[155] Eom HJ, Roca CP, Roh JY, Chatterjee N, Jeong JS, Shim I, et al. A systems toxicology approach on the mechanism of uptake and toxicity of MWCNT in *Caenorhabditis elegans*. Chemico-Biological Interactions 2015;239:153–63. https://doi.org/10.1016/j.cbi.2015.06.031.
[156] Kim DH, Kim JH, Park JH, Ewane EB, Lee DH. Correlation between above-ground and below-ground biomass of 13-year-old Pinus densiflora S. et Z. planted in a post-fire area in Samcheok. Forest Science and Technology 2016;12(3):115–24. https://doi.org/10.1080/21580103.2015.1107627.
[157] Dharni S, Sanchita, Unni SM, Kurungot S, Samad A, Sharma A, Patra DD. In vitro and in silico antifungal efficacy of nitrogen-doped carbon nanohorn (NCNH) against Rhizoctonia solani. Journal of Biomolecular Structure and Dynamics 2016;34(1):152–62. https://doi.org/10.1080/07391102.2015.1018841.
[158] Yadav T, Mungray AA, Mungray AK. Effect of multi-walled carbon nanotubes on UASB microbial consortium. Environmental Science and Pollution Research International 2016;23(5):4063–72. https://doi.org/10.1007/s11356-015-4385-y.
[159] Shrestha B, Anderson TA, Acosta-Martinez V, Payton P, Cañas-Carrell JE. The influence of multi-walled carbon nanotubes on polycyclic aromatic hydrocarbon (PAH) bioavailability and toxicity to soil microbial communities in alfalfa rhizosphere. Ecotoxicology and Environmental Safety 2015;116:143–9. https://doi.org/10.1016/j.ecoenv.2015.03.005.
[160] Kim YR, Park SH, Lee JK, Jeong J, Kim JH, Meang EH, et al. Organization of research team for nano-associated safety assessment in effort to study nanotoxicology of zinc oxide and silica nanoparticles. International Journal of Nanomedicine 2014;9(Suppl. 2):3–10. https://doi.org/10.2147/ijn.s57915.
[161] Oberdorster G, Maynard A, Donaldson K, Castranova V, Fitzpatrick J, Ausman K, et al. Principles for characterizing the potential human health effects from exposure to nanomaterials: elements of a screening strategy. Particle and Fibre Toxicology 2005;2:8. https://doi.org/10.1186/1743-8977-2-8.
[162] Colvin VL. The potential environmental impact of engineered nanomaterials. Nature Biotechnology 2003;21(10):1166–70. https://doi.org/10.1038/nbt875.
[163] Mogharabi M, Abdollahi M, Faramarzi MA. Toxicity of nanomaterials; an undermined issue. Daru Journal of Pharmaceutical Sciences 2014;22:59. https://doi.org/10.1186/s40199-014-0059-4.
[164] Mukherjee B, Maji R, Roychowdhury S, Ghosh S. Toxicological concerns of engineered nanosize drug delivery systems. American Journal of Therapeutics 2016;23(1):e139–150. https://doi.org/10.1097/01.mjt.0000433947.16654.75.
[165] Donaldson K, Stone V, Tran CL, Kreyling W, Borm PJ. Nanotoxicology. Occupational and Environmental Medicine 2004;61(9):727–8. https://doi.org/10.1136/oem.2004.013243.

[166] Xia T, Zhao Y, Sager T, George S, Pokhrel S, Li N, et al. Decreased dissolution of ZnO by iron doping yields nanoparticles with reduced toxicity in the rodent lung and zebrafish embryos. ACS Nano 2011;5(2): 1223–35. https://doi.org/10.1021/nn1028482.

[167] Ahmadi E, Zarghami N, Jafarabadi MA, Alizadeh L, Khojastehfard M, Yamchi MR, Salehi R. Enhanced anticancer potency by combination chemotherapy of HT-29 cells with biodegradable, pH-sensitive nanoparticles for co-delivery of hydroxytyrosol and doxorubicin. Journal of Drug Delivery Science and Technology 2019;51:721–35. https://doi.org/10.1016/j.jddst.2019.03.003.

[168] Cheong S, Ferguson P, Feindel KW, Hermans IF, Callaghan PT, Meyer C, et al. Simple synthesis and functionalization of iron nanoparticles for magnetic resonance imaging. Angewandte Chemie International Edition 2011;50(18):4206–9.

[169] Davidson DC, Derk R, He X, Stueckle TA, Cohen J, Pirela SV, et al. Direct stimulation of human fibroblasts by $nCeO_2$ in vitro is attenuated with an amorphous silica coating. Particle and Fibre Toxicology 2016;13(1): 23. https://doi.org/10.1186/s12989-016-0134-8.

[170] Wang X, Xia T, Duch MC, Ji Z, Zhang H, Li R, et al. Pluronic F108 coating decreases the lung fibrosis potential of multiwall carbon nanotubes by reducing lysosomal injury. Nano Letters 2012;12(6):3050–61.

[171] Chanana M, Rivera_Gil P, Correa-Duarte MA, Liz-Marzán LM, Parak WJ. Physicochemical properties of protein-coated gold nanoparticles in biological fluids and cells before and after proteolytic digestion. Angewandte Chemie International Edition 2013;52(15): 4179–83.

[172] Osmond-McLeod MJ, Osmond RI, Oytam Y, McCall MJ, Feltis B, Mackay-Sim A, et al. Surface coatings of ZnO nanoparticles mitigate differentially a host of transcriptional, protein and signalling responses in primary human olfactory cells. Particle and Fibre Toxicology 2013;10(1):54.

[173] Lee J, Park EY, Lee J. Non-toxic nanoparticles from phytochemicals: preparation and biomedical application. Bioprocess and Biosystems Engineering 2014;37(6): 983–9. https://doi.org/10.1007/s00449-013-1091-3.

[174] Ramamurthy C, Sampath K, Arunkumar P, Kumar MS, Sujatha V, Premkumar K, Thirunavukkarasu C. Green synthesis and characterization of selenium nanoparticles and its augmented cytotoxicity with doxorubicin on cancer cells. Bioprocess and Biosystems Engineering 2013; 36(8):1131–9.

[175] Arunachalam KD, Annamalai SK, Hari S. One-step green synthesis and characterization of leaf extract-mediated biocompatible silver and gold nanoparticles from Memecylon umbellatum. International journal of nanomedicine 2013;8:1307.

[176] Dhar S, Reddy EM, Shiras A, Pokharkar V, Prasad BeLeV. Natural gum reduced/stabilized gold nanoparticles for drug delivery formulations. Chemistry–A European Journal 2008;14(33):10244–50.

[177] Shukla R, Nune SK, Chanda N, Katti K, Mekapothula S, Kulkarni RR, . . . Katti KV. Soybeans as a phytochemical reservoir for the production and stabilization of biocompatible gold nanoparticles. Small 2008;4(9): 1425–36.

[178] Zhang L, Yu F, Cole AJ, Chertok B, David AE, Wang J, Yang VC. Gum arabic-coated magnetic nanoparticles for potential application in simultaneous magnetic targeting and tumor imaging. The AAPS journal 2009; 11(4):693.

[179] Lee J, Kim HY, Zhou H, Hwang S, Koh K, Han D-W, Lee J. Green synthesis of phytochemical-stabilized Au nanoparticles under ambient conditions and their biocompatibility and antioxidative activity. Journal of Materials Chemistry 2011;21(35):13316–26.

[180] Hainfeld J, Slatkin D, Focella T, Smilowitz H. Gold nanoparticles: a new X-ray contrast agent. The British Journal of Radiology 2006;79(939):248–53.

[181] Nune SK, Chanda N, Shukla R, Katti K, Kulkarni RR, Thilakavathy S, . . . Katti KV. Green nanotechnology from tea: phytochemicals in tea as building blocks for production of biocompatible gold nanoparticles. Journal of Materials Chemistry 2009;19(19):2912–20.

[182] Chanda N, Shukla R, Zambre A, Mekapothula S, Kulkarni RR, Katti K, . . . Boote EJ. An effective strategy for the synthesis of biocompatible gold nanoparticles using cinnamon phytochemicals for phantom CT imaging and photoacoustic detection of cancerous cells. Pharmaceutical research 2011;28(2):279–91.

[183] Kattumuri V, Katti K, Bhaskaran S, Boote EJ, Casteel SW, Fent GM, . . . Katti KV. Gum arabic as a phytochemical construct for the stabilization of gold nanoparticles: in vivo pharmacokinetics and X-ray-contrast-imaging studies. Small 2007;3(2):333–41.

CHAPTER 10

Biomedical Applications of Nanobots

DEEPANJALEE DUTTA, PHD • SUNIL KUMAR SAILAPU, PHD

1 INTRODUCTION

The evolution of life has been realized with many developments, in which motion (or movement) remained an integral feature. This characteristic phenomenon is observed widely in nature and is crucial to happening of biological processes and cellular activities regulated by mobile entities such as motor proteins. Stimulating research has been devoted in developing such synthetic micro/nanomachines to offer applications in various fields including medicine. The area of study has attained a portion of its motivation from motor proteins like kinesin and myosin, that perform intriguing multifaceted biological tasks in living cells with unexpectedly high efficiency and accuracy. Motor proteins are structures made up of biomolecules bestowed with the capacity to convert chemical energy into valuable mechanical form in a cyclic manner. Biological motors are one of the outstanding results of development in terms of evolution. Several efforts have been directed to develop autonomously propelling synthetic machines using these motor proteins, however, due to the characteristic unstable nature of the protein in the diverse environmental milieu, their applications were limited [1]. Nevertheless, the proficiency of the biological motors has driven researchers from a multitude of backgrounds to fabricate artificial counterparts that have inherent motion to achieve feats such as transportation of cargo, drug/gene delivery, surgeries, biosensing, bioimaging, and environmental remediation. The prominence of the field was encouraged by the recent award of Nobel Prize in physics to Arthur Ashkin in 2018 for his work on optical tweezers; and in chemistry to Sauvage, Stoddard, and Feringa in 2016, for the engineering and construction of molecular machines. In addition to exhibiting an actuation at the atomic or molecular level, clusters of these molecular machines and the subsequent hierarchic assemblies may exhibit numerous effects in a macro domain. These ultrasmall self-propelled micro/nanomachines, regarded in various forms as motors, engines, jets, rockets, robots, swimmers, embarks remarkable potential with rich applicability and operational scope in various fields, such as catalysis, biomedical, energy harvesting, and electronic sensing, along with some fictional yet probable areas to be explored in near future.

A substantial objective in the stride toward the development of tiny vehicles/sensors is to bestow them with the ability to monitor cellular activity and design/deliver targeted therapeutics inside living cells. As a result, considerable research has been aimed at developing autonomous nanomachines with the capability to penetrate cells and probe the cellular activities. In addition, interest has also been devoted toward exploring novel methods and techniques to accomplish tasks such as transportation of payload to subcellular organelles in body fluids and conduct surgical procedures. In spite of considerable progress in biological applications, there are important challenges in terms of achieving sensitive detection, controlled delivery and release, improved biocompatibility, efficient propulsion, mechanisms for safe exclusion, and so on. Nevertheless, these micro/nanomachines with their ultrasmall size and propulsion are promising to provide a wide range of biomedical applications with unprecedented accuracy and efficiency in a way that would possibly mimic the natural systems. Further, achieving such high-level functionality to manipulate and target at the cellular level would be highly fascinating and moreover would enable localized diagnosis and disease treatment in a simple and efficient way.

2 CHALLENGES TO MOTION AT NANOSCALE

Although the exciting prospects of micro/nanomotors rightly encourage their immediate fabrication and utilization, there lies certain challenges and hurdles in their realization that needs to be addressed. Here, although natural motors provide blueprints to mimic the propulsion mechanisms, it is not very easy to directly emulate such functionality and motion perfected over centuries

Intelligent Nanomaterials for Drug Delivery Applications. https://doi.org/10.1016/B978-0-12-817830-0.00010-2

of evolution. This may be said because the propulsion of small dimension entities in solution is governed by diverse and contradictory behavior of physical laws when compared to macroscale systems. Although the surface area of these small-scale particles tends to decrease only by ΔR^2 in the process of scaling by a factor of "R," the volume changes by ΔR^3. As a result, surface effects become more dominant with the proportional exposed area being relatively more in comparison to large-scale objects. Parameters dictated by surface area such as the rate of energy dissipation become considerably large than its ability to store the required fuel. This less fuel storage capability makes them submissive to motion that rapidly recedes due to quick exhaustion of on-board fuel. For this reason, they must harness energy from external sources (based on fuel or fuel-free approaches) for their propulsion. In addition, at these small-scale dimensions, the Brownian force is dominant, which forces random trajectories of motion. Electrostatic and van der Waals interactions also become significant due to the minuscule separation distances.

Movement at extremely small scales through a variety of fluids becomes particularly challenging owing to the prominent viscous drag relative to the associated inertial forces. As a consequence, a low Reynolds number system governs the motion that is very different from that observed in the macroscopic world. Therefore the pathway of movement implemented in the case of macroscale objects cannot be realized to impart movement to a micro/nanoscale particle. In its place, an uninterrupted motion is mandatory to detect a clear displacement. To accomplish this, the symmetry of the system needs to be disturbed through motor design, geometry, and surface chemistry to produce asymmetric surface gradients needed to generate the driving force for their propulsion. Further challenges to motion would also be experienced from the consequences of electric double-layer formation with its influence on the possible aggregation of small-scale particles.

The applicability of the micro/nanomotors for in vivo purposes such as drug delivery and cargo transport must overcome the immune system clearance, so as not to be destroyed as foreign material. Hence, besides augmenting the drug delivery and therapeutic efficiency, it is vital to first conquer the host's immune system. In addition, these motors must also deliver therapy avoiding unnecessary side effects with good levels of precision in targeting the specific site. This might expect the motor to reach out for remote locations of the host organism by penetrating through barriers offered by cells and tissues.

Finally, to make these motors commercially realistic, parameters like easiness in fabrication, scale-up, and cost would be crucial. Hence, to combat the earlier-mentioned challenges arising from scaling laws, physical and fluid parameters, environment, biological and commercial aspects, so as to achieve well-defined functional nano/microrobots, it is essential to adopt different approaches in design, composition, and swimming mechanisms.

3 CLASSIFICATION

As a result of rapidly progressing research, a novel combination of approaches, materials, and technological advancements, the realization of several artificial micro/nanomotors has been made possible in the past few years. To accomplish diverse biomedical applications, the design of micro/nano vehicles is particularly crucial so as to perform a given task. Thereby, their composition, shape, and source of propulsion must be carefully considered to successfully project them as powerful sensors or delivery systems for theranostic applications.

The self-propelled machines, with characteristic sizes in the range of nanometers to micrometers, have been fabricated in various shapes including tubular, spherical, conical, helical or asymmetric geometries. Based on the source of the driving force essential for the motion of nanomotors, these are classified into two groups: (1) those by the use of local chemical/biochemical fuel (fuel-based propulsion) and (2) through external stimuli such as magnetic field, electric field, light, or ultrasound (fuel-free propulsion).

Fuel-propelled motors operate in the presence of a particular fuel (aqueous solution) that would enable them to self-propel using surface reactions based on the composition of the motor. The mechanism of propulsion varies based on the reactions that may lead to generate local gradients of concentration, electrical potential, and bubbles. Although different types of motions can be achieved by fuel-based propulsion by varying the fuel type, composition, dimension, etc., they usually lack a defined trajectory. However, defined navigation could be achieved by involving external controlling factors. For example, by incorporating a magnetic component in the composition of a motor, it could be navigated using a magnetic field. External chemical sources used in the propulsion of motors cause hinder in biological environments due to their incomplete degradation. Hence, biocompatible fuels such as glucose, water, and acid are crucial for not only supporting the motor propulsion but to achieve it only in defined environments.

Fuel-free propelled motors operate through the energies provided by external stimuli like electric and magnetic fields, light, and ultrasound as driving sources. This propulsion is of high interest for particularly in vivo applications as the motors are not affected by the medium in which they operate, and thus can work efficiently in biological environments. By involving biocompatible external stimuli, fuel-free propulsion offers advantages such as on-demand motion control, biocompatibility, lifetime, flexibility, and sustainability. Examples of fuel-free propellers are magnetic helical swimmer that translates rotational motion to translational motion in the presence of low rotational magnetic fields, and nanowires that perform directional motion and spinning under acoustic signals.

The distinct ways in which motion of these micro/nanomotors is achieved affect their overall performance in terms of the operating environment, achieved speed/velocity, tuneability of motion, compatibility in diverse environments, operational lifetime, degradability, etc. The following section details about their design, composition, propulsion mechanism, limitations, and uses—presented in an order based on the source of driving force—fuel-based (chemical/biochemical energy) and fuel-free (magnetic, ultrasound, light, electric, organism-based).

4 FUEL-BASED MICRO/NANOMOTORS

Chemically powered motors are propelled from motion generated by converting chemical energy through the reaction between the fuel (in the fluid) and the motor. To achieve this, two components are constituted in motor design—a catalyst/active metal, an inert material to provide asymmetry required for propulsion. In this case, the source of power required for the propulsion is generally obtained through the decomposition of fuel by the catalytic material. Furthermore, depending on the type of reaction between fuel (H_2O_2, H_2O, glucose, acids, bases, etc.) and motor, propulsion can be through the generation of bubbles or established concentration and other gradients, etc.

In the presence of an external electric field, charged objects exhibit motion in liquid due to electrophoresis. However, without applying any external field, the motion of micro/nanoscale objects could be achieved. This is through the self-generated field in the system developed by chemical gradients from the reactions within the suspended fuel. This is termed as self-electrophoresis-based propulsion and relies on the formation of an indigenous electric field due to the asymmetric development of ionic products (Fig. 10.1A) [2]. Self-electrophoresis is characteristically displayed by bimetallic nanomotors, and the first of its kind was made up of Au/Pt, developed by Sen and coworkers. When the motor was placed in dilute H_2O_2 (fuel), an asymmetric decomposition happened at the ends of the metallic rod. Although the oxidation of fuel occurred at the anodic segment (Pt), reduction took place at the cathodic part (Au). As a result of this, a proton gradient was established along the axis of the metallic rod and the negatively charged rod reacted by moving in the direction of the anode at approximately 5–10 μm/s [3]. The speed of these motors has been drastically improved by adopting a tubular design and using Au/Ag alloy instead of Au in the cathodic

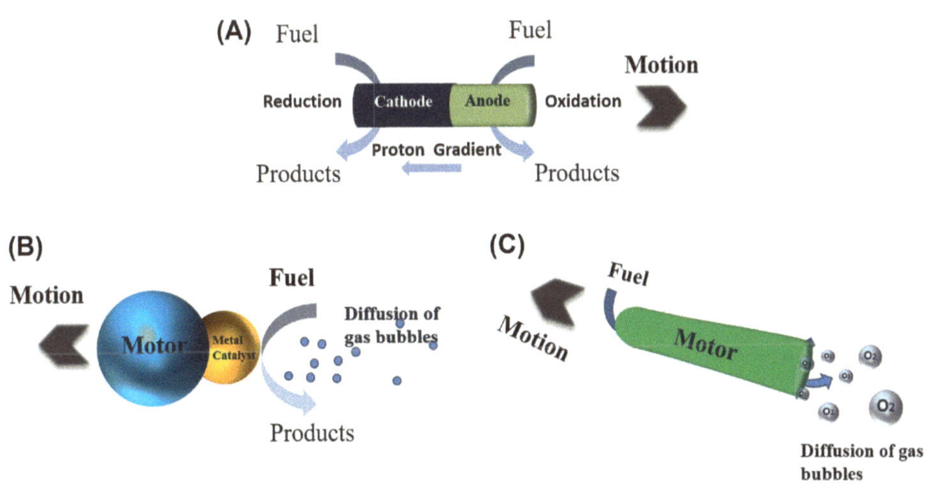

FIG. 10.1 Schematic of propulsion mechanisms of micro/nanomotors. **(A)** Self-electrophoresis, **(B)** Diffusiophoresis, **(C)** Bubble propulsion.

segment. These tubular microrobots developed by the Wang group could move at a speed of 150 μm/s in biorelevant fuels and high ionic-strength media [4]. They further improved the speed to as high as 200 μm/s by fabrication of a nanowire-based motor involving carbon nanotubes (Au/Pt-CNT) [5]. The Wang group also developed methods to regulate the motion of the tubular motors by modulating the temperature and electrical potential of the platform over which the motor propels [6,7]. The Sen group has explored other fuels such as bromide and iodide to create self-propulsion of Cu/Pt-based nanomotors [8]. In addition to the earlier-mentioned shapes, spherical geometry-based motors were also fabricated. Spherical Au/Pt Janus micromotor was developed by Posner et al. to self-propel in H_2O_2. Their study suggested that the speed of the spherical motor was dependent on the diameter and fuel concentration [9].

Synthetic motors have also been developed to propel based on diffusiophoresis mechanisms, namely electrolyte or nonelectrolyte diffusiophoresis. Some types of Janus particles (Pt/SiO_2) experience motion via neutral self-diffusiophoresis (Fig. 10.1B) [10–12]. The catalytic part of the particle is responsible for the generation of reaction products. These reaction products establish a concentration gradient, which further promotes asymmetric net repulsive force to propel the particle away from the catalyst. In another approach, based on self-diffusiophoresis, reversible assemblies of AgCl microparticles in both space and time have been achieved by the Sen group. This reversible collective schooling behavior of particles owing to developed gradients was observed in the presence of H_2O_2 and UV light. Janus particles made of Ir/SiO_2 are also shown to be driven by self-diffusiophoresis, in a similar mechanism [13]. In contrast to Pt/SiO_2 particles reaching speeds about 12 μm/s in 10% H_2O_2, the particles of Ir/SiO_2 can attain a speed of 20 μm/s speed in only 0.001% N_2H_4, owing to its superior catalytic behavior.

Another propulsion approach that is frequently implemented by a comparatively large number of spherical and tubular motors is the bubble-induced motion [14,15]. One of the approaches is based on the development of gas bubbles, like those of H_2 or O_2, on one side of the motor (near to the catalyst) due to the decomposition of the fuel by the catalyst in the motor. The discharge of these gas bubbles delivers a strong thrust for movement. Thereby, a continuous formation and release of bubbles drives the motor in the direction away from the ejection of bubbles or catalysts (Fig. 10.1C). Motors whose function is mediated through a bubble propulsion route can function irrespective of the ionic concentration of the fuel media that is a vital concern when utilizing these modules in a biological environment such as serum or blood. Several bubble propulsion based micro/nanomotors have been developed since the Pt-based motor was presented by the Whitesides group. The Whitesides motor propelled as a result of the bubble generated by the decomposition of H_2O_2 into water and O_2 [16]. Different geometries of motors in the form of spherical Janus motors, micro/nanotubular motors, micro/nanorods, micro/nanoshells, etc. have been explored for bubble propulsion. A bubble-propelled nanorocket based on biopolymers chitosan and alginate, with Pt nanoparticles (NPs) in the inner layer, was developed by the He group [17]. Due to the decomposition of H_2O_2, the nanorocket exhibited motion in straight, circular, and curved trajectories by means of force generated through the release of the O_2 bubble from one end of the rocket. Highly efficient polyaniline/Pt-based microtubular engines have also been shown to self-propel at a speed of 3 mm/s in low concentrations of 0.2% H_2O_2 fuel (Fig. 10.2A) [18]. Furthermore, Pt, Ag, and MnO_2 based motors have also been fabricated to achieve bubble propulsion in H_2O_2 [22]. Based on the bubble-mediated propulsion, chemotaxis and pH taxis have been also demonstrated. Through "active diffusion," the chemotactic behavior of Pt–Au microrods was shown, where the directional movement was observed toward higher concentrations of H_2O_2 [23]. Alternatively, the Chattopadhyay group developed different propulsion strategies in the form of buoyancy-driven motion from micro/nanorobots based on the formation of bubbles during the catalytic decomposition of H_2O_2 [24,25]. They demonstrated pH taxis of a Pd NP containing polymer microparticle towards increasing pH gradient by means of differential catalytic decomposition of H_2O_2. By modulating the size of the motor and imposed pH gradient, the microbot could achieve random walk, translation, vertical, hopping, and pulsed motion. All the earlier-mentioned motors mainly rely on the use of H_2O_2 for their propulsion. However, at high concentrations, H_2O_2 is not compatible with biological environments.

As an alternative to H_2O_2, a water-driven microsphere composed of Al–Ga binary alloy was developed by the Wang group to move by means of bubble propulsion [26]. Here, on the contact of the sphere with water, H_2 bubbles were ejected from the exposed Al–Ga alloy hemisphere side creating the thrust for its motion. These bubbles generated as a result of the reaction

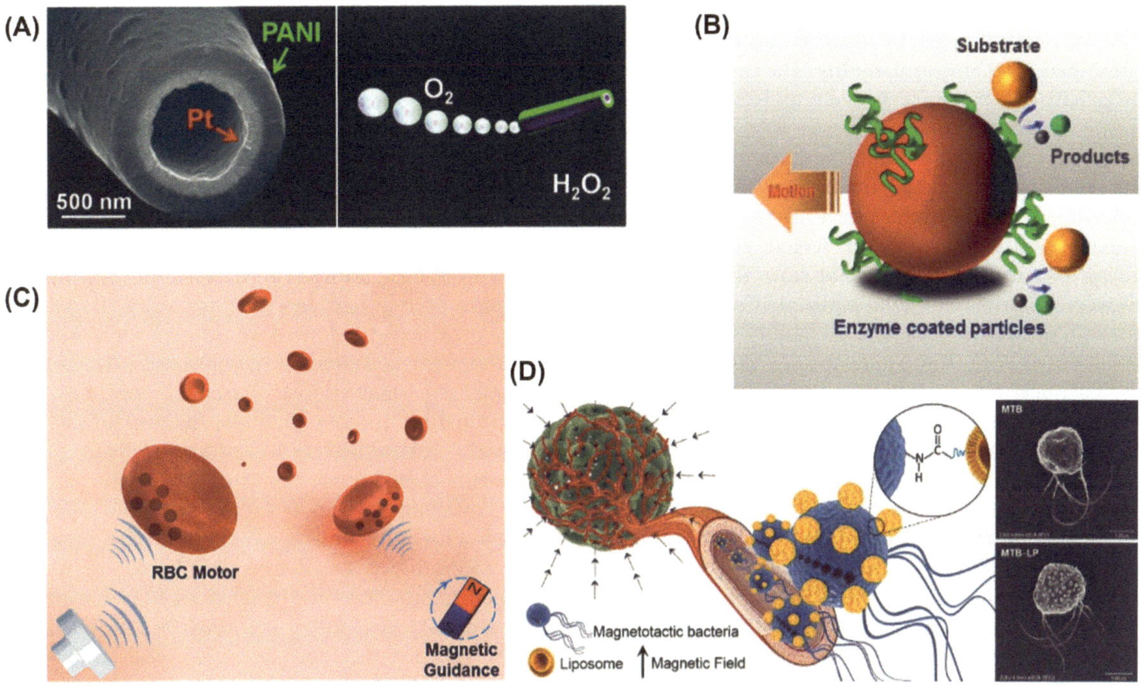

FIG. 10.2 **Different types of micro/nanomotors. (A)** Catalytic microtube engines driven by bubble-mediated propulsion. **(B)** Enzyme-functionalized microparticles as autonomous motors. **(C)** Ultrasound-propelled RBC micromotors. **(D)** Magnetotactic bacteria-based motor. ((A) Reproduced with permission [18]. Copyright 2011, American Chemical Society. (B) Reproduced with permission [19]. Copyright 2015, American Chemical Society. (C) Reproduced with permission [20]. Copyright 2014, American Chemical Society. (D) Reproduced with permission from Ref. [21] Copyright 2014, American Chemical Society.)

between Al and water propelled at a speed of 3 mm/s. Importantly, the motion was further shown in different biological media like human serum. In addition to water, acids were also investigated as potential fuels for bubble-mediated propulsion. Tubular polyaniline/Zn microrockets were developed to self-propel using acid as fuel through H_2 bubbles generated due to redox reaction at the inner Zn layer [27]. This is particularly interesting as acids are commonly found in biosystems. Al/Pd Janus microsphere-based motor was shown to be capable of self-propelling in three different fuels namely, acid, base, and H_2O_2, with a high speed of up to 200 μm/s in alkaline media [28]. The motion of these Al/Pd micromotors in acids and bases was a result of the formation of H_2 bubbles in the course of the reaction of Al with acids or bases. In addition, the motors were also capable of propelling in H_2O_2 as an outcome of the thrust generated from the catalytic break down of H_2O_2 mediated by Pd. The capability of the Janus microspheres to self-propel in more than one fuel source is highly beneficial so that it could suit for applications related to multiple fields. Metal-organic framework (MOF)-dependent materials present another exclusive way to fabricate catalytic motors [29]. These are micro/nanoporous materials with adaptable properties and have been applied in various fields namely drug delivery, purification, and catalysis. MOFs composed of zeolite imidazolate frameworks [30] and (UiO)-type scaffolds [31] have exhibited motion in the presence of H_2O_2, and this motion can be tuned according to the active metal content of the MOF design.

The most significant prerequisite for the application of catalytic nanomotors in biomedical applications is the cytocompatibility of the fuels used. The essential strategy is to enhance the properties of presently available motor systems by converting them into a hybrid system. Keeping this in mind, many artificial nanomotors were fabricated to propel in biological fluids or has been loaded with enzymes that are biological catalysts. Enzymes are biomolecules that process the catalysis of numerous biochemical reactions in an extremely specific and proficient way. The application of enzymes in the construction of artificial micro/nanomotors is encouraged by their extraordinary activity combined with biocompatibility. The improvement of diffusivity of a particular enzyme molecule submerged in the substrate solution enhances with an increase in substrate concentration that confirms the point that the individual enzyme molecules are capable of behaving as nanomotors. When in contact with a stationary surface, enzymes can behave as micropumps giving rise to a potential fluid flow that is capable enough to drive colloidal particles. Catalase, a bioenzyme present in cells, has the ability to proficiently break down H_2O_2 into water and O_2, thereby converting chemical energy into mechanical energy and can be viewed as a biological alternative to Ag, Pt, Ni, or MnO_2. The immobilization of the synthetic enzyme on SiO_2 microparticles was achieved via covalent coupling of carboxylate ligand to the aminopropyl modified SiO_2 microparticles [32]. The motor achieved propulsion with a speed of around 35 μm/s in 5% H_2O_2. In similar lines, a new approach was implemented for the immobilization of catalase inside the electrodeposited polypyrrole/Au microtubes. Schmidt et al. demonstrated that the microtubes constitute a multipurpose platform for the functionalization of enzymes as a substitute for Pt [33]. The motor was developed by covalently functionalizing catalase on the Au layer of the Ti/Au microtube, which was modified with 3-mercaptopropionic acid self-assembled monolayers (SAMs). This motor was capable of propelling even in less concentrations, that is, 1.5% of H_2O_2. Urease immobilized micron-sized particles exhibit improved diffusion in the presence of urea solutions, creating a platform for fully biocompatible enzyme-based motors (Fig. 10.2B) [19]. Recently a controlled validation of chemophoretic movement of Janus hollow mesoporous silica particles immobilized with urease has been reported [34]. These investigations propose an enormous potential of urease in the propulsion of synthetic motors of discrete geometrical symmetries. However, the immobilized enzymes are susceptible to eventual activity loss. In addition, prolonged exposure to high H_2O_2 levels can affect the catalase activity due to structural modifications induced by peroxide. For combating this issue, encapsulation of the enzymes inside shells of polymers has been achieved. Proteins and peptides are essential constituents existing in all living organisms. Another encouraging direction toward the development of artificial motors is the application of proteases like trypsin/pepsin, as fuel to drive the motors. Trypsin-immobilized Janus particles have been reported to show an improved diffusivity with increased instantaneous velocity in a solution of model fuel peptide [35]. Plant-tissue based motors using materials such as potato tubers, mushrooms, carrot roots, and millet seeds were also showed to self-propel due to the presence of catalase, which is a natural catalyst known to catalyze the breakdown of H_2O_2 into water and O_2 responsible for the movement of the motor [36]. Asymmetric distribution of the catalase is essential within the plant tissue for proficient propulsion, and hence they were partly covered with an inert layer of materials such as chitosan or epoxy resins. These effectively provided the desired asymmetry for viable propulsion with considerably good speeds. These plant tissue-based nanomotors presented low cost and biocompatible motors. Recently, Bandopadhyay et al. showed chemotherapeutic drug delivery using magnetite-coated mushroom-based nanomotors based on its catalytic and magnetic properties [37]. However, there is a lot of impending scope toward demonstrating the applicability of these plant-based motors in biomedical applications. In terms of fully biocompatible nanomotors, Sen et al. demonstrated the likelihood of applying enzymes, such as catalase itself as nanomotors [38]. The autonomous motion of the enzymes was revealed to be completely dependent on the substrate with a fuel gradient-based control of the movement. The H_2O_2 fuel gradient was generated at the specific site with the help of enzyme glucose oxidase and glucose. The catalase was migrated in the direction of a higher concentration of glucose oxidase giving rise to a chemically interrelated organization. This result is a pioneering step in the direction of biocompatible nanomotors for applications in a biomedical field like catalyst isolation and drug delivery.

5 FUEL-FREE MICRO/NANOMOTORS

As seen in the previous section, the field of fuel-based motors has been progressed greatly. However, certain issues related to concentrations of fuels used, biocompatibility, in vivo toxicity, and oxidation capacity create a hurdle for their applications, particularly in healthcare. For instance, the available fuel concentration levels (such as of H_2O_2) in in vivo systems are too low to generate the motion and required speed in the case of some motors. Thereby, besides addressing these issues related to fuel-driven motors, promising alternative approaches have been sought to design fuel-free motors, thus eliminating fuel dependency parameters. In addition, a high degree of freedom in terms of rotation and orientation of micro/nanorobot, tuneability of speed, collective operation at destined locations can be easily achieved through fuel-free motors. These further offer the capabilities in device integration, better compatibility, and lifetime.

5.1 Magnetic Field-Based Motors

In general, magnetic motors constitute a magnetic material (such as compounds of iron, nickel, and cobalt) and have the potential to be propelled competently depending on the sort of applied magnetic field. The magnetically compelled motors have a number of features that can be of great interest to biological systems. In addition, given the harmless nature of low strength magnetic fields toward cells and tissues, this propulsion technique has been widely employed in fuel-free motors.

Magnetic surface walkers are a kind of magnetically propelled micro/nanorobots that propel by breaking the spatial symmetry through hydrodynamic interaction with the nearby solid boundaries [39]. Based on this phenomenon, Tierno et al. developed anisotropic doublets composed of streptavidin-coated paramagnetic colloidal particles to achieve controlled propulsion above a flat plate when subjected to a suitable magnetic field [40]. In the development of walkers, inspiration has been taken from natural species, like cilia. Without involving complex fabrication methods, Alexander-Katz et al. developed artificial cilia that constituted a chain of self-assembled superparamagnetic particles. These colloidal rotors propel along surfaces with their shape and motion being regulated by the application of a rotating magnetic field [41]. Furthermore, Zhang et al. used Ni nanowires to demonstrate propulsion and precise steering with cargo transport capabilities in microchannel [42].

Inspired from natural species like spermatozoa, flexible magnetic micro/nano swimmers have been designed to achieve propulsion through asymmetric deformation of flexible filaments incorporated in the swimmer [43]. Exploiting the concept of bacterial movement, the Mishra group reported a magnetic field-based sperm flagella-driven microbot consisting of a bovine spermatozoon encapsulated inside Ti/Fe nanomembranes and established comprehensive motion control under the impact of controlled magnetic field lines [44]. A flexible artificial flagellum-based swimmer with a DNA-linked chain of paramagnetic beads was demonstrated by Dreyfus et al. In an external magnetic field, the swimmer achieved propulsion through bending wave propagation along the flagella and drew red blood cells (RBCs) attached to it [45]. Flexible metallic nanorods with polyelectrolyte hinges, fabricated through electrodeposition and chemical etching approaches, have been actuated by bending induced about the hinge through applied magnetic field [46]. The Wang group formulated flagella-like nanoswimmers using Au–Ag–Ni segment nanowires, so as to achieve a structure comprising Au head and Ni tail with a flexible Ag bridge. The swimmer achieved propulsion through rotation of Ni segment encouraging the motion of nanowires under rotating magnetic fields [47]. Drug-delivery capabilities were demonstrated by magnetically driving anticancer drug doxorubicin (DOX)-loaded magnetic poly(D,L-lactic-co-glycolic acid) microparticle by flexible Ni–Ag nanomotors [48].

Magnetic helical swimmers, inspired by helical bacterial flagella, generally constitute a magnetic head and a rigid helical body. These propel through rotation of their bases in a cock screw-like mechanism under a continuously rotating magnetic field [49]. By using the self-scrolling technique, Nelson and coworkers fabricated synthetic bacterial flagella made up of a ferromagnetic head harboring Cr/Ni/Au layers and a helical tail assembled from InGaAs/GaAs or InGaAs/GaAs/Cr [50]. Under the reversible magnetic field, the rotation of the head brought about magnetically was transformed into translational motion of the swimmer. It can move forward or backward by reversing the direction of the rotating magnetic field. Propulsion speeds about 18 mm/s that are analogous to those achieved by bacteria were obtained. The glancing angle deposition technique was employed by Fisher and Ghosh to prepare helical magnetic nanoswimmers for controlled propulsion in liquid [51]. To address large-scale preparation, a low-cost method based on membrane template-assisted electrochemical deposition was demonstrated by Li et al. [52] Another promising method for the above objective is based on biotemplated synthesis. This cost-effective approach was used in helical motors preparation by deriving spiral water-conducting vessels of vascular plants [53]. Furthermore, other biotemplates like *Spirulina platensis* was also used to achieve cytocompatible swimmers with controlled propulsion in a low strength magnetic field. Magnetically triggered ciliary microrobots were designed by the Choi group using a Ni/Ti bilayer covered cilia part so as to achieve magnetic control and biocompatibility [54]. The microrobots were operated using an electromagnetic coil that produces a magnetic field in steps to trigger the cilia via nonreciprocal motion.

5.2 Ultrasound-Based Motors

Ultrasound motors are driven by converting the acoustic waves (whose frequencies are higher than the upper human hearing limit) into motion. Similar to the magnetic field, ultrasound is biocompatible and widely used in ultrasonography, noninvasive imaging, and therapy. It, therefore, serves as a potential external stimulus that can be applied for remotely propelling nanomotors.

Micron-scale Au—Ru bimetallic rods, fabricated through template electrodeposition technique, have been showed to display fast axial motion (200 μm/s) under a continuous or pulsed ultrasound wave of MHz frequencies [55]. The microrods harbored a concave or convex end giving them an asymmetric nature responsible for their axial propulsion. This asymmetric shape resulted in a pressure gradient at the microrod's surface and propelled the rods in one direction. Thereby, a self-acoustophoresis mechanism was suggested as the driving force for the motion. An approach toward turning RBCs into useful micromotors was implemented by the Wang group, wherein the micromotors were propelled with the assistance of ultrasound propulsion and a magnetic field (Fig. 10.2C) [20]. The asymmetric geometry of RBC and the distribution of iron oxide NPs inside it helped to achieve the motor's magnetic guidance and acoustic propulsion. Wang et al. also demonstrated a multifunctional nanomotor comprising three segments (Au—Ni—Au), which was driven by ultrasound. The Ni element helped to further achieve guided motion directed by the applied magnetic field [56]. The motor showed capabilities of picking and transporting magnetic beads in a predetermined path. In further research, the Au surface was decorated with receptors such as lectins and antiprotein (A) antibody through a binary SAM to capture *E. coli*/*S. aureus* bacteria selectivity during its propulsion. Of late, ultrasound triggered tubular microbullets have been quite encouraging for meeting the challenges of targeted drug delivery and surgery. The force for the propulsion of these bullets was achieved through the vaporization of a loaded perfluorocarbon on an ultrasound trigger exhibiting a velocity of 6.3 m/s [57]. Additionally, these ultrasound propelled motors in comparison to the Pt-catalysis-based nanomotors can deal proficiently with the corrosive behavior of solutions with high ionic strength. Ultrasound powered motors functionalized by the RBC membrane reported by Wu et al. were utilized for efficient exclusion of toxin melittin with high efficiency [58].

5.3 Light-Based Motors

A photoactive element constituting particle can be powered through light by a mechanism, generally involving optical tweezers, or photocatalytic or photoisomerization or photothermal-mediated propulsions. Light offers biocompatibility and achieve deeper penetration as some wavelengths are not easily absorbed and harmful for biological components. Optical tweezers have been an exciting development for precise trapping and manipulation of micro/nanoscale particles using a strongly focused light. This technique is widely adopted to trap different nanostructures [59,60], and to transport and manipulate biological constituents [61]. Arthur Ashkin was awarded the Nobel Prize in Physics for his work on optical tweezers in 2018.

Light-induced photocatalytic decomposition-based propulsion has been demonstrated by the Sen group, where the photoreactive AgCl microparticles have been showed to achieve propulsion under UV irradiation in deionized water [62]. The UV illumination-induced break down of AgCl to protons and chloride ions. Due to the difference in proton diffusion speed (in the opposite direction of AgCl) from chloride ions, an electrolyte gradient resulted in the autonomous movement of particles at a high velocity of 100 mm s^{-1}. Duan et al. showed that in the presence of UV light, nanoscale AgCl were able to move autonomously in liquid [63]. They observed three diverse diffusive behaviors in the form of ballistic motion, enhanced diffusion and subdiffusion with the help of four distinct AgCl particles, namely inert isolated, active isolated, active coupled, and active schooled. Titanium dioxide (TiO_2) is a proficient photocatalyst with remarkable properties such as less cost, high stability, and refractive index as well as good biocompatibility. The Sen group showed the application of TiO_2 in the development of micromotors that were powered by light [64]. When placed under UV light, the micromotor achieved a speed of about 10 μm/s owing to the self-diffusiophoresis of TiO_2. Furthermore, the capabilities of TiO_2 were exploited in bimetallic colloids made up of hematite/TiO_2, which were showed to decompose H_2O_2 photocatalytically leading to an induced phoretic effect and osmotic self-propulsion [65]. Janus particles of TiO_2/Au were also showed to experience fast motion attaining a speed of 25 times their body length s^{-1} in water under UV light. The speed of this motor can be attained by the addition of a small quantity of H_2O_2 that increases the electrophoretic effect of TiO_2/Au particles [66].

Photothermal propulsion induced by light was demonstrated in Au-coated Janus particles by employing a laser light. The motion occurred as a result of radiation-pressure force and photophoretic pull on the Janus sphere. Under directed light, a temperate difference of 7 K could be established between the hemispheres, with the Au-coated side on the higher value. This difference induced a phoretic pull of approximately 0.1 pN on the particle [67]. The effect of light-induced thermophoresis has also been used to generate rotatory motion. Microfabricated gears with laser lithography technique and coated with carbon were showed to develop temperature gradient upon illumination with incoherent light. Rotational motion with a maximum speed of 300 rpm was observed as a result of considerable torque over inside parts of gear contour [68]. Near-infrared (NIR) light with deeper penetration capability has also been explored as a driving force and is particularly exciting considering that it is not easily absorbed by biological components. NIR light-propelled nanorocket was developed to achieve triggered self-propulsion through induced self-thermophoresis and demonstrated the potential for photothermal cancer cell therapy. These Au nanoshell-loaded polymer multilayer nanorockets were able to achieve a propulsion speed of about 160 μm/s due to the asymmetric thermal gradients induced by Au nanoshells through plasmonic absorption of NIR light [69].

5.4 Electric Field-Based Motors

Propulsion of electrically driven micro/nanorobots can be manipulated using direct or/and alternating currents. Velev et al. observed that semiconductor diodes of millimeter-scale lengths can propel autonomously under the influence of an external alternating electric field [70]. The motion of the diodes was directed by the local electroosmotic flow in proximity to the diodes. Nourhani et al. reported a flux-based motion analysis of electrocatalytic nanomotors that are spherical in nature based on an electrokinetic model with an overall distribution of cation flux on the motor surface [71]. To promote biomedical applications, the Wang group made an attempt to shrink the dimensions of these diodes to microscale by a membrane template growth method [72]. They fabricated nanowires of poly(pyrrole)-cadmium (PPy-Cd), CdSe–Au–CdSe and detected directional motion of the system at a speed of 17.2 mm/s due to the occurrence of electroosmotic flux in the presence of a spatially uniform alternating electric field. However, the ionic strength of the solution became a hindrance to the speed of the diodes. Recently, the Fan group has demonstrated a controllable and meticulous approach in employing catalytic nanomotors under three dimensional (3-D) electric fields for capture, delivery, and release of cargos to targeted locations, and to achieve integration with nanomechanical devices for powering the actuation [73]. Here, in the applied combined alternating and direct electric fields, alternating currents were used to align catalytic nanomotors, and the direct current helped in regulating their speed.

Electrical tweezers composed of multisegment nanowires (Au/Ni/Au) as rotors and nanomagnets (Au/Ni/Cr) as bearings were developed by Fan et al. The tweezer achieved rotation at a speed of over 18,000 rpm with the Ni segment acting as an anchor on the nanomagnet under an applied alternating electric field [74]. An alternative approach of using electrically triggered polarization across a micron-sized glassy carbon sphere was developed by Khun et al. to achieve its propulsion through asymmetric bubble generation initiated through spatially separated two electrochemical reactions. The speed and motion can be regulated by the applied redox chemistry and construction of the system [75].

5.5 Organism-Based Motors

Organism-based motors or biohybrid microbots are a combination achieved through the integration of mobile natural species like microorganisms and synthetic components. This particular combination gives the opportunity to make use of the organism's efficient propulsion and its ability to respond to external stimuli (Fig. 10.2D).

The Leonardo group recently reported rapid as well as tunable biohybrid micromotors, which can be developed by capturing specific bacteria into synthetic components that possess precise arrays of microchambers to accommodate it. The captured bacterial cells contribute maximally to the applied torque of the system [76]. Due to the light-responsive nature of bacteria, the control of rotational speeds was achieved by using a spatial light modulator. Another motile microorganism like sperms comprising lengthy flagella was used by the Schmidt group to achieve a biohybrid system. The system constituting a sperm cell inside a magnetic microtube was propelled using the motile cells as a driving force and could be guided remotely by an external magnetic field [77]. This strategy was a promising approach toward the targeted delivery of a specific sperm to an ovum for fertilization. In another approach, to access targeted locations through capillaries in humans, the Pouponneau group combined magnetotactic bacterium and flagellated nanomotors to develop biohybrid nanorobots [78]. However, for the functioning of the biohybrid systems, it is important to address the challenge of maintaining physiological conditions and nutrients so as to keep the involved biological component viable. This is still an upcoming area of research with significant developments to unfold in the near future.

6 IN VITRO AND IN VIVO APPLICATIONS OF MICRO/NANOMOTORS

6.1 Drug/Gene/Protein Delivery

Limited release of therapeutic moieties preferentially to unhealthy cells has been a serious limitation in conventional drug-delivery modules. Autonomous micro/nanomotors open up possibilities in this direction for active delivery to the site of interest by virtue of their directional movement and tissue retention/penetration capacity. Generally, to achieve this objective, the procedure involves encapsulation of drug/gene/protein to micro/nanorobots, propels them to the determined destination, and triggers their release using physiological or external parameters.

Wang and coworkers established an initial instance of applying an artificial catalytic micromotor for drug delivery and release. This exciting study was demonstrated by capturing Fe_3O_4 encapsulated drug carriers using self-propelled shuttles of catalytic nanowire through an applied magnetic field. Here, they demonstrated the pick-up, transport, and release of the drug molecule to predestined targeted locations [81]. The He group developed multilayer tubular nanorockets with incorporated Pt, Fe_3O_4 NPs and anticancer drug DOX to perform loading, transport, and release of drugs in cancer cells. While the rocket self-propelled through generated bubbles by catalytic decomposition of H_2O_2, the magnetic field was used to move autonomously and navigate to target cells. Upon reaching the target, the drug encapsulated rocket got attached to HeLa cells. Through the application of ultrasound, the DOX has been released to cancer cells [17]. Although earlier systems were promising, the amount of H_2O_2 used in their propulsion raises biocompatible issues. In this context, the research by Wu et al. established multilayer polyelectrolyte microtube machines modified by Pt NPs on the inside and a thin Au shell on the outside. With tumor-targeted peptide-functionalized to the Au layer, the microtubes can be triggered by NIR light to induce motion, either in the presence or absence of H_2O_2. Targeted detection and the consequent destruction of cancer cells by utilizing the benefits of photothermal activity were demonstrated using these engines [82]. To eliminate H_2O_2 completely, means of propulsion in other fuels were pursued. Janus micromotors of Mg/Pt-PNIPAM were identified for temperature-controlled delivery of drugs in simulated body fluid as well as blood plasma (Fig. 10.3A) [79]. Stomatocyte-based polymeric nanomotors composed of soft self-assembled block copolymers embedded with Pt-NPs were applied for loading DOX and subsequently its active release into cancer cells [83].

Zn- and Mg-based motors have presented proficient biocompatibility in fuel use, self-degradation, and generation of nontoxic by-products. Micromotors of PEDOT/Zn was the first report of in vivo application of artificial motors for drug delivery in mouse stomach [84]. Wang et al. described a PEDOT/Au microtubes loaded with Mg-microparticle covered with an enteric polymer for accurate placement and propulsion in the gastrointestinal tract of mouse [85]. The Kastrup group has applied microparticles constituting carbonate and tranexamic acid to achieve propulsion through bubble production for transporting therapeutic entities via blood. These were also showed to halt hemorrhage in an animal model when loaded with thrombin [86].

Gene/protein therapy is a popular form of treatment wherein the DNA or the genetic material/protein is applied as a therapeutic moiety to be transferred into a diseased cell for treatment purposes. Micromotor-based fuel independent gene delivery is an upcoming affordable as well as biocompatible methodology. Electric field-dependent motors made up of protein functionalized Au nanowire were showed to deliver tumor necrosis factor alpha to a target cell to combat cellular inflammation processes [87]. Au nanowires-mediated ultrasound propulsion-based siRNA gene delivery was demonstrated by Wang's group through rapid intracellular movement [88]. In addition, a magnetic micromotor decorated with lipoplexes has been employed in single-cell gene delivery to the human embryonic kidney [89].

6.2 Precision Surgery

Self-propelling nanotools have tremendous potential to offer surgical solutions at remote locations of an organism. In vitro applications in model organisms and various cells, tissues have been achieved using self-propelled micro/nanorobots. With further advancements, these microdrillers or nanojets are envisaged to propel to the desired location and deliver minimally invasive surgeries in the distant future.

A detailed and restricted incision is indispensable for successful surgical procedure especially for carrying out negligibly invasive surgery. Asymmetric microjets of InGaAs/GaAs/(Cr)Pt consisting of sharp tips and with diameters around 280–600 nm exhibited motion in hydrogen peroxide at 180 μm/s in a corkscrew fashion,

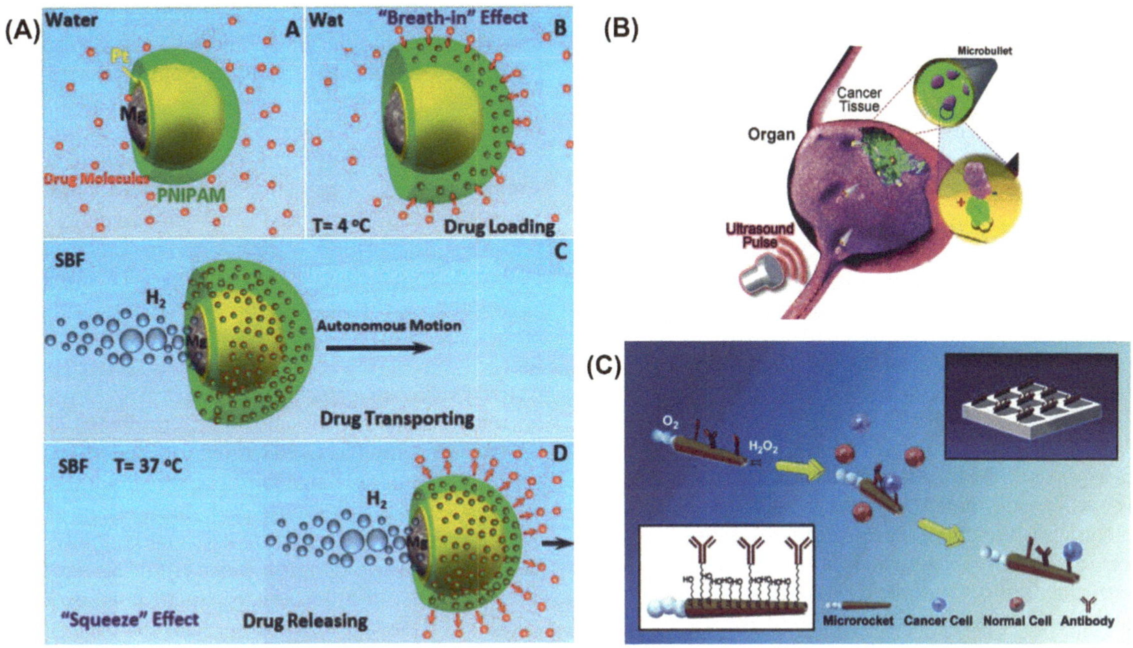

FIG. 10.3 **In vitro and in vivo applications of micro/nanomotors.** **(A)** Mg/Pt-Poly(N-isopropyl acrylamide) Janus micromotors for temperature-controlled drug delivery. **(B)** Microbullets for targeted tissue penetration and deformation. **(C)** Micromotor enabled capture and release of cancer cells. ((A) Reproduced with permission [79]. Copyright 2014, American Chemical Society. (B) Reproduced with permission [57]. Copyright 2012, John Wiley and Sons. (C) Reproduced with permission [80]. Copyright 2011, John Wiley and Sons.)

and thereby was able to achieve drilling into fixed HeLa cells [90]. Ti/Cr/Fe microtubes were explored for the execution of drilling processes under the influence of the external magnetic field eradicating the need for toxic fuels like H_2O_2/hydrazine. These microtubes have been activated in a rotating magnetic field and the applied frequency altered its alignment in a way to be made suitable for drilling. These microtubes were shown to drill holes in porcine liver tissue in the ex vivo model system [91]. Ultrasound-activated microbullet was applied in biolistic circulation and precision nanosurgery [57]. These Ti/Ni/Au composite microbullets contained perfluorocarbon, which undergoes vaporization upon applying an ultrasound field and established movement of microbullets at a speed of 6.3 m/s permitting in-depth penetration into the lamb kidney tissues for possible applications in nanosurgery. These developments demonstrate the control achieved in the propulsion of the micro/nanomotors and signify a noteworthy step in the realization of prominent nanosurgery tools (Fig. 10.3B).

6.3 Biosensing and Bioimaging

Nanomotors have the capacity to sense and segregate biological entities from a mixture of samples. Exploiting the antigen-antibody, donor-receptor, oligonucleotides complementary pairing, nanomotors that seize specific molecules or cells have been reported. The Wang group has developed a system for detecting DNA and bacterial ribosomal RNA based on the speed dependency of a catalytic motor on Ag ion concentration. The sensing concept relied on monitoring the speed variations of nanomotors caused during DNA hybridization assay by the dissolution of captured Ag NP tags [92]. An in vitro strategy was also developed by the Wang group based on the specific interaction of mAb-loaded microrockets so as to achieve isolation of cancer cells (Fig. 10.3C) [80]. These rockets contained Pt that helped in catalytic decomposition of H_2O_2 to generate bubble-mediated propulsion. On the course of motion, cancer cell containing targeting ligand was selectively isolated in complex media. Lectin, anti-CEA antibody, as well as aptamers functionalized microengines were some of the realized platforms for biosensing. The extremely specific and accurate detection, as well as the segregation potential of such motors, can be utilized in the fabrication of self-powered analytical platforms.

Micromachines were found to be applicable for magnetic resonance imaging wherein the Pouponneau group demonstrated the capacity of magnetotactic bacteria-based nanorobots for steering to remote locations in humans and be tracked using MRI imaging to obtain comprehensive information about the miniature microvascular system [78]. Micro/nanomotor's exclusive behavior being sensitive to pathological factors such as water content, temperature, and H_2O_2, they can be imaged in biological tissues. As a result of this sensitivity, they are evolving as a novel class of proficient candidates for nominally invasive diagnostics. Mattrey et al. confirmed the application of catalase-coated silica nanosphere converters toward ultrasound imaging based on microbubble formation when inoculated in the area of a bacterial inflammation [93].

6.4 Cargo Transport

Biological motor proteins that are equipped for cargo transport achieve intracellular transportation of vesicles as well as organelles in microtubule filaments. This encouraged the initial use of synthetic micro/nanomotors in the transportation and delivery of nano or microparticles. Motors harboring recognition entities or guided via external control possess the capacity for transportation and specific identification of targeted cargoes. Microtubular motors have been successfully applied in numerous targeted as well as nontargeted applications of payload transport. In addition, the transportation of microtubules across surfaces has been achieved in vitro using the motor protein, kinesin [94].

Pt/Au nanorods adapted to load polystyrene (PS)–amidine and PS–streptavidin microspheres facilitated the transport of different sizes of particles with speeds inversely proportional to the size via self-electrophoresis pathway [95]. Moreover, to achieve on-demand loading and release of cargo, a helical microswimmer with microholder at one end was developed by the Nelson group [96]. Constructed with non-cytotoxic materials, this helical swimmer exhibited propulsion through corkscrew mechanism in deionized water as well as fetal bovine serum (FBS) with abilities to pick up, transport, and release colloidal particles using microholder. Magnetotactic bacteria have also been explored as carriers of therapeutic cargo [21].

6.5 In Vivo Applications

The most important application of these micro/nanomotors could be their use in vivo applications. This may be said based on the enormous complexity involved and steep challenges to overcome. In vivo applications demand extremely small size motors for better penetrating capacities. Furthermore, it is of utmost importance to address the biocompatibility issues related to artificial motors.

As discussed earlier, fuel-based motors based on Zn and Mg-based tubular microrobots have been explored in drug delivery related to in vivo applications [84,85]. Organism-based propulsion systems were also developed for in vivo applications. Magnetotactic bacteria were used in the transportation of drug-containing liposomes near regions of tumor in mice model based on aerotactic mechanism and magnetic guidance [97]. Improved therapeutic response of nanocarriers was observed using the propulsion of magnetotactic bacteria.

Based on fuel-free propulsion, thrombolysis therapy employing magnetically powered Ni-microrod motors for delivery of tissue plasminogen activator was reported in C57/BL6 mice by Cheng et al. [98] Artificial bacterial flagella functionalized NIR probe-based fluorescence imaging has been carried out under the influence of a rotating magnetic field for applications in peritoneal cavity of a mouse [99]. Applications of optical tweezers for nanoparticle/cell manipulation have been reported for zebrafish larva [100]. Nevertheless, in vivo application of micro/nanomotors is in the budding phase, and relentless efforts have been dedicated to steering, transport, imaging, and tracing the micro/nanomotors in vivo.

7 MISCELLANEOUS APPLICATIONS

Micro/nanomotors have been applied in numerous other interesting applications. Singh et al. devised a strategy employing Janus Pt-black/Ti motors to not only achieve autonomous motion but also to produce useable H_2 gas [102]. The propulsion of the micromotor led to improved mixing and transport of the fuel, $NaBH_4$, in which it was suspended, which eventually led to the rapid production of H_2. Furthermore, the practical utility of the generated H_2 was demonstrated by powering a fuel cell car. In another application, the motion of an autonomously propelling microbot was used to generate electrical energy. Sailapu et al. demonstrated the induction of electromotive force (emf) into a Faraday coil through the motion of a magnetic microbot (Fig. 10.4A) [25]. The microbot constituted catalytic Pd NPs and a micron-sized magnet. While the microbot propelled in H_2O_2 through buoyancy-driven vertical motion due to the generated O_2 bubbles, the micromagnet was responsible for the induction of emf during its journey in and out of a copper coil. The so generated electrical signal during the flight of the microbot could illuminate light-emitting diodes following appropriate amplification, thus extending the application of micro/nanobots in energy harvesting.

Micromotors were also employed by researchers to examine and localize at the microscopic cracks for

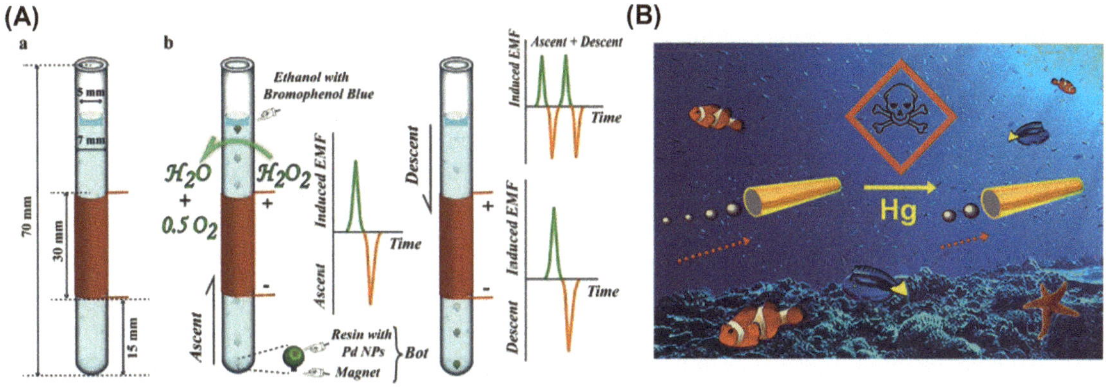

FIG. 10.4 **Miscellaneous applications of micro/nanorobots. (A)** Electromotive force induction from an autonomously moving magnet bot. **(B)** Enzyme-powered microrobot for water-quality testing. ((A) Reproduced with permission [25]. Copyright 2014, John Wiley and Sons. (B) Reproduced with permission [101]. Copyright 2013, American Chemical Society.)

refurbishing microelectronic circuits inspired by the self-healing capacity of living organisms [103]. The destruction of the environment due to the release of toxic pollutants possesses a severe threat to the ecosystem. Nanotechnology and material science offer innovative alternatives for the detection and removal of pollutants owing to their excellent photophysical and catalytic properties. The upcoming environment remediation technologies in the future intend to apply low-cost and energy-efficient methods for the removal of pollutants. Micro/nanomotors have shown promising developments in environmental remediation applications including sensing and removal of the pollutants (Fig. 10.4B) [101]. Through highly sensitive direct visualization of variations in propulsion behavior, enzyme-powered biocompatible polymeric (PEDOT)/Au-catalase tubular microengine offers a real-time assessment of water quality in the presence of common pollutants. Smart and functional microsupercapacitor microbots that could self-propel to a certain part of an electrical circuit to enhance the capacitance significantly was developed using WS_2 NPs-PANI/Pt microtubes [104]. Nanomotors have also found application in lithography as a new nanopatterning method to construct submicron/nanometer-scale features [105].

8 FUTURE PROSPECTS AND CHALLENGES

It can be inferred from all the earlier insights that noteworthy progress has been achieved in the field of micro/nanomotors to cover a wide range of applications related to carrying a payload, immunoassays, drug delivery, environmental remediation/sensing as well as energy production and storage, etc. During this remarkable journey of micro/nanorobots, commendable works have been documented overcoming several challenges in their realization. For these to further expand their reach into bigger practical uses, particularly related to medical applications, it becomes necessary to consider some vital elements.

For instance, to realize in vivo applications, the size and geometrical aspect of the micro/nanomotors is to be regulated and should be made significantly smaller so as to facilitate their delivery into cells and vessels of narrow diameters. In addition, optimization of the shape and surface can be carried out to attain required speeds by minimizing the protein adhesion. The driving force must be made adequate for their convenient propulsion, transfer or delivery of payloads, task execution in biological fluids that are generally viscous as well as have low Reynolds number. Improvements in energy conversion and propulsion efficiencies must be achieved for prolonged operation in challenging environments. Alongside, the drug retention and delivery efficiencies should also be made higher with optimum loading concentrations. Besides achieving optimal efficiencies, other critical aspects would be related to their removal from the body. Here, motors that can self-degrade, biocompatible are though very much encouraged, other options to remotely expel them from the body could also be considered.

In spite of the intrinsic advantages of remotely governed micro as well as nanodevices, micro/nanomotors that can achieve self-functionality with navigational capabilities are more interesting. In this regard, combinatorial approaches of propulsion mechanisms, careful design optimizations, material compositions become critical. The improvement in versatility and sophistication of such motors for incorporating multiple functions is necessary for functionalization, self-regulation, environmental monitoring, and coordination. Finally, practical bio- and industry-related applications should be explored. The activity of enzyme-driven nanomotors in viscous environments is to be established for their practical biomedical applications. Moreover, the corrosive behavior of metal-based motors is to be taken care of by improving their long-term stability, and their industrial-scale applications should be taken into consideration. Biocompatibility and environmentally friendly aspects must be considered cautiously, both in terms of the materials used in micro/nanomotors fabrication procedures and the operating fuels. Motors that could be self-decomposed into nontoxic products after executing a task would be extremely advantageous. Alternatively, means for the motor isolation after the completion of the assigned task need to be studied.

With dedicated efforts and considerable developments, in the future, it can be envisaged that these micro/nanorobots in a way could largely mimic the natural counterparts so as to conduct high-end functions to revolutionize in vivo and in vitro biomedical applications in a way to transform life toward better living standards.

REFERENCES

[1] van den Heuvel MGL, Dekker C. Motor proteins at work for nanotechnology. Science 2007;317(5836):333–6.

[2] Wang Y, Hernandez RM, Bartlett DJ, Bingham JM, Kline TR, Sen A, Mallouk TE. Bipolar electrochemical mechanism for the propulsion of catalytic nanomotors in hydrogen peroxide solutions. Langmuir 2006; 22(25):10451–6.

[3] Paxton WF, Sen A, Mallouk TE. Motility of catalytic nanoparticles through self-generated forces. Chemistry 2005;11(22):6462–70.

[4] Demirok UK, Laocharoensuk R, Manesh KM, Wang J. Ultrafast catalytic alloy nanomotors. Angewandte Chemie 2008;47(48):9349–51.

[5] Laocharoensuk R, Burdick J, Wang J. Carbon-nanotube-induced acceleration of catalytic nanomotors. ACS Nano 2008;2(5):1069–75.

[6] Balasubramanian S, Kagan D, Manesh KM, Calvo-Marzal P, Flechsig G-U, Wang J. Thermal modulation of nanomotor movement. Small 2009;5(13):1569–74.

[7] Calvo-Marzal P, Manesh KM, Kagan D, Balasubramanian S, Cardona M, Flechsig G-U, Posner J, Wang J. Electrochemically-triggered motion of catalytic nanomotors. Chemical Communications 2009;30:4509–11.

[8] Liu R, Sen A. Autonomous nanomotor based on copper–platinum segmented nanobattery. Journal of the American Chemical Society 2011;133(50):20064–7.

[9] Wheat PM, Marine NA, Moran JL, Posner JD. Rapid fabrication of bimetallic spherical motors. Langmuir 2010;26(16):13052–5.

[10] Brown A, Poon W. Ionic effects in self-propelled Pt-coated Janus swimmers. Soft Matter 2014;10(22):4016–27.

[11] Gibbs J, Zhao Y. Catalytic nanomotors: fabrication, mechanism, and applications. Frontiers of Materials Science 2011;5(1):25–39.

[12] Gibbs JG, Zhao Y-P. Design and characterization of rotational multicomponent catalytic nanomotors 2009; 5(20):2304–8.

[13] Gao W, Pei A, Dong R, Wang J. Catalytic iridium-based Janus micromotors powered by ultralow levels of chemical fuels. Journal of the American Chemical Society 2014;136(6):2276–9.

[14] Gibbs JG, Zhao Y-P. Autonomously motile catalytic nanomotors by bubble propulsion 2009;94(16): 163104.

[15] Mei Y, Solovev AA, Sanchez S, Schmidt OG. Rolled-up nanotech on polymers: from basic perception to self-propelled catalytic microengines. Chemical Society Reviews 2011;40(5):2109–19.

[16] Ismagilov RF, Schwartz A, Bowden N, Whitesides GM. Autonomous movement and self-assembly. Angewandte Chemie 2002;41(4):652–4.

[17] Wu Z, Wu Y, He W, Lin X, Sun J, He Q. Self-propelled polymer-based multilayer nanorockets for transportation and. Drug Release 2013;52(27):7000–3.

[18] Gao W, Sattayasamitsathit S, Orozco J, Wang J. Highly efficient catalytic microengines: template electrosynthesis of polyaniline/platinum microtubes. Journal of the American Chemical Society 2011;133(31):11862–4.

[19] Dey KK, Zhao X, Tansi BM, Méndez-Ortiz WJ, Córdova-Figueroa UM, Golestanian R, Sen A. Micromotors powered by enzyme catalysis. Nano Letters 2015;15(12): 8311–5.

[20] Wu Z, Li T, Li J, Gao W, Xu T, Christianson C, Gao W, Galarnyk M, He Q, Zhang L, Wang J. Turning erythrocytes into functional micromotors. ACS Nano 2014; 8(12):12041–8.

[21] Taherkhani S, Mohammadi M, Daoud J, Martel S, Tabrizian M. Covalent binding of nanoliposomes to the surface of magnetotactic bacteria for the synthesis of self-propelled therapeutic agents. ACS Nano 2014; 8(5):5049–60.

[22] Wang H, Zhao G, Pumera M. Beyond platinum: bubble-propelled micromotors based on Ag and MnO_2 catalysts. Journal of the American Chemical Society 2014;136(7):2719–22.

[23] Hong Y, Blackman NMK, Kopp ND, Sen A, Velegol D. Chemotaxis of nonbiological colloidal rods. Physical Review Letters 2007;99(17):178103.

[24] Dey KK, Bhandari S, Bandyopadhyay D, Basu S, Chattopadhyay A. The pH taxis of an intelligent catalytic microbot. Small 2013;9(11):1916–20.

[25] Sailapu SK, Chattopadhyay A. Induction of electromotive force by an autonomously moving magnetic bot. Angewandte Chemie 2014;53(6):1521–4.

[26] Gao W, Pei A, Wang J. Water-driven micromotors. ACS Nano 2012;6(9):8432–8.

[27] Gao W, Uygun A, Wang J. Hydrogen-bubble-propelled zinc-based microrockets in strongly acidic media. Journal of the American Chemical Society 2012;134(2):897–900.

[28] Gao W, D'Agostino M, Garcia-Gradilla V, Orozco J, Wang J. Multi-fuel driven Janus micromotors. Small 2013;9(3):467–71.

[29] Furukawa H, Cordova KE, O'Keeffe M, Yaghi OM. The chemistry and applications of metal-organic frameworks. Science 2013;341(6149).

[30] Tan TTY, Cham JTM, Reithofer MR, Andy Hor TS, Chin JM. Motorized Janus metal organic framework crystals. Chemical Communications 2014;50(96): 15175–8.

[31] Li J, Yu X, Xu M, Liu W, Sandraz E, Lan H, Wang J, Cohen SM. Metal–Organic frameworks as micromotors with tunable engines and brakes. Journal of the American Chemical Society 2017;139(2):611–4.

[32] Vicario J, Eelkema R, Browne WR, Meetsma A, La Crois RM, Feringa BL. Catalytic molecular motors: fuelling autonomous movement by a surface bound synthetic manganese catalase. Chemical Communications 2005;(31):3936–8.

[33] Sanchez S, Solovev AA, Mei Y, Schmidt OG. Dynamics of biocatalytic microengines mediated by variable friction control. Journal of the American Chemical Society 2010;132(38):13144–5.

[34] Ma X, Jannasch A, Albrecht U-R, Hahn K, Miguel-López A, Schäffer E, Sánchez S. Enzyme-powered hollow mesoporous Janus nanomotors. Nano Letters 2015; 15(10):7043–50.

[35] Schattling PS, Ramos-Docampo MA, Salgueiriño V, Städler B. Double-fueled Janus swimmers with magnetotactic behavior. ACS Nano 2017;11(4):3973–83.

[36] Gu Y, Sattayasamitsathit S, Kaufmann K, Vazquez-Duhalt R, Gao W, Wang C, Wang J. Self-propelled chemically-powered plant-tissue biomotors. Chemical Communications 2013;49(66):7307–9.

[37] Bhuyan T, Singh AK, Dutta D, Unal A, Ghosh SS, Bandyopadhyay D. Magnetic field guided chemotaxis of iMushbots for targeted anticancer therapeutics. ACS Biomaterials Science & Engineering 2017;3(8):1627–40.

[38] Sengupta S, Dey KK, Muddana HS, Tabouillot T, Ibele ME, Butler PJ, Sen A. Enzyme molecules as nanomotors. Journal of the American Chemical Society 2013;135(4):1406–14.

[39] Erb RM, Martin JJ, Soheilian R, Pan C, Barber JR. Actuating soft matter with magnetic torque. Advanced Function Materials 2016;26(22):3859–80.

[40] Tierno P, Golestanian R, Pagonabarraga I, Sagués F. Controlled swimming in confined fluids of magnetically actuated colloidal rotors. Physical Review Letters 2008; 101(21):218304.

[41] Sing CE, Schmid L, Schneider MF, Franke T, Alexander-Katz A. Controlled surface-induced flows from the motion of self-assembled colloidal walkers. Proceedings of the National Academy of Sciences 2010;107(2):535.

[42] Zhang L, Petit T, Lu Y, Kratochvil BE, Peyer KE, Pei R, Lou J, Nelson BJ. Controlled propulsion and cargo transport of rotating nickel nanowires near a patterned solid surface. ACS Nano 2010;4(10):6228–34.

[43] Pak OS, Gao W, Wang J, Lauga E. High-speed propulsion of flexible nanowire motors: theory and experiments. Soft Matter 2011;7(18):8169–81.

[44] Khalil ISM, Magdanz V, Sanchez S, Schmidt OG, Misra S. Biocompatible, accurate, and fully autonomous: a sperm-driven micro-bio-robot. Journal of Micro-Bio Robotics 2014;9(3):79–86.

[45] Dreyfus R, Baudry J, Roper ML, Fermigier M, Stone HA, Bibette J. Microscopic artificial swimmers. Nature 2005; 437:862.

[46] Mirkovic T, Foo ML, Arsenault AC, Fournier-Bidoz S, Zacharia NS, Ozin GA. Hinged nanorods made using a chemical approach to flexible nanostructures. Nature Nanotechnology 2007;2:565.

[47] Gao W, Sattayasamitsathit S, Manesh KM, Weihs D, Wang J. Magnetically powered flexible metal nanowire motors. Journal of the American Chemical Society 2010;132(41):14403–5.

[48] Gao W, Kagan D, Pak OS, Clawson C, Campuzano S, Chuluun-Erdene E, Shipton E, Fullerton EE, Zhang L, Lauga E, Wang J. Cargo-towing fuel-free magnetic nanoswimmers for targeted drug delivery. Small 2012;8(3): 460–7.

[49] Peyer KE, Tottori S, Qiu F, Zhang L, Nelson BJ. Magnetic helical micromachines. Chemistry 2013;19(1):28–38.

[50] Zhang L, Abbott JJ, Dong L, Kratochvil BE, Bell D, Nelson BJ. Artificial bacterial flagella: fabrication and magnetic control. Applied Physics Letter 2009;94(6): 064107.

[51] Ghosh A, Fischer P. Controlled propulsion of artificial magnetic nanostructured propellers. Nano Letters 2009;9(6):2243–5.

[52] Li J, Sattayasamitsathit S, Dong R, Gao W, Tam R, Feng X, Ai S, Wang J. Template electrosynthesis of tailored-made helical nanoswimmers. Nanoscale 2014; 6(16):9415–20.

[53] Gao W, Feng X, Pei A, Kane CR, Tam R, Hennessy C, Wang J. Bioinspired helical microswimmers based on vascular plants. Nano Letters 2014;14(1):305–10.

[54] Kim S, Lee S, Lee J, Nelson BJ, Zhang L, Choi H. Fabrication and manipulation of ciliary microrobots with non-reciprocal magnetic actuation. Scientific Reports 2016;6: 30713.

[55] Wang W, Castro LA, Hoyos M, Mallouk TE. Autonomous motion of metallic microrods propelled by ultrasound. ACS Nano 2012;6(7):6122–32.

[56] Garcia-Gradilla V, Orozco J, Sattayasamitsathit S, Soto F, Kuralay F, Pourazary A, Katzenberg A, Gao W, Shen Y, Wang J. Functionalized ultrasound-propelled

magnetically guided nanomotors: toward practical biomedical applications. ACS Nano 2013;7(10): 9232–40.
[57] Kagan D, Benchimol MJ, Claussen JC, Chuluun-Erdene E, Esener S, Wang J. Acoustic droplet vaporization and propulsion of perfluorocarbon-loaded microbullets for targeted tissue penetration and deformation. Angewandte Chemie 2012;51(30): 7519–22.
[58] Wu Z, Li T, Gao W, Xu T, Jurado-Sánchez B, Li J, Gao W, He Q, Zhang L, Wang J. Cell-membrane-coated synthetic nanomotors for effective biodetoxification. Advanced Function Materials 2015;25(25):3881–7.
[59] Bosanac L, Aabo T, Bendix PM, Oddershede LB. Efficient optical trapping and visualization of silver nanoparticles. Nano Letters 2008;8(5):1486–91.
[60] Ploschner M, Čižmár T, Mazilu M, Di Falco A, Dholakia K. Bidirectional optical sorting of gold nanoparticles. Nano Letters 2012;12(4):1923–7.
[61] Pang Y, Song H, Kim JH, Hou X, Cheng W. Optical trapping of individual human immunodeficiency viruses in culture fluid reveals heterogeneity with single-molecule resolution. Nature Nanotechnology 2014;9:624.
[62] Ibele M, Mallouk TE, Sen A. Schooling behavior of light-powered autonomous micromotors in water. Angewandte Chemie 2009;48(18):3308–12.
[63] Duan W, Ibele M, Liu R, Sen AJTEPJE. Motion analysis of light-powered autonomous silver chloride nanomotors. The European Physical Journal E 2012;35(8):77.
[64] Hong Y, Diaz M, Córdova-Figueroa UM, Sen A. Light-driven titanium-dioxide-based reversible microfireworks and micromotor/micropump systems. Advanced Function Materials 2010;20(10):1568–76.
[65] Palacci J, Sacanna S, Kim S-H, Yi G-R, Pine DJ, Chaikin PM. Light-activated self-propelled colloids 2014;372(2029).
[66] Dong R, Zhang Q, Gao W, Pei A, Ren B. Highly efficient light-driven TiO2–Au Janus micromotors. ACS Nano 2016;10(1):839–44.
[67] Qian B, Montiel D, Bregulla A, Cichos F, Yang H. Harnessing thermal fluctuations for purposeful activities: the manipulation of single micro-swimmers by adaptive photon nudging. Chemical Science 2013;4(4):1420–9.
[68] Maggi C, Saglimbeni F, Dipalo M, De Angelis F, Di Leonardo R. Micromotors with asymmetric shape that efficiently convert light into work by thermocapillary effects. Nature Communications 2015;6:7855.
[69] Wu Z, Si T, Gao W, Lin X, Wang J, He Q. Superfast near-infrared light-driven polymer multilayer rockets. Small 2016;12(5):577–82.
[70] Chang ST, Paunov VN, Petsev DN, Velev OD. Remotely powered self-propelling particles and micropumps based on miniature diodes. Nature Materials 2007;6:235.
[71] Nourhani A, Lammert PE, Crespi VH, Borhan A. A general flux-based analysis for spherical electrocatalytic nanomotors. Physics of Fluids 2015;27(1):012001.
[72] Calvo-Marzal P, Sattayasamitsathit S, Balasubramanian S, Windmiller JR, Dao C, Wang J. Propulsion of nanowire diodes. Chemical Communications 2010;46(10):1623–4.
[73] Guo J, Gallegos JJ, Tom AR, Fan D. Electric-field-guided precision manipulation of catalytic nanomotors for cargo delivery and powering nanoelectromechanical devices. ACS Nano 2018;12(2):1179–87.
[74] Kim K, Xu X, Guo J, Fan DL. Ultrahigh-speed rotating nanoelectromechanical system devices assembled from nanoscale building blocks. Nature Communications 2014;5:3632.
[75] Loget G, Kuhn A. Electric field-induced chemical locomotion of conducting objects. Nature Communications 2011;2:535.
[76] Vizsnyiczai G, Frangipane G, Maggi C, Saglimbeni F, Bianchi S, Di Leonardo R. Light controlled 3D micromotors powered by bacteria. Nature Communications 2017;8:15974.
[77] Magdanz V, Sanchez S, Schmidt OG. Development of a sperm-flagella driven micro-bio-robot. Advanced Materials 2013;25(45):6581–8.
[78] Martel S, Mohammadi M, Felfoul O, Lu Z, Pouponneau P. Flagellated magnetotactic bacteria as controlled MRI-trackable propulsion and steering systems for medical nanorobots operating in the human microvasculature. The International Journal of Robotics Research 2009;28(4):571–82.
[79] Mou F, Chen C, Zhong Q, Yin Y, Ma H, Guan J. Autonomous motion and temperature-controlled drug delivery of Mg/Pt-poly(N-isopropylacrylamide) Janus micromotors driven by simulated body fluid and blood plasma. ACS Applied Materials & Interfaces 2014;6(12): 9897–903.
[80] Balasubramanian S, Kagan D, Jack Hu C-M, Campuzano S, Lobo-Castañon MJ, Lim N, Kang DY, Zimmerman M, Zhang L, Wang J. Micromachine-enabled capture and isolation of cancer cells in complex media. Angewandte Chemie 2011;50(18):4161–4.
[81] Kagan D, Laocharoensuk R, Zimmerman M, Clawson C, Balasubramanian S, Kang D, Bishop D, Sattayasamitsathit S, Zhang L, Wang J. Rapid delivery of drug carriers propelled and navigated by catalytic. Nanoshuttles 2010;6(23):2741–7.
[82] Wu Z, Lin X, Wu Y, Si T, Sun J, He Q. Near-infrared light-triggered "On/Off" motion of polymer multilayer rockets. ACS Nano 2014;8(6):6097–105.
[83] Tu Y, Peng F, André AAM, Men Y, Srinivas M, Wilson DA. Biodegradable hybrid Stomatocyte nanomotors for drug delivery. ACS Nano 2017;11(2):1957–63.
[84] Gao W, Dong R, Thamphiwatana S, Li J, Gao W, Zhang L, Wang J. Artificial micromotors in the Mouse's stomach: a step toward in vivo use of synthetic motors. ACS Nano 2015;9(1):117–23.
[85] Li J, Thamphiwatana S, Liu W, Esteban-Fernández de Ávila B, Angsantikul P, Sandraz E, Wang J, Xu T, Soto F, Ramez V, Wang X, Gao W, Zhang L, Wang J. Enteric micromotor can selectively position and spontaneously propel in the gastrointestinal tract. ACS Nano 2016;10(10):9536–42.

[86] Baylis JR, Yeon JH, Thomson MH, Kazerooni A, Wang X, St John AE, Lim EB, Chien D, Lee A, Zhang JQ, Piret JM, Machan LS, Burke TF, White NJ, Kastrup CJ. Self-propelled particles that transport cargo through flowing blood and halt hemorrhage. Science Advances 2015;1(9):e1500379.
[87] Fan D, Yin Z, Cheong R, Zhu FQ, Cammarata RC, Chien CL, Levchenko A. Sub-cellular resolution delivery of a cytokine via precisely manipulated nanowires. Nature Nanotechnology 2010;5(7):545–51.
[88] Esteban-Fernández de Ávila B, Angell C, Soto F, Lopez-Ramirez MA, Báez DF, Xie S, Wang J, Chen Y. Acoustically propelled nanomotors for intracellular siRNA delivery. ACS Nano 2016;10(5):4997–5005.
[89] Qiu F, Fujita S, Mhanna R, Zhang L, Simona BR, Nelson BJ. Magnetic helical microswimmers functionalized with lipoplexes for targeted gene delivery. Advanced Function Materials 2015;25(11):1666–71.
[90] Solovev AA, Xi W, Gracias DH, Harazim SM, Deneke C, Sanchez S, Schmidt OG. Self-propelled nanotools. ACS Nano 2012;6(2):1751–6.
[91] Xi W, Solovev AA, Ananth AN, Gracias DH, Sanchez S, Schmidt OG. Rolled-up magnetic microdrillers: towards remotely controlled minimally invasive surgery. Nanoscale 2013;5(4):1294–7.
[92] Wu J, Balasubramanian S, Kagan D, Manesh KM, Campuzano S, Wang J. Motion-based DNA detection using catalytic nanomotors. Nature Communications 2010;1:36.
[93] Olson ES, Orozco J, Wu Z, Malone CD, Yi B, Gao W, Eghtedari M, Wang J, Mattrey RF. Toward in vivo detection of hydrogen peroxide with ultrasound molecular imaging. Biomaterials 2013;34(35):8918–24.
[94] Brunner C, Wahnes C, Vogel V. Cargo pick-up from engineered loading stations by kinesin driven molecular shuttles. Lab on a Chip 2007;7(10):1263–71.
[95] Sundararajan S, Lammert PE, Zudans AW, Crespi VH, Sen A. Catalytic motors for transport of colloidal cargo. Nano Letters 2008;8(5):1271–6.
[96] Tottori S, Zhang L, Qiu F, Krawczyk KK, Franco-Obregón A, Nelson BJ. Magnetic helical micromachines: fabrication, controlled swimming, and cargo transport. Advanced Materials 2012;24(6):811–6.
[97] Felfoul O, Mohammadi M, Taherkhani S, de Lanauze D, Xu YZ, Loghin D, Essa S, Jancik S, Houle D, Lafleur M, Gaboury L, Tabrizian M, Kaou N, Atkin M, Vuong T, Batist G, Beauchemin N, Radzioch D, Martel S. Magneto-aerotactic bacteria deliver drug-containing nanoliposomes to tumour hypoxic regions. Nature Nanotechnology 2016;11(11):941–7.
[98] Cheng R, Huang W, Huang L, Yang B, Mao L, Jin K, ZhuGe Q, Zhao Y. Acceleration of tissue plasminogen activator-mediated thrombolysis by magnetically powered nanomotors. ACS Nano 2014;8(8):7746–54.
[99] Servant A, Qiu F, Mazza M, Kostarelos K, Nelson BJ. Controlled in vivo swimming of a swarm of bacteria-like microrobotic flagella. Advanced Materials 2015;27(19):2981–8.
[100] Johansen PL, Fenaroli F, Evensen L, Griffiths G, Koster G. Optical micromanipulation of nanoparticles and cells inside living zebrafish. Nature Communications 2016;7:10974.
[101] Orozco J, García-Gradilla V, D'Agostino M, Gao W, Cortés A, Wang J. Artificial enzyme-powered microfish for water-quality testing. ACS Nano 2013;7(1):818–24.
[102] Singh VV, Soto F, Kaufmann K, Wang J. Micromotor-based energy generation. Angewandte Chemie 2015;54(23):6896–9.
[103] Li J, Shklyaev OE, Li T, Liu W, Shum H, Rozen I, Balazs AC, Wang J. Self-propelled nanomotors autonomously seek and repair cracks. Nano Letters 2015;15(10):7077–85.
[104] Mayorga-Martinez CC, Moo JGS, Khezri B, Song P, Fisher AC, Sofer Z, Pumera M. Self-propelled supercapacitors for on-demand circuit configuration based on WS2 nanoparticles micromachines. Advanced Materials 2016;26(36):6662–7.
[105] Li J, Gao W, Dong R, Pei A, Sattayasamitsathit S, Wang J. Nanomotor lithography. Nature Communications 2014;5:5026.

Index

Note: Page numbers followed by "f" indicate figures and "t" indicate tables.

Kolibri 360 GmbH
Ferdinand-Jühlke-Straße 7
99095 Erfurt, Deutschland
produktsicherheit@kolibri360.de